MONOGRAPHS ON STATISTICS AND APPLIED PROBABILITY

General Editors

D.R. Cox, D.V. Hinkley, N. Reid, D.B. Rubin and B.W. Silverman

1 Stochastic Population Models in Ecology and Epidemology *M.S. Bartlett* (1960)
2 Queues *D.R. Cox and W.L. Smith* (1961)
3 Monte Carlo Methods *J.M. Hammersley and D.C. Handscomb* (1964)
4 The Statistical Analysis of Series of Events *D.R. Cox and P.A.W. Lewis* (1966)
5 Population Genetics *W.J. Ewens* (1969)
6 Probability, Statistics and Time *M.S. Barlett* (1975)
7 Statistical Inference *S.D. Slivey* (1975)
8 The Analysis of Contingency Tables *B.S. Everitt* (1977)
9 Multivariate Analysis in Behavioural Research *A.E. Maxwell* (1977)
10 Stochastic Abundance Models *S. Engen* (1978)
11 Some Basic Theory for Statistical Inference *E.J.G. Pitman* (1978)
12 Point Processes *D.R. Cox and V. Isham* (1980)
13 Identification of Outliers *D.M. Hawkins* (1980)
14 Optimal Design *S.D. Silvey* (1980)
15 Finite Mixture Distributions *B.S. Everitt and D.J. Hand* (1981)
16 Classification *A.D. Gordon* (1981)
17 Distribution-free Statistical Methods *J.S. Maritz* (1981)
18 Residuals and Influence in Regression *R.D. Cook and S. Weisberg* (1982)
19 Applications of Queueing Theory *G.F. Newell* (1982)
20 Risk Theory, 3rd edition *R.E. Beard, T. Pentikainen and E. Pesonen* (1984)
21 Analysis of Survival Data *D.R. Cox and D. Oakes* (1984)
22 An Introduction to Latent Variable Models *B.S. Everitt* (1984)
23 Bandit Problems *D.A. Berry and B. Fristedt* (1985)
24 Stochastic Modelling and Control *M.H.A. Davis and R. Vinter* (1985)
25 The Statistical Analysis of Compositional Data *J. Aitchison* (1986)
26 Density Estimation for Statistical and Data Analysis *B.W. Silverman*
27 Regression Analysis with Application *G.B. Wetherill* (1986)
28 Sequential Methods in Statistics, 3rd edition *G.B. Wetherill* (1986)
29 Tensor Methods in Statistics *P. McCullagh* (1987)
30 Transformation and Weighting in Regression *R.J. Carroll and D. Ruppert* (1988)

31 Asymptotic Techniques for Use in Statistics *O.E. Barndorff-Nielson and D.R. Cox* (1989)

32 Analysis of Binary Data, 2nd edition *D.R. Cox and E.J. Snell* (1989)

33 Analysis of Infectious Disease Data *N.G. Becker* (1989)

34 Design and Analysis of Cross-Over Trials *B. Jones and M.G. Kenward* (1989)

35 Empirical Bayes Method, 2nd edition *J.S. Maritz and T. Lwin* (1989)

36 Symmetric Multivariate and Related Distributions *K.-T. Fang S. Kotz and K. Ng* (1989)

37 Generalized Linear Models, 2nd edition *P. McCullagh and J.A. Nelder* (1989)

38 Cyclic Designs *J.A. John* (1987)

39 Analog Estimation Mehtods in Econometrics *C.F. Manski* (1988)

40 Subset Selection in Regression *A.J. Miller* (1990)

41 Analysis of Repeated Measures *M. Crowder and D.J. Hand* (1990)

42 Statistical Reasoning with Imprecise Probabilities *P. Walley* (1990)

43 Generalized Additive Models *T.J. Hastie and R.J. Tibshirani* (1990)

44 Inspection Errors for Attributes in Quality Control *N.L. Johnson, S. Kotz and X. Wu* (1991)

45 The Analysis of Contingency Tables, 2nd edition *B.S. Everitt* (1992)

47 Longitudinal Data with Serial Correlation: A State-space Approach *R.H. Jones* (1993)

48 Differential Geometry and Statistics *M.K. Murray and J.W. Rice* (1993)

49 Markov Models and Optimization *M.H.A. Davis* (1993)

50 Networks and Chaos – Statistical and Probalistic Aspects *O.E. Barndorff-Nielson, J.H. Jensen and W.S. Kendall* (1993)

51 Number Theoretic Methods in Statistics *K.-T. Fang and Y. Wang* (1993)

52 Inference and Asymptotics *D.R. Cox and O.E. Barndorff-Nielson* (1994)

53 Practical Risk Theory for Actuaries *C.D. Daykin, T. Pentikäinen and M. Pesonen* (1993)

54 Statistical Concepts and Applications in Medicine *J. Aitchison and I.J. Lauder* (1994)

55 Predictive Inference *S. Geisser* (1993)

56 Model Free Curve Estimation *M. Tartar and M. Lock* (1994)

57 An Introduction to the Bootstrap *B. Efron and R. Tibshirani* (1993)

58 Nonparametric Regression and Generalized Linear Models *P.J. Green and B.W. Silverman* (1994)

(Full details concerning this series are available from the publisher)

Number-theoretic Methods in Statistics

K.-T. FANG

Hong Kong Baptist College

AND

Y. WANG

Institute of Mathematics
Academia Sinica
Beijing, China

CHAPMAN & HALL

London · Glasgow · New York · Tokyo · Melbourne · Madras

Published by Chapman & Hall, 2-6 Boundary Row, London SE1 8HN, UK

Chapman & Hall, 2-6 Boundary Row, London SE1 8HN, UK

Blackie Academic & Professional, Wester Cleddens Road, Bishopbriggs, Glasgow G64 2NZ, UK

Chapman & Hall Inc., One Penn Plaza, 41st Floor, New York, NY 10119, USA

Chapman & Hall Japan, Thomson Publishing Japan, Hirakawacho Nemoto Building, 6F, 1-7-11 Hirakawa-cho, Chiyoda-ku, Tokyo 102, Japan

Chapman & Hall Australia, Thomas Nelson Australia, 102 Dodds Street, South Melbourne, Victoria 3205, Australia

Chapman & Hall India, R. Seshadri, 32 Second Main Road, CIT East, Madras 600 035, India

First edition 1994

Printed in Great Britain by Clays Ltd, St Ives Plc, Bungay, Suffolk

ISBN 0 412 46520 5

A catalogue record for this book is available from the British Library
Library of Congress Cataloging-in-Publication Data available

Contents

Preface

This book contains some of the research work conducted by the authors in past seventeen years. Our research started in 1976. At that time, the first author was studying a problem of alloy steel standardization, where a series of probabilities of the 5-dimensional normal distribution needed to be calculated. Since the usual methods for numerical evaluation of multiple definite integral were all ineffective, he asked the second author for input on the number-theoretic method for numerical integration. As a result, his problem was solved by using the number-theoretic method. The essence of the number-theoretic method (NTM) is to find a set of points that is uniformly scattered over an s-dimensional unit cube, and sometimes this set can be used instead of random number in Monte Carlo method. The authors then understood that this idea can apply also to some other problems, for instance, problems in experimental design.

In 1978, a Chinese industrial agency proposed a problem of experimental designs to the authors, where six factors with at least 12 levels for each factors should be considered. Since the experiment was quite expensive, we were asked to provide an efficient experimental design within 50 experiments. This lead us to study the problem on how to apply the number-theoretic method to arrange the experiments, and we proposed a so-called "uniform design" for designing experiments. Applying the uniform design to our problem, 31 experiments were arranged only, each factor has 31 levels, and a satisfactory result was achieved. From then on, the uniform design was gradually popularized inside China in agriculture, the textile industry, the watch industry, science researches, military sciences and so on.

In 1987, another Chinese industry department asked us a typical geometrical probabilistic problem. It was difficult to get an analytic solution on the one hand, and on the other hand the precision was not high if the problem was simulated by a Monte Carlo method. Hence we tried a number-theoretic method. As a result, the computer time was reduced by 1/100 and an extra digit of precision was gained when compared with the result obtained by the Monte Carlo method. We were excited by these fruitful outcomes, and decided to organize a small research group "Applications of Number Theory to Statistics" in Academia Sinica, Beijing. Some young graduate students Gang Wei, Ke-Hai Yuan and Jin-Ting Zhang, Research Associate Bi-Qi Fang and Professor Yao-Chen Zhu are members of our group, and their contributions are included in this book.

This book illustrates the idea of number-theoretic methods and their applications in statistics. Since we are mainly concerned with applications, proofs of certain useful theorems are only given in Appendix B of the book. For more details on number-theoretic related results, the reader is referred to the monograph of Hua and Wang (1981). This book assumes a basic knowledge in calculus and a background in statistics at the graduate level.

In Chapter 1 we describe a series of problems in applied statistics that may be considered as the background of the typical "applications of number theory to statistics". All of these problems can be satisfactorily solved by applying number-theoretic methods. In addition, we give the essence of number-theoretic method and the methods for finding the uniformly scattered sets of points over s-dimensional cube, rectangle, sphere, ball and simplex. In Chapter 2, we illustrate the numerical methods for calculating the probabilities and moments of various multivariate statistical distributions, including the distributions over a sphere or a simplex which are required by the analysis of compositional data. In Chapter 3, we introduce a so-called SNTO algorithm for finding the global maximum and a maximum point of a continuous function over a closed and bounded domain. The SNTO can serve for optimization of a multiple modal function without requirements of continuous derivatives. Many applications of SNTO to maximum likelihood estimation, likelihood ratio statistics, nonlinear regression, robust regression and so on will be given also. A continuous multivariate distribution is often represented by a finite set of points carefully chosen to capture certain statistical properties of the distribution. This set is usually called the set of representative points of the distribution. The F-discrepancy and the mean square error have been used for measuring the accuracy of the representation of the points. Using these measures we give in Chapter 4 some methods for finding the representative points of many well-known multivariate distributions including classes of elliptically symmetric, multivariate l_1-norm symmetric, and multivariate Liouville distributions. In this chapter, we also give the applications of number-theoretic methods to the problems of statistical simulation, thus some difficult geometric probabilistic problems can be solved in this fashion. In Chapter 5, we introduce the idea and method of uniform designs. This method can also be applied to experiments with mixtures where the factors are dependent. In the last chapter, we give a combination of number-theoretic and projection pursuit methods, and then derive some new methods of statistical inference such as a test of multivariate normality and the robust estimation of a mean vector. Some results in our book are new and have not been published previously. To help the reader understand the theory of number-theoretic methods and their applications, we give in Appendix A the generating vectors of *glp* sets within dimensions 18, the tables of uniform designs, and the proofs of some theoretical results.

We wish to express our sincere thanks to the following institutions for their financial support or their computer and other facilities: Chinese National Science Foundation; Institutes of Mathematics and Applied Mathematics, Academia Sinica; the Hong Kong UPGC grant; Sino-British Exchange Scholarship of Hong Kong University; Wei Lun Foundation and United College Foundation of Chinese University of Hong Kong; Science Foundation of Hong Kong Baptist College; the Glorious Sun Fellowship of the CEEC of State University of New York at Stony Brook, and Grant DA01070 from the US Public Health Service. Some materials in this book were presented by the authors at the First Asian Mathematical Conference in 1990, the First Conference on Recent Development in Statistical Research organized by International Chinese Statistical Association in 1990 and the International Symposium on Multivariate Analysis and Its Applications in 1992 as an invited talk; University of North Carolina at Chapel Hill; University of Florida; University of California, Los Angeles; Chinese University of Hong Kong; and University of Hong Kong. We are grateful to many attendants for their helpful suggestions. We are grateful to Professors T.W. Anderson and I. Olkin of Stanford University, Professor P.M. Bentler of University of California, Los Angeles, Professor P.A.S. Fraser of University of Toronto, Professor Y.L.

Tong of Georgia Institute of Technology and Professors S. Kotz and Grace Yang of University of Maryland for their help and encouragement. We also would like to thank Dr. F.J. Hickernell and Dr. K.W. Ng for their valuable comments and helps. The authors are grateful to Professors H.H. Tsang and C.F. Ng, the Vice President and the Dean of Science Faculty of Hong Kong Baptist College for their support and encouragement. Our special thanks are due to Sir David R. Cox who brought our work to the Monographs on Statistics and Applied Probability. Finally, we thank Mr. Jin-Rong Wu and Ms Catherine Shum for the excellent typing in TEXand Mr. Hoi-Lam Wong for the graph drawing in MATLAB.

We are thankful to Ms Elizabeth Johnston (Senior Editor) and Ms Nicki Dennis (Commissioning Editor) of Chapman and Hall for their capable, experienced and patient guidance related to editorial and technical matters.

Kai-Tai Fang
Department of Mathematics
Hong Kong Baptist College
224 Waterloo Road, Kowloon
Hong Kong
and
Institute of Applied Mathematics, Academia Sinica
P.O. Box 2734
100080 Beijing, China

Yuan Wang
Institute of Mathematics, Academia Sinica
100080 Beijing, China

CHAPTER 1

Introduction to the number-theoretic method

1.1 Statistical problems

Statistics has developed greatly in the twentieth century, and many efficient statistical methods have become powerful instruments in the study of every science. In many complicated statistical applications the underlying problem may involve some of the following properties: (a) multiple variates, (b) non-normality, (c) nonlinearity, (d) multiple modes, (e) the presence of outliers and (f) missing data. The treatment of such problems requires more powerful statistical methods which require more tools and results from pure mathematics. In the past fifteen years differential geometry and group theory have been found useful in statistics; see for example Efron (1975), Amari (1985, 1987), Eaton (1989) and Diaconis (1989). This book introduces the so-called **number-theoretic method** (NTM) that can be used to solve a variety of problems in statistics. Some of these problems take a comparatively long time to solve using other methods. The NTM or **quasi-Monte Carlo method** is a special method which represents a combination of number theory and numerical analysis. Like many mixed breeds, it has its fascinations and attractions. The widest range of applications and indeed the historical origin of this method is found in numerical integration, but related problems such as interpolation and the numerical solutions of integral equations and differential equations can also be dealt with successfully. Korobov (1963), Hua and Wang (1961, 1963, 1981) and Niederreiter (1978b, 1988a, 1988b, 1992) give a comprehensive review in a bibliographic setting.

Although there is a close relationship between the NTM and the Monte Carlo method, in the past only a few statisticians have studied the NTM and its applications in statistics. The first applications of the NTM in statistics were naturally in evaluating probabilities and moments of a multivariate distribution (for example, Fang and Wu (1979), Zhang and Fang (1982)). Fang (1980) and Wang and Fang (1981) were the first to apply the NTM idea to experimental design, and they proposed a new design called the uniform design. This is perhaps the first application of NTM in statistics after integration. Since then many results have been obtained by uniform design in various fields in China. Shaw (1988) gave a detailed discussion on the applications of NTM to Bayesian statistics, mainly for numerical computation of posterior distributions. Recently Wang and Fang (1990a, 1990b, 1992), Fang and Wang (1990, 1991) and Fang, Yuan and Bentler (1990, 1992) have systematically studied the applications of NTM in statistics. This direction has attracted more statisticians and others who use statistics. Many facts illustrate that the NTM is a powerful tool.

The reader may have the following questions: What is the number-theoretic method? And how is it applied to various problems in statistics? In this section we shall introduce some statistical problems to be discussed in this book and the main ideas for solving these problems. We shall illustrate that all problems can be reduced to one key problem: *How to find a set of points which are uniformly scattered in the s-dimensional unit cube C^s?* Note that a "uniformly scattered" set of points stated here means roughly that the set has a small discrepancy (defined in section 1.2), not a set which is uniformly distributed in the usual statistical sense.

1.1.1 *Evaluation for the probabilities and moments of a multivariate distribution.*

Suppose that the s-dimensional random vector $\boldsymbol{x}$ has a continuous **probability density function** (p.d.f.) $p(\boldsymbol{x})$, and its probability on a domain D is

$$p = \int_D p(\boldsymbol{x})d\boldsymbol{x}. \tag{1.1.1}$$

D is usually an s-dimensional rectangle $[\boldsymbol{a}, \boldsymbol{b}] = [a_1, b_1] \times \cdots \times [a_s, b_s]$, so

$$p = \int_{a_1}^{b_1} \cdots \int_{a_s}^{b_s} p(x_1, \cdots, x_s)dx_1 \cdots dx_s \tag{1.1.2}$$

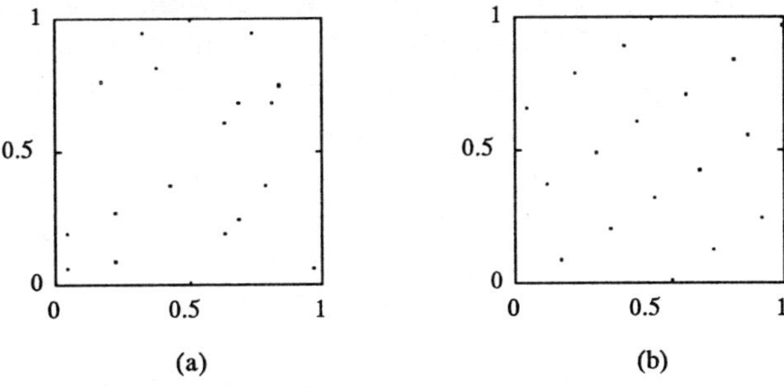

Figure 1.1 Plots of points generated by Monto Carlo method and NTM

which can be reduced to a canonical form

$$I(f) = \int_{C^s} f(\boldsymbol{x})d\boldsymbol{x} = \int_0^1 \cdots \int_0^1 f(x_1, \cdots, x_s)dx_1 \cdots x_s, \quad (1.1.3)$$

where $f(\boldsymbol{x})$ is a continuous function on the unit cube C^s. One often uses the sample mean method in the Monte Carlo methods to approximate $I = I(f)$ if the analytic expression of I can not be obtained, namely,

$$I \approx \frac{1}{n} \sum_{k=1}^{n} f(\boldsymbol{y}_k), \quad (1.1.4)$$

where $\{\boldsymbol{y}_k\}$ is a random sample from the uniform distribution on C^s, i.e. $\boldsymbol{y}_1, \cdots, \boldsymbol{y}_n$ are independently identically distributed (i.i.d.) as $U(C^s)$, the uniform distribution on C^s. However the efficiency of the Monte Carlo method is not high, i.e. a good approximation can be obtained by (1.1.4) only when n is very large. It will be pointed out in Chapter 2 that when $I(f^2) < \infty$ the rate of convergence of the Monte Carlo method is, in average, $O(1/\sqrt{n})$ in the sense of probability and is no case worse than $O(\sqrt{\ln(\ln(n))/n})$.

Why is the efficiency of the Monte Carlo method low? The key reason is that the set $\{\boldsymbol{y}_k\}$ is not scattered very uniformly on C^s. For example, Figure 1.1 (a) is a plot of a sample of size 17 taken from $U(C^2)$ using the Monte Carlo method, and Figure 1.1 (b) is a plot of 17 points obtained by the NTM. It is visually seen that the latter are scattered more uniformly on C^2 than the former. The precise definition of "uniformly scattered" set will be given in section 1.2.

Let $\mathcal{P} = \{\boldsymbol{c}_1, \cdots, \boldsymbol{c}_n\}$ be a set of n points uniformly scattered on C^s obtained by the NTM. Since they are scattered more uniformly than the points $\{\boldsymbol{y}_k\}$ obtained by random numbers, they can be used instead of $\{\boldsymbol{y}_k\}$ in (1.1.4). This yields

$$I \approx I(f, n) = \frac{1}{n} \sum_{k=1}^{n} f(\boldsymbol{c}_k). \tag{1.1.5}$$

It will be shown in Chapter 2 that $|I(f,n) - I| = O(n^{-1}(\log n)^{s-1})$ for this set $\mathcal{P}$ which is better than the order $O(n^{-\frac{1}{2}})$ in the sense of probability by the Monte Carlo method (1.1.4).

It is obvious that we can apply the NTM for numerical integration to evaluate functions of a multivariate distribution, such as moments, posterior probabilities, and some related moments in Bayesian statistics.

The key problem in numerical approximation of a multiple integral over C^s by the NTM approach is to choose a uniformly scattered set of points on C^s. Effective NTMs for this purpose have been proposed by many authors, such as Korobov (1959a, 1963), Hua and Wang (1960, 1964, 1965), Hlawka (1962) and Halton (1960), and we shall illustrate these methods in section 1.3.

Suppose that a random vector $\boldsymbol{x}$ is distributed on a domain D of R^s. D might be an s-dimensional sphere, ball or simplex. Then the problem of evaluating probabilities and moments over D is reduced to the problem of numerical approximation of the integrals of the following type

$$I(f, D) = \int_D f(\boldsymbol{x}) dv, \tag{1.1.6}$$

where dv denotes the volume element of D. For example, if D is the s-dimensional unit sphere, the study of distributions on D is the statistical basis of **directional data analysis** (Mardia (1972), Watson (1984)). If D is the simplex

$$T_s = \{(x_1, \cdots, x_s) : \ x_i \geq 0, i = 1, \cdots, s, \ x_1 + \cdots + x_s = 1\}, \tag{1.1.7}$$

then an observation from a distribution over D is called a **composition**. Aitchison (1986) gives a comprehensive discussion of the statistical analysis of compositional data.

Wang and Fang (1990a) proposed the following methods for approximating $I(f, D)$: **the indicator function method, transformation method** and **direct method**. Similar to the miss-hit method in Monte Carlo methods, we can develop the so-called **NT-hit-miss method** (section 4.7). The key requirement of these methods is to find uniformly scattered sets on C^s or D. We shall discuss these methods in Chapter 2 in detail and also in Chapter 4.

1.1.2 Optimization and statistics

There is a close relationship between optimization and statistics. There are many problems in statistics, such as maximum likelihood estimation, parameter estimation in regression analysis and optimal experimental design that all can be regarded as optimization problems. Suppose that D is a closed and bounded domain in R^s, and $f(\boldsymbol{x})$ is a continuous function on D. We want to find a point $\boldsymbol{x}^* \in D$ such that

$$M = f(\boldsymbol{x}^*) = \max_{\boldsymbol{x} \in D} f(\boldsymbol{x}). \tag{1.1.8}$$

We now discuss a few optimization problems in statistics.

(a) The maximum likelihood estimate (MLE)
Suppose that $\boldsymbol{x}_1, \cdots, \boldsymbol{x}_N$ is a sample from a p.d.f. $g(\boldsymbol{x}, \boldsymbol{\theta})$, where the parameter $\boldsymbol{\theta} \in \Theta \subset R^s$, and Θ is called the **parameter space**. Then the MLE of $\boldsymbol{\theta}$ is $\hat{\boldsymbol{\theta}}$ such that the **likelihood function**

$$L(\boldsymbol{\theta}) = \prod_{i=1}^{N} g(\boldsymbol{x}_i, \boldsymbol{\theta}) \tag{1.1.9}$$

attains its maximum at $\hat{\boldsymbol{\theta}}$, i.e.,

$$L(\hat{\boldsymbol{\theta}}) = \max_{\boldsymbol{\theta} \in \Theta} L(\boldsymbol{\theta}). \tag{1.1.10}$$

When the population is normally distributed $\hat{\boldsymbol{\theta}}$ has an analytic expression. In general $\hat{\boldsymbol{\theta}}$ has no analytic expression, for example, if the underlying distribution is the Weibull, beta or Cauchy distribution. Thus $\hat{\boldsymbol{\theta}}$ can only be obtained by means of numerical methods. In general, the likelihood function is not unimodal. As a result, numerical methods are frequently trapped into local maxima of the likelihood function that are not the MLE of $\boldsymbol{\theta}$.

(b) Likelihood Ratio Statistics
Consider the case of hypothesis testing

$$H_0: \quad \boldsymbol{\theta} \in \omega \subset \Theta, \quad H_1: \boldsymbol{\theta} \notin \omega, \quad \boldsymbol{\theta} \in \Theta \tag{1.1.11}$$

for the parameter of a p.d.f. $g(\boldsymbol{x}, \boldsymbol{\theta})$. The **likelihood ratio criterion** is

$$\lambda = \frac{\sup_{\boldsymbol{\theta} \in \omega} L(\boldsymbol{\theta})}{\sup_{\boldsymbol{\theta} \in \Theta} L(\boldsymbol{\theta})}, \tag{1.1.12}$$

where the numerator and denominator of the expression both involve optimization.

(c) Robust regression and non-linear regression
Consider the regression model

$$EY = h(\boldsymbol{x}, \boldsymbol{\theta}), \quad \boldsymbol{\theta} \in \Theta \subset R^s, \tag{1.1.13}$$

where $\boldsymbol{\theta}$ is the vector of regression parameters or regression coefficients and h is some function. One of the main tasks of regression analysis is to estimate $\boldsymbol{\theta}$ by observations $\{(Y_i, \boldsymbol{x}_i) : i = 1, \cdots, N\}$. Let

$$Q(\boldsymbol{\theta}) = \sum_{i=1}^{N} (Y_i - h(\boldsymbol{x}_i, \boldsymbol{\theta}))^2. \tag{1.1.14}$$

Then the widely used **least squares estimator** $\hat{\boldsymbol{\theta}}$ of $\boldsymbol{\theta}$ satisfies

$$Q(\hat{\boldsymbol{\theta}}) = \min_{\boldsymbol{\theta} \in \Theta} Q(\boldsymbol{\theta}). \tag{1.1.15}$$

The LSE $\hat{\boldsymbol{\theta}}$ has an analytic expression if h is a linear function of $\boldsymbol{\theta}$. Otherwise we must usually use a numerical method in nonlinear optimization for finding the least squares estimator $\hat{\boldsymbol{\theta}}$ (Ratkowsky (1983), Nash and Walker-Smith (1987)).

Robust regression has been drawn more attention in recent years, and the function $Q(\boldsymbol{\theta})$ in (1.1.14) is often replaced by a more **robust function**

$$Q_1(\boldsymbol{\theta}) = \sum_{i=1}^{N} |Y_i - h(\boldsymbol{x}_i, \boldsymbol{\theta})| \tag{1.1.16}$$

or more generally,

$$Q_2(\boldsymbol{\theta}) = \sum_{i=1}^{N} u(Y_i, h(\boldsymbol{x}_i, \boldsymbol{\theta})), \tag{1.1.17}$$

where $u(x, y) > 0$ is a function of x, y with certain properties (Huber(1972), Hampel (1986)). Since the absolute value operation appears in the expression of $Q_1(\boldsymbol{\theta})$, there will be difficulties when using Newton-like methods for finding the **robust estimate** $\hat{\boldsymbol{\theta}}^*$ of $\boldsymbol{\theta}$ such that

$$Q_1(\hat{\boldsymbol{\theta}}^*) = \min_{\boldsymbol{\theta} \in \Theta} Q_1(\boldsymbol{\theta}). \tag{1.1.18}$$

Usually, the parameter space Θ is R^s or a rectangle $[\boldsymbol{a}, \boldsymbol{b}]$, where $\boldsymbol{a}, \boldsymbol{b} \in R^s$. Sometimes $\Theta \subset R^s_+$, where

$$R^s_+ = \{(x_1, \cdots, x_s) : x_i \geq 0, i = 1, \cdots, s\}. \tag{1.1.19}$$

In this case, there exists no analytic solution for either the least squares estimate or the robust estimate of $\boldsymbol{\theta}$. Several authors (Boot (1964), Waterman (1974), Fang, Wang and Wu (1982) and Fang and He (1985)) have suggested different approaches to this problem.

All the above statistical problems are special cases of optimization (1.1.8). There are several numerical methods in optimization theory which can be applied to this problem, for example, the Newton-Gauss method, the simplex method, the truncated Newton method and the conjugate direction methods. For these methods, the function f should be unimodal, otherwise only a local optimum can be reached in general. However many functions arising in statistics are not unimodal. Furthermore most conventional methods assume that f has one or two continuous derivatives, which is not always the case if "max" "min" and "$|x|$" are contained in the expression for $f(\boldsymbol{x})$. Recently, many diverse algorithms for solving a wide variety of multiextremal global optimization problems have been developed (Horst and Tuy (1990)). On the other hand, in the past twenty years there has been considerable activity related to Monte Carlo optimization (Rubinstein (1986)). However, we may use the NTM for optimization and find its applications in statistics. These compete well with other optimization methods.

The following algorithm is quite simple. Let $\mathcal{P}_n = \{\boldsymbol{x}_1, \cdots, \boldsymbol{x}_n\}$ be a set of n points which are uniformly scattered over D. Let

$$M_n = \max_{1 \le k \le n} f(\boldsymbol{x}_k) = f(\boldsymbol{x}_{k_n}) \tag{1.1.20}$$

be the maximum of f on $\mathcal{P}_n$ and $\boldsymbol{x}_{k_n}$ be a point of $\mathcal{P}_n$ such that f attains its maximum at $\boldsymbol{x} = \boldsymbol{x}_{k_n}$ on $\mathcal{P}_n$. It is obvious that $M_n \to M$ as $n \to \infty$. Hence $M_n \cong M$ and $\boldsymbol{x}_{k_n} \cong \boldsymbol{x}^*$ if n is sufficiently large.

The NTM for optimization has many advantages. First of all, it only requires that the function f is continuous. There is no assumption that f is unimodal or differentiable. So this method suits, in particular, the problem of finding the maximum of a function with many local maxima. Secondly, the NTM requires only the values of the function f on $\mathcal{P}_n$ and not the derivatives of f. For example, we could use the NTM to maximise Q_2 to obtain the robust estimate as well as for Q. Hence the difference for the algorithms of the above two problems lies only on the expressions of Q and Q_2. Thirdly, the NTM does not depend on the choice of initial guess as conventional numerical methods do.

Unfortunately, theoretical and practical results show that the rate of convergence of M_n to M is not fast. In order to enhance the efficiency of NTM, we shall propose a sequential algorithm in section 3.2, called SNTO to reduce the amount of calculation FOR the case where D is a rectangle. In Chapter 3, many applications of SNTO to the MLE of parameters, the likelihood ratio of hypothesis testing and various estimates of regression parameters will be given.

When D is not a rectangle but can be mapped from C^t by some transformation, we can also use the versions of SNTO to find the approximations of the maximum M and the corresponding $\boldsymbol{x}^*$ of a function f over D. From SNTO on a rectangle we shall obtain its versions on ball, sphere, simplex, and give various applications in sections 3.5 – 3.8.

1.1.3 Representative points of a continuous distribution

The uniformly scattered set of points on C^s obtained by NTM is usually called a set of **quasi random numbers** or a **number-theoretic net (NT-net)**, since it may be used instead of random numbers in many statistical problems. Actually, the NT-net can be defined as **a set of representative points (rep-points)** of the uniform distribution $U(C^s)$. Therefore in this book we also call them rep-points of $U(C^s)$ or rep-points for simplicity. The definition of rep-points for a given distribution will be given in section 1.2. The rep-points have some advantages over random numbers. Therefore, we want to get the rep-points for various univariate and multivariate statistical distributions. By mixing the NTM and the Monte Carlo method we can generate rep-points for various distributions. The rep-points of the uniform distribution on the unit sphere as well as those on the unit simplex are extremely important for generating rep-points of some other distributions. These details will be discussed in section 4.2.

There are another kind of rep-points in the literature, especially in information theory, cluster analysis and experimental design. Let $F(x)$ be a given **cumulative distribution function** (c.d.f.). We may assume without loss of generality that the variance of $F(x)$ is 1. For a given positive integer s, we want to find s numbers $x_1, \cdots, x_s$, such that they give the best information about $F(x)$ under the following sense. We often use the **mean square error**

$$\mathrm{MSE}(\boldsymbol{x}) = \mathrm{MSE}(x_1, \cdots, x_s) = \int_{-\infty}^{\infty} \min_i (x_i - x)^2 p(x) dx \quad (1.1.21)$$

to measure the representation of $\{x_i\}$, where $p(x)$ is the p.d.f. of $F(x)$. We may assume without loss of generality that $x_1 < \cdots < x_s$. Let

$$A_s(R) = \{(x_1, \cdots, x_s) : -\infty < x_1 < \cdots < x_s < \infty\}. \quad (1.1.22)$$

Suppose that $\boldsymbol{x}^* \in A_s(R)$ such that MSE($\boldsymbol{x}^*$) attains the minimum of MSE($\boldsymbol{x}$) on $A_s(R)$, i.e.

$$\mathrm{MSE}(\boldsymbol{x}^*) = \min_{\boldsymbol{x}\in A_s(R)} \mathrm{MSE}(\boldsymbol{x}). \tag{1.1.23}$$

Then we call $\boldsymbol{x}^*$ the **mse-rep-points** of size s of $F(x)$.

For the uniform distribution on $(0,1)$, normal distribution, Rayleigh distribution and Laplacian distribution, the mse-rep-points are known. It is worth mentioning that we can give a universal algorithm for obtaining various mse-rep-points with high precision by the NTM in Chapters 4.

If $F(\boldsymbol{x})$ is a multivariate c.d.f, we may define similarly its mse-rep-points, but there is an essential difficulty in finding them compared with the univariate case just stated, so it is still an open problem in statistics. However we shall propose a numerical method for finding approximately the mse-rep-points of a continuous multivariate distribution in Chapter 4.

1.1.4 Experimental and uniform design

Experimental design has wide applications in science research, manufacturing industries, agricultural experiments and quality management. As usual, experimental design should satisfy the following rules:

(a) Uniformity: The experimental points should be scattered uniformly on the domain for experimentation such that these points are a good representation.

(b) Regularity: The experiment points satisfy some regularity condition so that it is convenient to do the analysis of variance for the experimental data.

The **orthogonal array design** and **BIB design** are methods produced according to these principles, but sometimes too many experiments are required in order to satisfy rule (b). For example, if the number of levels of a factor in an experiment is 12, then we must arrange at least $144 = 12^2$ experiments by the orthogonal array design. This is too much for most laboratories. Is it possible to reduce the number of experiments whilst still ensuring a good result? In fact, the purpose of experimental design is to find a finite number of points in the experimental domain so that the results of experiments on these points have a good representation. Hence our purpose completely coincides with the problems of numerical approximation of multiple integrals and of finding the global optimum of a function on a domain. If we drop principle (b) and only keep principle (a), then we can use a uniformly scattered set of points on the experimental domain D, which is obtained by the NTM, to arrange the experiments. We call this method for experimental design the **uniform design** (UD) (Wang and Fang (1981, 1990a)). The number of experiments required by UD is considerably less than for other known experimental designs if the number of levels is large. Many nice results were obtained by UD in the areas of watch making, textiles, military study and agriculture in China. We shall introduce the UD and the corresponding data analysis in Chapter 5.

1.1.5 Geometric probability and simulation

We often meet the following problem in geometric probability and simulation: Suppose that D is a bounded and closed domain in R^s, for instance, D is an s-dimensional unit ball. How do we find a set $\mathcal{P}$ of N points in D to stand for D? Suppose T is a statistic on D. Then M/N is regarded as a sample of T, where M is the number of points in $\mathcal{P}$ which are covered in the processes of simulation according to the definition of T, i.e. M/N denotes approximately a sample value of T. In Figure 1.2 D is a unit disk and is covered by two random disks. The T is the covered area in D. One often uses an **equi-distribution set** (or **equi-lattice points**) to represent D if D is a rectangle. It is obvious that this method is not always acceptable when D is not a rectangle. For example, if D is a ball, and the natural method is using a circumscribed cube C of D and to define the set $\mathcal{P}$ for D by those points of C inside D (Figure 1.3). This method is simple but has many disadvantages: (a) The computational effort is large if we use the equi-distribution set on C. (b) The points that are uniformly scattered on C are usually not very uniform on D. (c) The points near the boundary of D often cause a big error in the calculation. (d) If D is a manifold of lower dimension, for example, D is a unit sphere U_s in R^s, the above method is useless, since it is likely that no point in C falls on U_s, i.e., D.

If we use the NTM to get a uniformly scattered set $\mathcal{P}$ of n points on D, and use $\mathcal{P}$ to stand for D (Figure 1.4), then the above troubles can be avoided. So the NTM is very useful for this type of

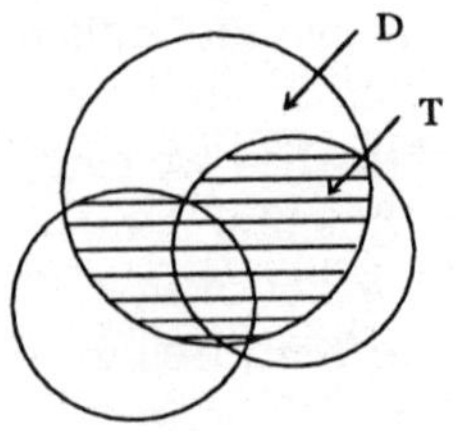

Figure 1.2 Geometric probability

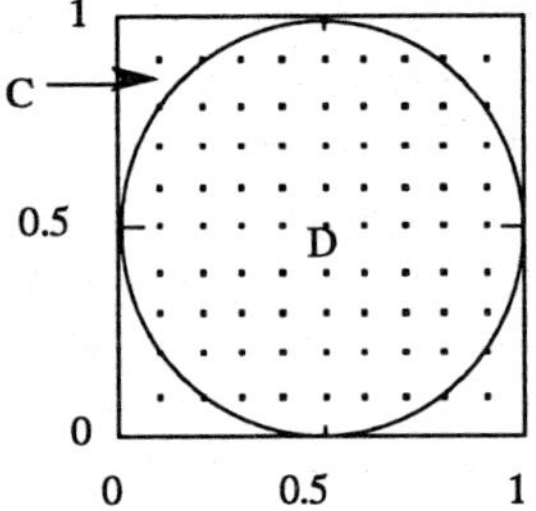

Figure 1.3 An NT-net on B_2 induced from that on C^2

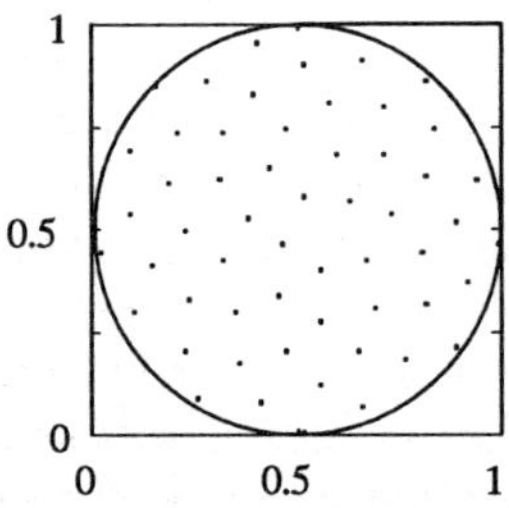

Figure 1.4 An NT-net on B_2

problem in geometric probability and simulation. We shall introduce in sections 4.8 and 4.9 two case studies which have been solved satisfactorily using the NTM.

1.1.6 *Miscellaneous*

We often use the method of projection in multivariate analysis, for instance, principle component analysis, canonical correlation analysis and projection pursuit. Fortunately, one can find analytic expressions for principal component analysis and canonical correlation analysis. But there are no analytic solutions in many other cases involving projection pursuit, so we have to use numerical methods. Let $\boldsymbol{x}$ be an s-dimensional random vector with a one-dimensional projection $\boldsymbol{a}'\boldsymbol{x}, \boldsymbol{a} \in R^s$. We may suppose without loss of generality that $\boldsymbol{a}'\boldsymbol{a} = 1$, i.e. $\boldsymbol{a}$ lies on the unit-sphere U_s of R^s. Let $t(u)$ be a continuous scale function and $t(\boldsymbol{a}'\boldsymbol{x})$ be the designated statistic. In many methods of projection pursuit we are required to find an $\boldsymbol{a}_0 \in U_s$ such that $t(\boldsymbol{a}'\boldsymbol{x})$ attains its optimum at $\boldsymbol{a} = \boldsymbol{a}_0$. By a uniformly scattered set of points on U_s and SNTO, we may obtain approximations of $\boldsymbol{a}_0$ and $t(\boldsymbol{a}'_0\boldsymbol{x})$.

For instance, it is well-known that testing the multinormality of $\boldsymbol{x} \in R^s$ is equivalent to testing one-dimensional normality for $\boldsymbol{a}'\boldsymbol{x}$, $\boldsymbol{a} \in U_s$ (Anderson (1982)). For each $\boldsymbol{a} \in U_s$, suppose that $t(\boldsymbol{a}'\boldsymbol{x})$ is a statistic for testing one-dimensional normality, and that $\boldsymbol{a}'\boldsymbol{x}$ is less normal when $t(\boldsymbol{a}'\boldsymbol{x})$ is large. If $\boldsymbol{a}_0$ is a vector such that

$$t(\boldsymbol{a}_0'\boldsymbol{x}) = \max_{\boldsymbol{a} \in U_S} t(\boldsymbol{a}'\boldsymbol{x})$$

then testing multinormality is equivalent to testing normality of $\boldsymbol{a}_0'\boldsymbol{x}$. The NTM can help us finding approximations of $\boldsymbol{a}_0$ and $t(\boldsymbol{a}_0'\boldsymbol{x})$. We shall discuss this problem in sections 6.3 and 6.4. In section 6.5 we shall treat the more general problem of projecting R^s into a lower dimensional space R^l, $1 \leq l \leq s$ by the NTM.

In fact, the methods introduced in this book are not only useful for the problems in statistics but also for some other fields. For example, the SNTO may be used for solving a system of nonlinear equations with constraints or without constraints, and for finding a fixed point of a continuous mapping which maps a domain onto itself. As for numerical integration, it is useful in almost every field in the natural sciences.

1.2 Discrepancy and F-discrepancy

In the previous section we have introduced many problems in statistics whose numerical solutions depend on generating a uniformly scattered set of points on C^s. What is the meaning of "uniformly scattered" on C^s ? First we need to give a measurement for the uniformity of the set on C^s. In fact, this is a special case of the following problem.

Let $F(\boldsymbol{x})$ be a continuous multivariate distribution in R^s and n be a given integer. We want to find a set of n points $\boldsymbol{x}_1, \cdots, \boldsymbol{x}_n$ in R^s such that they have a good representation for $F(\boldsymbol{x})$. What is the meaning of representation? Let us consider a measurement of representation.

Definition 1.1
Let $\boldsymbol{x}_1, \cdots, \boldsymbol{x}_n$ be any n points in R^s. Then the function

$$F_n(\boldsymbol{x}) = \frac{1}{n} \sum_{i=1}^{n} I\{\boldsymbol{x}_i \leq \boldsymbol{x}\} \tag{1.2.1}$$

is called the **empirical distribution** of $\boldsymbol{x}_1, \cdots, \boldsymbol{x}_n$, where all inequalities are understood with respect to the componentwise order of R^s and $I\{A\}$ is the **indicator function** of A, i.e.

$$I\{A\} = \begin{cases} 1, & \text{if } A \text{ occurs}, \\ 0, & \text{otherwise}. \end{cases}$$

The empirical distribution can be defined by another equivalent way: Let $\boldsymbol{x}$ be a random vector of R^s such that

$$P(\boldsymbol{x} = \boldsymbol{x}_i) = \frac{1}{n}, \quad i = 1, \cdots, n.$$

Then the c.d.f. of $\boldsymbol{x}$, $F_n(\boldsymbol{x})$, is called the empirical distribution of $\boldsymbol{x}_1, \cdots, \boldsymbol{x}_n$.

Definition 1.2

Let $F(\boldsymbol{x})$ be a c.d.f. in R^s and $\mathcal{P} = \{\boldsymbol{x}_k,\ k = 1, \cdots, n\}$ be a set of points on R^s. Then

$$D_F(n, \mathcal{P}) = \sup_{\boldsymbol{x} \in R^s} |F_n(\boldsymbol{x}) - F(\boldsymbol{x})| \tag{1.2.2}$$

is called the **F-discrepancy** of $\mathcal{P}$ with respect to $F(\boldsymbol{x})$, where $F_n(\boldsymbol{x})$ is the empirical distribution of $\boldsymbol{x}_1, \cdots, \boldsymbol{x}_n$.

Remark 1.1

Obviously, the F-discrepancy is a measure of the representation of $\mathcal{P}$ with respect to $F(\boldsymbol{x})$. Consider testing the hypothesis H_0 : the underlying distribution is $F(\boldsymbol{x})$. Then $D_F(n, \mathcal{P})$ is just the **Kolmogorov-Smirnov statistic** for the goodness of fit test of $F(\boldsymbol{x})$.

When $F(\boldsymbol{x})$ is the uniform distribution on $C^s = [0, 1]^s$, denoted by $U(C^s)$, the F-discrepancy reduces into the common discrepancy in the literature (Hua and Wang (1981)), in this case we shall use $D(n, \mathcal{P})$ instead of F-discrepancy. It was Weyl (1916) who gave the concept of discrepancy to measure the uniformity of a set of points on C^s. The more general concept given by Definition 1.2 was suggested by Wang and Fang (1990a).

Remark 1.2

In the literature of NTM the discrepancy is defined as follows: Let $\mathcal{P} = \{\boldsymbol{x}_k,\ k = 1, \cdots, n\}$ be a set of points on C^s. For any $\boldsymbol{\gamma} \in C^s$, let $N(\boldsymbol{\gamma}, \mathcal{P})$ be the number of points satisfying $\boldsymbol{x}_k \le \boldsymbol{\gamma}$. Then

$$D(n, \mathcal{P}) = \sup_{\boldsymbol{\gamma} \in C^s} \left| \frac{N(\boldsymbol{\gamma}, \mathcal{P})}{n} - v([\mathbf{0}, \boldsymbol{\gamma}]) \right| \tag{1.2.3}$$

is called the **discrepancy of** $\mathcal{P}$, where $v([\mathbf{0}, \boldsymbol{\gamma}]) = \gamma_1 \cdots \gamma_s$ denotes the volume of the rectangle $[\mathbf{0}, \boldsymbol{\gamma}]$. For simplicity, we sometimes denote $D(n, \mathcal{P})$ by $D(n)$.

Figure 1.5 gives an illustration of the discrepancy of a set. If the absolute value of ratio of the number of points lying in the rectangle $[\mathbf{0}, \boldsymbol{\gamma}]$ and the total number of points of the set minus the volume of the rectangle $[\mathbf{0}, \boldsymbol{\gamma}]$ is small, then the set of points

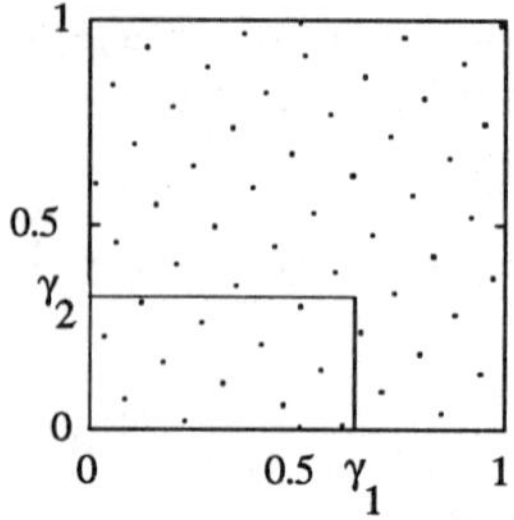

Figure 1.5 The illustration of discrepancy

is scattered uniformly. This is why we use discrepancy to measure the uniformity for a set of points.

Now we proceed to find the set of points with lowest discrepancy. This problem is solved for the case $s = 1$.

Example 1.1

Let n be an integer ≥ 1, and let

$$\mathcal{Q} = \left\{ \frac{2i-1}{2n}, \quad i = 1, \cdots, n \right\}. \tag{1.2.4}$$

Now we shall show that $\mathcal{Q}$ has the lowest discrepancy $\frac{1}{2n}$ among all sets of n points in $C^1 = [0, 1]$.

Let $[x]$ and $\{x\}$ denote respectively the integral part and fractional part of x. For any $\gamma \in [0, 1]$, let $x = 2n\gamma$. Then

$$\begin{aligned} \frac{N(\gamma, \mathcal{Q})}{n} - \gamma &= \frac{N(\frac{x}{2n}, \mathcal{Q})}{n} - \frac{x}{2n} \\ &= \begin{cases} \frac{[x]+1}{2n} - \frac{[x]+\{x\}}{2n} = \frac{1-\{x\}}{2n}, & \text{if } 2 \not| \, [x], \\ \frac{[x]}{2n} - \frac{[x]+\{x\}}{2n} = -\frac{\{x\}}{2n}, & \text{if } 2 \mid [x], \end{cases} \end{aligned}$$

where $a|b$ denotes that b is a multiple of a. Otherwise we denote by $a \not| b$. Hence $D(n, \mathcal{Q}) \leq \frac{1}{2n}$. If we take $\gamma = \frac{2i-1}{2n}$, then

$$\frac{N(\gamma, \mathcal{Q})}{n} - \gamma = \frac{i}{n} - \frac{2i-1}{2n} = \frac{1}{2n}.$$

Hence $D(n, \mathcal{Q}) = 1/(2n)$.

Now let $\mathcal{P} = \{x_i,\ i = 1, \cdots, n\}$ be a subset of $[0,1]$. Without loss of generality, we may suppose that

$$0 \leq x_1 \leq \cdots \leq x_n \leq 1.$$

If $D(n, \mathcal{P}) \leq 1/(2n)$ then it follows from (1.2.2) that for any γ

$$\left| \frac{N(\gamma, \mathcal{Q})}{n} - \gamma \right| \leq \frac{1}{2n},$$

$$-\frac{1}{2} + \gamma n \leq N(\gamma, \mathcal{P}) \leq \frac{1}{2} + \gamma n.$$

For any γ such that $(2i-1)/(2n) < \gamma < (2i+1)/(2n)$ the above inequality implies that

$$i - 1 < -\frac{1}{2} + \gamma n \leq N(\gamma, \mathcal{P}) \leq \frac{1}{2} + \gamma n < i + 1,$$

so $N(\gamma, \mathcal{P}) = i$. This restriction on $N(\gamma, \mathcal{P})$ means that $\mathcal{P}$ must be the same as $\mathcal{Q}$ defined in (1.2.4). Hence we have proved that if $\mathcal{P}$ is not Q, then $D(n, \mathcal{P}) > 1/(2n)$. □

It is very difficult to find a set with the smallest discrepancy for the case $s \geq 2$, since the distributions of n points in C^s may be very complicated. Hence we want to find sets with asymptotically small discrepancies.

A conjecture in number theory states that for any given set $\mathcal{P}$ of $n (\geq 2)$ points and every $s \geq 2$ we have

$$D(n, \mathcal{P}) \geq c(s) n^{-1} \log^{s-1} n.$$

Hereafter we use $c(f, \cdots, g)$ to denote a positive constant depending on $f, \cdots, g$, but not the same value in different occurrences. This conjecture was proved by Schmidt(1972) for the case $s = 2$. In general we have the following theorem of Roth (1954):

$$D(n, \mathcal{P}) > 2^{-2(s+2)} (s-1)^{-(s-1)/2} n^{-1} (\log_2 n)^{(s-1)/2}. \qquad (1.2.5)$$

Therefore if we can find a sequence of sets $\mathcal{P}_n$, where $\mathcal{P}_n$ has n elements such that the order of $D(n, \mathcal{P}_n)$ is similar to the right-hand side of (1.2.5), then $\mathcal{P}_n$ can be regarded as a set of points which are well uniformly scattered on C^s. More precisely we have the following:

Definition 1.3

Let $F(\boldsymbol{x})$ be a s-dimensional continuous c.d.f. and $\boldsymbol{N}$ denote an infinite subset of natural numbers and $\{\mathcal{P}_n, n \in \boldsymbol{N}\}$ be a sequence of sets of R^s, where $\mathcal{P}_n$ has n points with a certain structure. If

$$D_F(n, \mathcal{P}_n) = o(n^{-1/2}), \quad \text{as} \quad n \to \infty, \tag{1.2.6}$$

then $\{\mathcal{P}_n\}$ is called a set of **representative points** (for simplicity **rep-points**) of $F(\boldsymbol{x})$. When $F(\boldsymbol{x})$ is the c.d.f. of $U(D)$ where D is a closed and bounded domain, $\{\mathcal{P}_n\}$ is said to be **uniformly scattered** on D or an **NT-net** on D. If for any given $\varepsilon > 0$, we have

$$D_F(n, \mathcal{P}_n) = O(n^{-1+\varepsilon}), \tag{1.2.7}$$

then $\{\mathcal{P}_n\}$ is called to be **well uniformly scattered** on D.

Remark 1.3

In the literature of the NTM most authors say that $\mathcal{P}_n$ is "uniformly distributed" on C^s if $D(n, \mathcal{P}_n) = o(1)$, as $n \to \infty$ (e.g. Kuipers and Niederreiter (1974)). But for our purpose the order $o(1)$ is not good enough. Besides, the term "uniformly distributed" in statistics has the exact meaning which is different from the present sense, and therefore we recommend the term "uniformly scattered".

From the convergence point of view, the convergence rate of $D(n, \mathcal{P}_n)$ does not change if we change a finite number of sets in the sequence $\mathcal{P}_n$. For example, we can put some sets with high discrepancy into the sequence. Obviously, it is not allowed for practical use and we must keep the same structure for all the $\mathcal{P}_n$.

Example 1.2
Let m be an integer > 1 and $n = m^s$. We call the set

$$\mathcal{E} = \Big\{\Big(\frac{2l_1 - 1}{2m}, \cdots, \frac{2l_s - 1}{2m}\Big), \ 1 \le l_i \le m, \ i = 1, \cdots, m\Big\} \quad (1.2.8)$$

to be the set of **equi-distribution** or the set of **equi-lattice points**. We can show that

$$c_1(s)n^{-1/s} \le D(n, \mathcal{E}) \le c_2(s)n^{-1/s} \qquad (1.2.9)$$

(Hua and Wang (1981)). Hence under Definition 1.3 the set $\mathcal{E}$ is not uniformly scattered on C^s when $s \ge 2$. This is the reason why $\mathcal{E}$ is not recommended for use in many problems in numerical analysis and statistics.

Example 1.3
Let $\boldsymbol{x}_1, \cdots, \boldsymbol{x}_n$ be i.i.d. according to the uniform distribution on C^s and $\mathcal{P}_n = \{\boldsymbol{x}_1, \cdots, \boldsymbol{x}_n\}$. Chung (1949) and Kiefer (1961) proved that

$$D(n, \mathcal{P}_n) = O(n^{-1/2}(\log\log n)^{1/2}) \qquad (1.2.10)$$

with probability one. This fact indicates that the sequence $\mathcal{P}_n$ generated by the Monte Carlo method is not uniformly scattered on C^s with probability one.

Example 1.4
Let $F(x)$ be a univariate continuous c.d.f. and $F^{-1}(y)$ be its inverse function. Then the set

$$\mathcal{Q}_F = \Big\{F^{-1}\Big(\frac{2i-1}{2n}\Big), \ \ i = 1, \cdots, n\Big\} \qquad (1.2.11)$$

is the rep-points of size n of $F(x)$ with lowest F-discrepancy $\frac{1}{2n}$.

In fact, let $\mathcal{Q}$ be the set defined by (1.2.4), $N(x, \mathcal{Q})$ the number of points satisfying $(2i - 1)/(2n) \le x$, and $N(x, \mathcal{Q}_F)$ the number of points such that $F^{-1}((2i-1)/(2n)) \le x$. We have

$$\frac{N(x,\mathcal{Q}_F)}{n} - F(x) = \frac{N(F(x),\mathcal{Q})}{n} - F(x)$$

and thus

$$\begin{aligned} D_F(n,\mathcal{Q}_F) &= \sup_{x\in R}\left|\frac{N(x,\mathcal{Q}_F)}{n} - F(x)\right| \\ &= \sup_{y\in[0,1]}\left|\frac{N(y,\mathcal{Q})}{n} - y\right| \\ &= D(n,\mathcal{Q}) = 1/2n. \end{aligned}$$

On the other hand, if $\mathcal{P} = \{x_1,\cdots,x_n\}$ has F-discrepancy$< \frac{1}{2n}$, then it follows from the above argument that the set $\mathcal{Q}_F = \{F^{-1}(x_1),\cdots,F^{-1}(x_n)\}$ has discrepancy $D(n,\mathcal{Q}_F) < \frac{1}{2n}$ which contradicts the conclusion of Example 1.1. □

The basis of Example 1.4 is the following theorem:

Theorem 1.1

Suppose that the random variable X has a continuous c.d.f. $F(x)$. Then the random variable $Y = F(X)$ is uniformly distributed on $[0,1]$, i.e. $Y \sim U[0,1]$. Conversely, if $Y \sim U[0,1]$, then X has c.d.f. $F(x)$.

PROOF For any $u \in [0,1]$, we have

$$P(Y \le u) = P(F(X) \le u) = P(X \le F^{-1}(u)) = F(F^{-1}(u)) = u.$$

Conversely, if $Y \sim U[0,1]$, then $P(X \le F^{-1}(u)) = u$, and thus the c.d.f. of X is $F(x)$. The theorem is proved. □

Example 1.5

Let $F(\boldsymbol{x})$ be a multivariate distribution and

$$F(\boldsymbol{x}) = \prod_{i=1}^{s} F_i(x_i), \tag{1.2.12}$$

where $F_i(x_i),\ i = 1, \cdots, n,$ are univariate continuous c.d.f.'s. Let $\{\boldsymbol{c}_k = (c_{k1}, \cdots, c_{ks}),\ k = 1, \cdots, n\}$ be a set of points on C^s with discrepancy d, then we may show similarly that the set

$$\{\boldsymbol{x}_k = (F^{-1}(c_{k1}), \cdots, F^{-1}(c_{ks})),\ k = 1, \cdots, n\} \tag{1.2.13}$$

has also F-discrepancy d with respect to $F(\boldsymbol{x})$. The proof is left to the reader.

Recently, Bundschuh and Zhu (1993) proposed a method for exact calculation of the discrepancy of low-dimensional finite point sets. Our computing program is based on their method.

1.3 NT-nets on C^s

In 1959-1964, Korobov (1959a), Hua and Wang (1960, 1964), Halton (1960) and Hlawka (1962) proposed various methods for obtaining sets of points on C^s with lower discrepancies, and their results are contained in Hua and Wang's book (1981). In this section we shall introduce several useful methods for generating uniformly scattered sets of points on C^s. These sets are also called **NT-nets** on C^s. Some theoretical results can be found in Hua and Wang's book (1981) or in Niederreiter's new book (1992).

1.3.1 The glp set

The set obtained by a so-called **good lattice point** modulo n is called the ***glp* set** which is often used in practice and is convenient for computation.

Definition 1.4
Let $(n; h_1, \cdots, h_s)$ be a vector with integral components satisfying $1 \le h_i < n,\ h_i \ne h_j\ (i \ne j),\ s < n$ and the **greatest common divisors** $(n, h_i) = 1,\ i = 1, \cdots, s$. Let

$$\begin{cases} q_{ki} \equiv kh_i (\bmod n) \\ x_{ki} = (2q_{ki} - 1)/2n, \end{cases} \qquad k = 1, \cdots, n,\ i = 1, \cdots, s, \tag{1.3.1}$$

where we use the usual multiplicative operation modulo n such that q_{ki} is confined by $1 \le q_{ki} \le n$. Then the set $\mathcal{P}_n = \{\boldsymbol{x}_k =$

$(x_{k1}, \cdots, x_{ks})$, $k = 1, \cdots, n\}$ is called the **lattice point set** of the **generating vector** $(n; h_1, \cdots, h_s)$. If the set $\mathcal{P}_n$ has the smallest discrepancy among all possible generating vectors, then the set $\mathcal{P}_n$ is called a ***glp* set**. It can be seen that x_{ki} defined in (1.3.1) can be easily calculated by

$$x_{ki} = \left\{\frac{2kh_i - 1}{2n}\right\}, \tag{1.3.2}$$

where $\{x\}$ stands for the fractional part of x.

Example 1.6
Take $n = 7, s = 3, h_1 = 1, h_2 = 3$ and $h_3 = 6$. We have

$$(q_{kj}) = \begin{pmatrix} 1 & 3 & 6 \\ 2 & 6 & 5 \\ 3 & 2 & 4 \\ 4 & 5 & 3 \\ 5 & 1 & 2 \\ 6 & 4 & 1 \\ 7 & 7 & 7 \end{pmatrix} \qquad \begin{array}{rcl} \boldsymbol{x}_1 & = & (1/14, 5/14, 11/14) \\ \boldsymbol{x}_2 & = & (3/14, 11/14, 9/14) \\ \boldsymbol{x}_3 & = & (5/14, 3/14, 7/14) \\ \boldsymbol{x}_4 & = & (7/14, 9/14, 5/14) \\ \boldsymbol{x}_5 & = & (9/14, 1/14, 3/14) \\ \boldsymbol{x}_6 & = & (11/14, 7/14, 1/14) \\ \boldsymbol{x}_7 & = & (13/14, 13/14, 13/14) \end{array}$$

and $\{\boldsymbol{x}_k, k = 1, \cdots, 7\}$ is the lattice point set of the generating vector $(7; 1, 3, 6)$.

The lattice points on C^s of a given $(n; h_1, \cdots, h_s)$ are usually not uniformly scattered; for example, Figure 1.6 shows the four distributions of $n = 21$ points with distinct $\{h_i\}$, where the points in (a) and (b) are uniformly scattered while those in (c) and (d) are not.

Korobov (1959) and Hlawka (1962) pointed out independently that for a given prime number p, we can choose $\{h_i\}$ such that the lattice points of $\{p; h_1, \cdots, h_s\}$ are well uniformly scattered on C^s. More precisely, they proved the following:

Theorem 1.2
For any given prime number p, there exists an integral vector $\boldsymbol{h}_p = (h_1, \cdots, h_s)$ such that the lattice point set of $(p; h_1, \cdots, h_s)$ has discrepancy

$$D(p) < c(s)p^{-1}(\log p)^s.$$

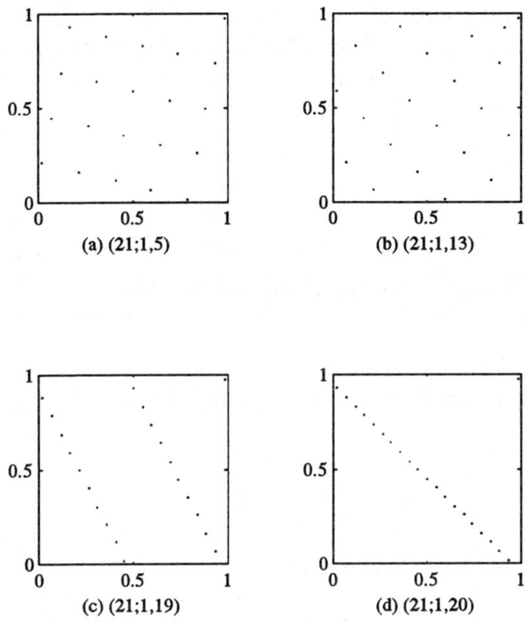

Figure 1.6 Comparisons between different generating vectors

Refer to Hua and Wang (1981) for the proof of Theorem 1.2. Finding the generating vector such that its corresponding lattice point set has lowest or lower discrepancy is our main problem. Many mathematicians contributed to this problem (Korobov (1963), Saltykov (1963), Maisonneuve (1972), Haber (1972), Moon (1974) and Wang, Xu and Zhang (1978) for the sets with comparatively large n, and Wang and Fang (1981) and Shaw (1988) for the sets with smaller n). Niederreiter (1978a) pointed out that p can be replaced by any composite number n in Theorem 1.2.

Denote the lattice point set of $(n; h_1, \cdots, h_s)$ by an $n \times s$ matrix

$$\boldsymbol{X} \equiv \boldsymbol{X}(h_1, \cdots, h_s) = (\boldsymbol{x}_1, \cdots, \boldsymbol{x}_n)'.$$

Definition 1.5
Two $n \times s$ matrices $\boldsymbol{X}$ and $\boldsymbol{Y}$ are said to be equivalent if they have the same discrepancy. In this case we denote $\boldsymbol{X} \stackrel{u}{=} \boldsymbol{Y}$.

We can easily find the following properties:

(a) Assume that $\boldsymbol{X}$ and $\boldsymbol{Y}$ are $n \times s$ matrices of two lattice points sets respectively. Then $\boldsymbol{X} \stackrel{u}{=} \boldsymbol{Y}$ if and only if there exist two permutation matrices $\boldsymbol{P} : n \times n$ and $\boldsymbol{Q} : s \times s$ such that $\boldsymbol{Y} = \boldsymbol{PXQ}$. A square matrix $\boldsymbol{P}$ is called a **permutation matrix** if $\boldsymbol{P}'\boldsymbol{P} = \boldsymbol{I}$ and the elements of $\boldsymbol{P}$ are all ones and zeros. In this case, $\boldsymbol{Y}$ can be obtained by exchanging rows and columns of $\boldsymbol{X}$

(b) Since each integer h $(0 < h < n)$ has an inverse h^{-1} modulo n such that $hh^{-1} = 1 \pmod n$ if $(n, h) = 1$, we have

$$\boldsymbol{X}(h_1, \cdots, h_s) \stackrel{u}{=} \boldsymbol{X}(1, h_1^{-1}h_2, \cdots, h_1^{-1}h_s) \pmod n. \quad (1.3.3)$$

Thus we can always assume $h_1 = 1$ for $(h_1, \cdots, h_s)$.

Property (b) will be proved in Chapter 5 and it can be used to save computing time for choosing the good generating vector $(n; h_1^*, \cdots, h_s^*)$. For instance, we have C_{p-1}^s choices of $\{h_1, \cdots, h_s\}$ among $\{1, 2, \cdots, p-1\}$ for the case of $n = p$ being prime number. Since we can take $h_1 = 1$, the number of choices of $\{1, h_2, \cdots, h_s\}$ reduces to C_{p-1}^{s-1}. Even so, when p and s are large the calculation is still heavy for obtaining a good generating vector $\{h_1^*, \cdots, h_s^*\}$. Hence, Korobov(1959b) suggested to consider $(h_1, \cdots, h_s)$ to be the form

$$(h_1, \cdots, h_s) \equiv (1, a, a^2, \cdots, a^{s-1}) \pmod p, \quad (1.3.4)$$

where $1 < a < p$, and he proved that there exists an integer $a \pmod p$ such that the discrepancy of the lattice points corresponding to $(1, a, \cdots, a^{s-1}) \pmod p$ is also $D(p) \le c(s)p^{-1}(\log p)^s$. Niederreiter (1977) pointed out that a can be chosen among the **primitive roots** modulo p, i.e. a satisfies

$$a^i \not\equiv a^j \pmod p, \quad 1 \le i < j < p.$$

For example, 3 is a primitive root modulo 7, and 2 is not, because

$$3 \equiv 3,\ 3^2 \equiv 2,\ 3^3 \equiv 6,\ 3^4 \equiv 4,\ 3^5 \equiv 5,\ 3^6 \equiv 1 \pmod 7$$

and

$$2 \equiv 2,\ 2^2 \equiv 4,\ 2^3 \equiv 1,\ 2^4 \equiv 2,\ 2^5 \equiv 4,\ 2^6 \equiv 1 \pmod 7.$$

Niederreiter proved that for any given prime p, there is a primitive root g such that the discrepancy of lattice points generated by $(1, g, \cdots, g^{s-1})$ is

$$D(p) \le c(s) p^{-1} (\log p)^s \log \log p.$$

Since the lattice point set generated by the vectors of the form (1.3.4) can be the well uniformly scattered set, we can find the best g among the primitive roots modulo p such that the corresponding lattice point set, which is still called the *glp* set, has the smallest discrepancy. When $n = 2, 4, p^l, 2p^l$, where p is an odd prime and $l \ge 1$, we know that there always exist primitive roots mod n, and that the number of primitive roots mod n is $\phi(\phi(n))$, where $\phi(n)$ is the **Euler ϕ-function** which denotes the number of m satisfying $1 \le m < n$, and $(m, n) = 1$. Note that, $\phi(\phi(p^l)) = p^{l-2}(p-1)\phi(p-1)$ (Hua (1956), Chap. 3). Most generating vectors in the literature are obtained in this way. When the generating vector $(n; h_1, \cdots, h_s)$ is obtained by another way, the condition $(n, h_i) = 1, i = 1, \cdots, s$ may be not satisfied, but the corresponding lattice point set is still called the *glp* set if its discrepancy is low. Since the *glp* set is convenient to use for lower discrepancy and *glp* sets have already been derived, we often use *glp* sets in practice and collect them in Appendix A of this book.

In the rest of the book when $(h_1, \cdots, h_s)$ has the form (1.3.4) we write $\boldsymbol{X}(a)$ instead of $\boldsymbol{X}(h_1, \cdots h_s)$. Furthermore, we have the following property:

(c) For given n and s, $\boldsymbol{X}(a) \stackrel{u}{=} \boldsymbol{X}(a^{-1}) \pmod n$ if $a^i \neq a^j \pmod n$, $1 \le i < j \le s-1$.

Remark 1.4

Let p be a prime. We can prove that for any $1 > \varepsilon > 0$, the number of integers a satisfying $1 \le a \le p$ such that the *glp* set generated by $(p; 1, a, \cdots, a^{s-1})$ has discrepancy

$$D(p) = O(p^{-1} \log^{2s-1} p)$$

is greater than $(1 - \varepsilon)p$ (Hua and Wang (1981), Theorem 3.2 and Lemma 7.8). Hence we may even choose the integer a randomly when p is very large.

1.3.2 The gp set.

The set obtained by a so-called **good point** is called a ***gp* set**.

Definition 1.6

Let $\boldsymbol{\gamma} = (\gamma_1, \cdots, \gamma_s) \in R^s$. If the first n terms $\mathcal{P}_n$ of the set of the form

$$\{(\{\gamma_1 k\}, \cdots, \{\gamma_s k\}), \quad k = 1, 2, \cdots\} \tag{1.3.6}$$

has discrepancy

$$D(n, \mathcal{P}_n) \le c(\boldsymbol{\gamma}, \varepsilon) n^{-1+\varepsilon}, \quad n = 1, 2, \cdots$$

then the set (1.3.6) is called a ***gp* set** and $\boldsymbol{\gamma}$ a **good point**.

Obviously, $\mathcal{P}_n$ is an NT-net on C^s by Definition 1.3. By Definition 1.6, the *gp* set is easily obtained if we have a good point $\boldsymbol{\gamma}$. Baker (1965) and Schmidt (1970) gave the existence of *gp* set. The following good points are recommended for practice:

(a) The **square root sequence**: We take

$$\boldsymbol{\gamma} = (\sqrt{p_1}, \cdots, \sqrt{p_s}), \tag{1.3.7}$$

where p_j's are different primes, and e.g. the first s primes.

(b) Let p be a prime and $q = p^{1/(s+1)}$. Take

$$\gamma = (q, q^2, \cdots, q^s). \tag{1.3.8}$$

(c) The **cyclotomic field method**: This method was suggested by Hua and Wang (1964) with

$$\gamma = \left(\left\{2\cos\frac{2\pi}{p}\right\}, \left\{2\cos\frac{4\pi}{p}\right\}, \cdots, \left\{2\cos\frac{2\pi s}{p}\right\}\right), \tag{1.3.9}$$

where p is a prime $\geq 2s+3$. For example, we have

$$\gamma_1 = \left\{2\cos\frac{2\pi}{7}\right\} \cong 0.2469796037,$$
$$\gamma_2 = \left\{2\cos\frac{4\pi}{7}\right\} \cong 0.5549581321,$$

for the case $p = 7$, and the respective *gp* set is

$$(0.247, 0.555), (0.494, 0.110), (0.741, 0.665), (0.988, 0.220),$$
$$(0.235, 0.775), (0.482, 0.330), (0.729, 0.885), (0.976, 0.440), \cdots$$

Using the rational approximations to $\gamma_i (1 \leq i \leq s)$ defined by (1.3.9), Hua and Wang (1964) proposed a method for obtaining the lattice point sets. Many generating vectors contained in the appendix of their book(1981) are obtained by their method. These sets have discrepancies $D(n) = O(n^{-\frac{1}{2}-\frac{1}{2(s-1)}+\varepsilon})$. Therefore, their method is good for high dimension only.

1.3.3 The H-set

Halton (1960) first proposed a generalization of the Van der Corput (1953) set of points on C^2 to a set on $C^s (s > 2)$. We call this set the ***H*-set**. Halton's method is based on the p-adic representation of natural numbers.

Let m be a natural number ≥ 2. Then any natural number k has a unique **m-digits representation**

$$\begin{aligned} &k = b_0 + b_1 m + b_2 m^2 + \cdots + b_r m^r, \\ &0 \leq b_i \leq m-1, \ i = 0, 1, \cdots, r, \end{aligned} \tag{1.3.10}$$

where $m^r \leq k < m^{r+1}$. For any $c \in (0,1)$ c has a unique m-digits representation

$$c = c_0 m^{-1} + c_1 m^{-2} + \cdots, \ 0 \leq c_i \leq m-1, \ i = 0, 1, 2, \cdots.$$

We write $k = b_r b_{r-1} \cdots b_1 b_0$ and $c = 0.c_0 c_1 \cdots$. It can be established a one-to-one correspondence between the positive integers and the rational numbers in $(0,1)$ as follows: For any integer $k \geq 1$ with representation (1.3.10), let

$$y_m(k) = b_0 m^{-1} + b_1 m^{-2} + \cdots + b_r m^{-r-1}. \tag{1.3.11}$$

Then $y_m(k) \in (0,1)$ is called the **radical inverse of k with base m**. Halton proposed the following set:

Let $p_i (1 \leq i \leq s)$ be s distinct prime numbers. Then

$$\boldsymbol{x}_k = (y_{p_1}(k), \cdots, y_{p_s}(k)), \ k = 1, 2, \cdots \tag{1.3.12}$$

is called an ***H*-set**.

Halton proved that the set formed by the first n ($> \max(p_1, \cdots, p_s)$) points of (1.3.12) has discrepancy

$$D(n) \leq n^{-1} \prod_{i=1}^{s} \frac{p_i \log(p_i n)}{p_i} = O(n^{-1} (\log n)^s). \tag{1.3.13}$$

Hence the H-set is well uniformly scattered on C^s.

Example 1.7
Take $m = 2$. Then the respective binary-representations of k and $y_2(k)$, for $k = 1, 2, \cdots$, are as following:

k	b_0	b_1	b_2	b_3	$y_2(k)$
1	1				0.5
2	0	1			0.25
3	1	1			0.75
4	0	0	1		0.125
5	1	0	1		0.625
6	0	1	1		0.375
7	1	1	1		0.875
8	0	0	0	1	0.0625

Take $p_1 = 2$, $p_2 = 3$ and $p_3 = 5$. Then the corresponding H-set is

$$
\begin{array}{ll}
(0.5000, 0.3333, 0.2000) & (0.8125, 0.7037, 0.2800) \\
(0.2500, 0.6667, 0.4000) & (0.1875, 0.1481, 0.4800) \\
(0.7500, 0.1111, 0.6000) & (0.6875, 0.4815, 0.6800) \\
(0.1250, 0.4444, 0.8000) & (0.4375, 0.8148, 0.8800) \\
(0.6250, 0.7778, 0.0400) & (0.9375, 0.2593, 0.1200) \\
(0.3750, 0.2222, 0.2400) & (0.0313, 0.5926, 0.3200) \\
(0.8750, 0.5556, 0.4400) & (0.5313, 0.9259, 0.5200) \\
(0.0625, 0.8889, 0.6400) & (0.2813, 0.0741, 0.7200) \\
(0.5625, 0.0370, 0.8400) & (0.7813, 0.4074, 0.9200) \\
(0.3125, 0.3704, 0.0800) & (0.1563, 0.7407, 0.1600) \quad \cdots
\end{array}
$$

Since the computations of (1.3.10) and (1.3.11) are quite heavy when n and s are sufficiently large, we may calculate the $y_p(k)$ $(k = 1, 2, \cdots)$ for the smaller primes $p = 2, 3, \cdots$ and then keep them on disk for the purpose of practical use.

Using the H-set, we may obtain finite sets with still lower discrepancies than that of H-set. Let $s \geq 2$ and $p_1, \cdots, p_{s-1}$ be $s - 1$ distinct prime numbers. Then the set that is called the **Hammersley set** (Hammersley (1960)),

$$
\boldsymbol{x}_k = \left(\frac{2k-1}{2n}, \; y_{p_1}(k), \cdots, y_{p_{s-1}}(k)\right), \; k = 1, 2, \cdots, n
$$

has discrepancy

$$D(n) \leq n^{-1} \prod_{i=1}^{s-1} \frac{p_i \log(p_i n)}{\log p_i}. \tag{1.3.14}$$

There are several variations of the H-set. One is the so-called **scrambled Halton set** or **scrambled H-set.** Let π_p be a permutation on the digits $(1, 2, \cdots, p)$. We define the scrambled radical inverse $y_{p,\sigma}(k)$ in analogy with $y_p(k)$:

$$y_{p,\sigma}(k) = \pi_p(b_0)/p + \pi_p(b_1)/p^2 + \cdots + \pi_p(b_r)/p^{r+1} \tag{1.3.15}$$

if k has the expansion $k = b_r \cdots b_1 b_0$ (base p). The scrambled Halton set is then given by

$$\boldsymbol{x}_k = (y_{p_1,\sigma_1}(k), \cdots, y_{p_s,\sigma_s}(k)), \quad k = 1, 2, \cdots \tag{1.3.16}$$

where σ_i is a permutation of $(1, 2, \cdots, p_i)$, $i = 1, \cdots, s$. Note that the H-set is just the special case in which each π_p is chosen to be the identity permutation. Braaten and Weller (1979) pointed out that the scrambled H-set can significantly improve the following behavior of the H-set. The sequence $y_p(k)$, $k = 1, 2, \cdots$ consists of cycles of length p of monotonically increasing numbers. We usually choose $p_1, \cdots, p_s$ to be the first s primes. For example, in the case $s = 8$, the last two coordinates are given by $y_{17}(k)$ and $y_{19}(k)$, and consist of 17 and 19 increasing numbers, respectively. This produces a strong correlation between the seventh and eighth coordinates of the sequence. Braaten and Weller recommended some permutations for the scrambled H-set for $p \leq 53$. They also pointed out that the scrambled H-set has lower discrepancy than that of the H-set.

Another way to extend the H-set is to define a **radical inverse p-adic addition** by setting:

$$\begin{aligned} x \oplus_p y =& 0.x_1 \cdots x_n \cdots \text{ regular } p\text{-adic expansion of } x \\ & \oplus_p 0.y_1 \cdots y_n \cdots \text{ regular } p\text{-adic expansion of } y \\ =& 0.\overline{x_1 + y_1}^p \cdots \overline{x_n + y_n}^p \cdots \end{aligned}$$

The result $x \oplus_p y$ is computed as in classical addition with carrying over, but from the left to the right (for example $0.123333\cdots \oplus_p 0.412777\cdots = 0.535011\cdots$). For every $p \geq 2$, $y = \sum_{k\geq 0} y_k/p^{k+1} \in [0,1] \cap Q_p$ (Q_p are the p-adic rationals), define

$$\Phi_{p,y}(x) = x \oplus_p y.$$

Now we can define a **Halton-like NT-net sequence** on C^s: If $p_1, \cdots, p_s$ are the first s prime numbers, $y_1, \cdots, y_s \in [0,1] \cap Q_p$ and $z_1, \cdots, z_s \in [0,1] \cap Q_p$, the sequence defined by

$$\boldsymbol{x}_k = (\Phi^{k-1}_{p_1,y_1}(z_1), \cdots, \Phi^{k-1}_{p_s,y_s}(z_s)), \quad k = 1, 2, \cdots$$

has a discrepancy in its first n terms which is bounded above by $O(n^{-1} \log^s n)$ (Lapeyre and Pagès (1989)).

Another version of the *H*-set is called the **Faure sequence** which has still lower discrepancy than that of the *H*-set (Faure (1982)). Niederreiter (1988) proposed some low-discrepancy sequences by the theory of (t, s)-sequences. The reader can find more detailed discussion in Niederreiter (1992).

1.3.4 Miscellaneous

There are many kinds of NT-nets on C^s which are not included in the above three methods. For example, the so-called **Haber sequence** (Haber (1970)) is defined by

$$\boldsymbol{x}_k = \left(\frac{k(k+1)}{2}\sqrt{p_1}, \cdots, \frac{k(k+1)}{2}\sqrt{p_s}\right) \pmod 1,$$

where $p_1, \cdots, p_s$ are the first s primes. Another sequence, defined in Sobol (1967), is based on the radical inverse also. For more details see Sobol (1967) and Niederreiter (1988b). The reader can find more methods for generating an NT-net on C^s in Hua and Wang (1981), Shaw (1988) and Pagès and Xiao (1991).

The above three main kinds of sets, the *glp* set, *gp* set and *H*-set have many variants and wide applications in practice (Hua and Wang (1981) and Shaw (1988)). The *glp* set is a finite set while the *gp* set and *H*-set have infinite number of elements. Each set has its advantages and also shortcomings, and the choice in practice should be in accordance with experience. We would give the following recommendations:

(a) If n is comparatively small, it is better to use the *glp* set. The other two sets are not usually uniformly scattered.
(b) Suppose one first generates an NT-net of n_1 points and then finds that in fact $n_2(> n_1)$ points are needed. With the *glp* set method one must start from scratch since the generating vectors for n_2 and n_1 are usually different. However if either the *gp* set or H-set method was used, then only an additional $n_2 - n_1$ points have to be generated.
(c) In these methods, the *H*-set method has the heaviest computational burden. When p_i and s are large, the number of points in the set must be quite large to insure a uniform scattering on C^s. Therefore the *H*-set is suitable only for small s.
(d) When s is large, for example $s > 10$, there are only a few generating vectors of lattice points contained in Appendix A of our book, and no generating vectors for $s > 18$. In the latter case, we suggest using the *gp* set method, in particular, the *gp* set obtained by the integral basis of cyclotomic field.
(e) When $s \leq 10$, the *glp* set is best for most practical cases.

Table 1.1 gives comparison among 7 NTM's, where *gp*-1 is the *gp*-set with γ (1.3.7), *gp*-2 the *gp*-set with γ (1.3.8), Hua-Wang the cyclotomic field method with γ (1.3.9). The values in the table are the discrepancies for different methods with $s = 2$ and various n. From the table we can see that the *glp*-set is the best for all cases, the Hammersley set is the next one in most cases and the Haber set is the worst in most cases. In section 1.4 we will give further comparisons among these methods by another criterion.

Table 1.1 *Comparisons among various methods*

n	34	89	144	233	610	1597	2584
gp-1	0.0770	0.0469	0.0249	0.0239	0.0152	0.0048	0.0028
gp-2	0.0693	0.0484	0.0253	0.0180	0.0086	0.0050	0.0037
Hua-Wang	0.1853	0.0786	0.0481	0.0350	0.0205	0.0072	0.0053
Haber	0.1644	0.0947	0.0838	0.0604	0.0327	0.0222	0.0217
Halton	0.1106	0.0432	0.0328	0.0207	0.0109	0.0052	0.0030
Hammersley	0.0864	0.0406	0.0251	0.0169	0.0073	0.0031	0.0020
glp	0.0642	0.0276	0.0181	0.0120	0.0050	0.0021	0.0013

1.4 Other measures of uniformity

Let $\mathcal{P} = \{\boldsymbol{x}_k, k = 1, \cdots, n\}$ be a set of points on C^s. There are several measurements of **uniformity** of the set $\mathcal{P}$ on C^s. We introduced one of the most important measures, the discrepancy, in section 1.2. Now we shall introduce some other measures.

1.4.1 The star discrepancy D^*

Note that only rectangles of the type $[\mathbf{0}, \boldsymbol{\gamma}]$ are considered in (1.2.3) for discrepancy. We may of course consider all the rectangles $[\boldsymbol{a}, \boldsymbol{b})(\mathbf{0} \leq \boldsymbol{a} \leq \boldsymbol{b} \leq \mathbf{1})$ in the definition of discrepancy.

Definition 1.7
Let $\mathcal{P} = \{\boldsymbol{x}_k\}$ be a set of n points on C^s and let $\mathcal{K}$ be a set of rectangles

$$\mathcal{K} = \{[\boldsymbol{a}, \boldsymbol{b}) : \mathbf{0} \leq \boldsymbol{a} \leq \boldsymbol{b} \leq \mathbf{1}\}. \tag{1.4.1}$$

Further let $N([\boldsymbol{a}, \boldsymbol{b}), \mathcal{P})$ be the number of points of $\mathcal{P}$ satisfying $\boldsymbol{x}_k \in [\boldsymbol{a}, \boldsymbol{b})$. Then

$$D^*(n, \mathcal{P}) = D^*(\mathcal{P}) = \sup_{[\boldsymbol{a},\boldsymbol{b}) \in \mathcal{K}} \left| \frac{N([\boldsymbol{a}, \boldsymbol{b}), \mathcal{P})}{n} - v([\boldsymbol{a}, \boldsymbol{b}))\right| \tag{1.4.2}$$

is called the **star discrepancy** of $\mathcal{P}$, where $v([\boldsymbol{a}, \boldsymbol{b})) = \prod_{i=1}^{s}(b_i - a_i)$ is the volume of $[\boldsymbol{a}, \boldsymbol{b})$.

The following theorem shows that these two definitions of discrepancy are essentially the same.

Theorem 1.3

$$D(n) \leq D^*(n) \leq 2^s D(n). \tag{1.4.3}$$

The discrepancy and star discrepancy can be expressed by the terminology of **measure theory**. Let ν and ν_n be the measures generated by $F(\boldsymbol{x})$ and $F_n(\boldsymbol{x})$ in s-dimensional Lebesgue-measurable space, respectively, where $F(\boldsymbol{x})$ is the c.d.f. of the uniform distribution $U(C^s)$, and $F_n(\boldsymbol{x})$ is the empirical distribution of $\{\boldsymbol{x}_k\}$. Then

$$D(n, \mathcal{P}) = \sup_{\gamma \in C^s} |\nu_n(B_\gamma) - \nu(B_\gamma)|, \tag{1.4.4}$$

where $B_\gamma = [\mathbf{0}, \gamma)$ and

$$D^*(n, \mathcal{P}) = \sup_{B \in \mathcal{K}} |\nu_n(B) - \nu(B)|. \tag{1.4.5}$$

In the next section we will find that these expressions can be easily extended to more general case.

1.4.2 The ℓ_p-discrepancy

Another measure for uniformity is called the **$\boldsymbol{\ell_p}$-discrepancy** defined by

$$T_p(n, \mathcal{P}) = \left[\int_{C^s} |\nu_n(B_\gamma) - \nu(B_\gamma)|^p d\gamma \right]^{1/p}. \tag{1.4.6}$$

When $p = 1, 2$ and ∞ we have **$\boldsymbol{\ell_1}$-, $\boldsymbol{\ell_2}$- and $\boldsymbol{\ell_\infty}$-discrepancy**, the latter is just the common discrepancy $D(n, \mathcal{P})$. Niederreiter (1978, p.971) pointed out that

$$C(s)(D^*(n, \mathcal{P}))^{(s+2)/2} \leq T_2(n, \mathcal{P}) \leq D(n, \mathcal{P}), \tag{1.4.7}$$

where $C(s) > 0$ is a constant depending only on s. Therefore, $D(n)$, $D^*(n)$ and $T_2(n, \mathcal{P})$ are essentially equivalent.

1.4.3 Dispersion

Definition 1.8

Let D be a closed and bounded domain and $\mathcal{P} = \{\boldsymbol{x}_k\}$ be a set of n points on D. Then

$$DP(\mathcal{P}, D) = \max_{\boldsymbol{x} \in D} \min_{1 \leq k \leq n} d(\boldsymbol{x}, \boldsymbol{x}_k) \tag{1.4.8}$$

is called the **dispersion** of $\mathcal{P}$ on D, where $d(\boldsymbol{x}, \boldsymbol{x}_k)$ denotes the Euclidean distance between $\boldsymbol{x}$ and $\boldsymbol{x}_k$. If $D = C^s$, then we write $DP(\mathcal{P}, D) = DP(\mathcal{P})$.

The concept of the dispersion of a set was introduced and studied by Niederreiter (1983). This idea is reasonable: For each $\boldsymbol{x} \in D$, we find a point in $\mathcal{P}$ to represent $\boldsymbol{x}$ which has the least distance with $\boldsymbol{x}$. If $\mathcal{P}$ is scattered uniformly on D, then the distance between $\boldsymbol{x}$ and its representation can be expected to be small. The computational burden calculating the dispersion is heavy, since we should find the maximum of $\min_{1 \le k \le n} d(\boldsymbol{x}, \boldsymbol{x}_k)$ for all $\boldsymbol{x} \in D$. We often use the NTM to estimate the dispersion, although sometimes we can compute $DP(\mathcal{P}, D)$ directly when s is small or when the distribution of the points of $\mathcal{P}$ is regular in D.

Example 1.8
Suppose that $D = C^s$ and $\mathcal{P}$ is the set of equi-distribution stated in Example 1.2. Then the dispersion of $\mathcal{P}$ is equal to the diameter of an s-dimensional cube with edge-length $(2m)^{-1}$, where $n = m^s$, and so

$$DP(\mathcal{P}, D) = \frac{\sqrt{s}}{2m} = \frac{\sqrt{s}}{2} n^{-1/s}. \tag{1.4.9}$$

The exact value of dispersion is in general not easy to obtain, but the estimations of its upper and lower bounds can be found. First of all, we introduce the following equivalent definition for dispersion:

Let $B(\boldsymbol{x}, r)$ denote the s-dimensional ball with centre $\boldsymbol{x}$ and radius r. Then

$$DP(\mathcal{P}, D) = \min\{r \ge 0 : \cup_{1 \le k \le n} B(\boldsymbol{x}_k, r) \supset D\}. \tag{1.4.10}$$

This means that dispersion is equal to the least radius r such that the union of the n balls with centers $\boldsymbol{x}_k (1 \le k \le n)$ and radius r forms a covering of D.

Theorem 1.4
Let $v(D)$ denote the volume of D. Then

$$DP(\mathcal{P}, D) \geq \left(\frac{v(D)}{\gamma_s}\right)^{1/s} n^{-1/s}, \tag{1.4.11}$$

where $\gamma_s = \pi^{s/2}/\Gamma(1+s/2)$ is the volume of the s-dimensional unit ball.

PROOF Let $r = DP(\mathcal{P}, D)$. Then by (1.4.10) we have

$$\cup_{1\leq k\leq n} B(\boldsymbol{x}_k, r) \supset D$$

and so

$$nr^s\gamma_s \geq v(\cup_{1\leq k\leq n} B(\boldsymbol{x}_k, r)) \geq v(D).$$

The theorem follows. □

The following theorem gives a relation between the dispersion and discrepancy of a set $\mathcal{P}$ on C^s, and hence it induces an upper estimation for $DP(\mathcal{P}, D)$ by the upper estimation of $D(n, \mathcal{P})$.

Theorem 1.5
If $D = C^s$, then we have

$$DP(\mathcal{P}, D) \leq 2\sqrt{s}\, D(n, \mathcal{P})^{1/s}. \tag{1.4.12}$$

The proof of Theorem 1.5 will be given in section 3.1. In particular, it follows from Theorem 1.5 that if $\mathcal{P}$ is well uniformly scattered on D, then for each $\varepsilon > 0$

$$DP(\mathcal{P}, D) = O(n^{-1/s+\varepsilon}). \tag{1.4.13}$$

Note that the Example 1.8 shows that the dispersion of the set of equi-distribution attains the order $O(n^{-1/s})$ which is better than the right hand side of (1.4.13). Therefore the distribution of the set of equi-distribution is uniform in accordance with the measurement of dispersion. This means that the different conclusions can

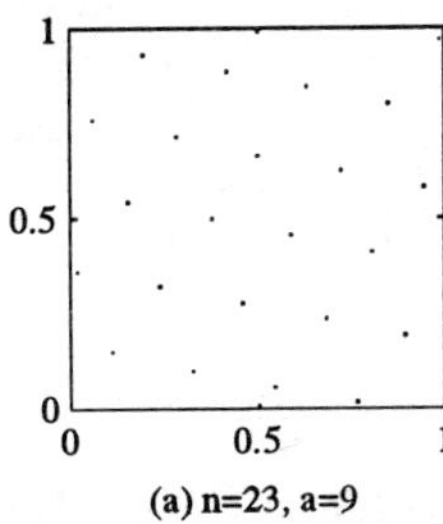

(a) n=23, a=9

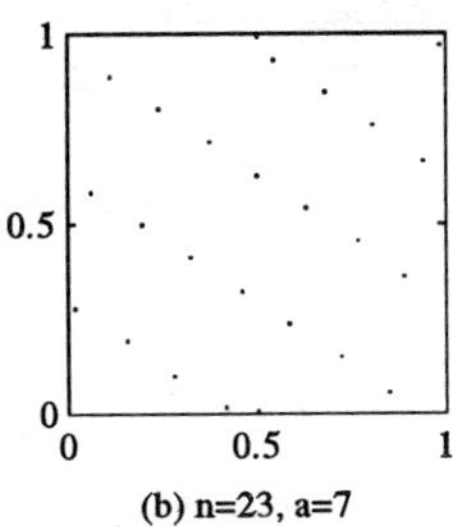

(b) n=23, a=7

Figure 1.7 The two *glp* sets

be obtained by the use of distinct measurements for uniformity. However the set of equi-distribution is still disadvantageous because the number of points of this set is of the form $n = m^s$ which increases fast with m increasing, in particular, when s is large.

1.4.4 Mean square error (MSE)

Suppose that $\mathcal{P} = \{\boldsymbol{x}_k, k = 1, \cdots, n\}$ is a set on C^s. Then for any $\boldsymbol{x} \in C^s$, we have $\boldsymbol{x}_{k(\boldsymbol{x})}$ such that

$$d(\boldsymbol{x}, \boldsymbol{x}_{k(\boldsymbol{x})}) = \min_{1 \leq j \leq n} d(\boldsymbol{x}, \boldsymbol{x}_j), \tag{1.4.14}$$

i.e., $\boldsymbol{x}_{k(\boldsymbol{x})}$ is the representative of $\boldsymbol{x}$. We use the integral of $d(\boldsymbol{x}, \boldsymbol{x}_{k(\boldsymbol{x})})^2$ over C^s to be the measurement for uniformity of $\mathcal{P}$ on C^s. This is called the **mean square error (MSE)** of $\mathcal{P}$:

$$\text{MSE}(\mathcal{P}) = \int_{C^s} \min_{1 \leq j \leq n} d(\boldsymbol{x}, \boldsymbol{x}_j)^2 d\boldsymbol{x}. \tag{1.4.15}$$

The quantity MSE has been widely used in the theory of quantization (section 4.4). MSE($\mathcal{P}$) can be calculated by the method introduced in the next chapter. In the past the estimation of discrepancy was obtained usually by neglecting some terms of lower orders, this may lead to a large error when n is small. For example, if we want to choose a *glp* set $\mathcal{P}$ of the type (1.3.4) for the case $s = 2$ and $n = 23$, then we obtain $a = 9$ by minimizing MSE($\mathcal{P}$) for all possible $\mathcal{P}$, and $a = 7$ by the minimum of discrepancies of all sets $\mathcal{P}$ (Fang, Wang and Yuan (1990)). If we calculate the exact values of discrepancy instead of their approximations we obtain $D(23,7) = 0.0903$ and $D(23,9) = 0.0846$. This shows that the approximate formula for discrepancy may lead to a big error. The plots of these two cases are given in Fig. 1.7.

1.4.5 Sample moments

Suppose that $\{\boldsymbol{y}_k, k = 1, \cdots, n\}$ is a sample of $U(C^s)$, the uniform distribution over C^s. Then its sample moments should be close to the corresponding moments of $U(C^s)$. Let $\{\boldsymbol{c}_k, k = 1, \cdots, n\}$ be an NT-net on C^s. Since the empirical distribution function generated by $\{\boldsymbol{c}_k\}$ is close to the c.d.f. of $U(C^s)$ (section 1.2), the sample moments defined by $\{\boldsymbol{c}_k\}$ should close also to the corresponding moments of $U(C^s)$.

We consider for simplicity the case $s = 1$. The r-th moment about the origin of $U(0,1)$ is

$$\mu_r = \frac{1}{r+1}, \quad r = 1, 2, \cdots, \tag{1.4.16}$$

and the r-th sample moment about the origin of $\{c_k\}$ is

$$m_r = \frac{1}{n} \sum_{j=1}^{n} c_j^r, \quad r = 1, 2, \cdots. \tag{1.4.17}$$

From Example 1.1, it follows that $\{c_k = \frac{2k-1}{2n}, k = 1, \cdots, n\}$ is the set on [0, 1] with the lowest discrepancy, and we obtain

$$m_r = \frac{1}{n^{r+1}} \sum_{k=1}^{n} (k - 0.5)^r, \quad r = 1, 2, \cdots.$$

Let $\{x_i, i = 1, \cdots, n\}$ be a random sample of $U(0,1)$. We use ν_r to denote the r-th sample moment of $\{x_i\}$. Figure 1.8 gives the comparison between $\{m_r\}$ and $\{\nu_r\}$ for $n = 30, 100$, where the dashed curve denotes the values of $\nu_r - \mu_r$ while the solid curve denotes the values of $m_r - \mu_r$. The figure shows that m_r is much

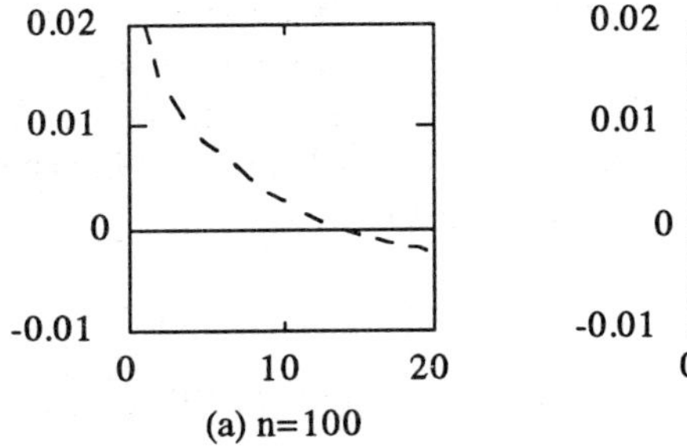

(a) n=100

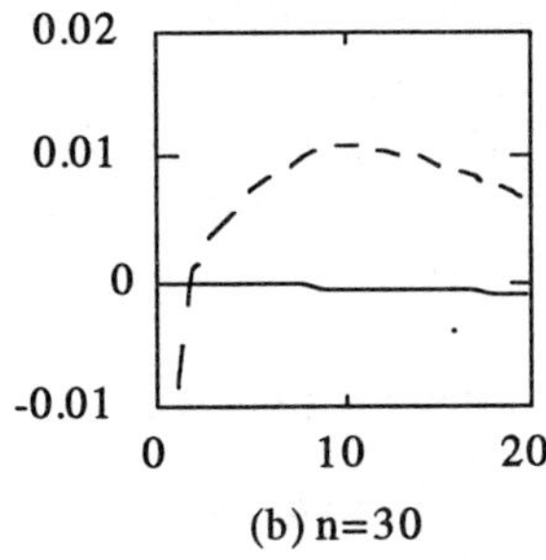

(b) n=30

Figure 1.8 Sample moments

closer to μ_r than ν_r, so that the set obtained by NTM is superior than those given by random numbers.

In Table 1.2 we give comparisons of the methods in Table 1.1 for $s = 3$ and various n. If $\boldsymbol{x} \sim U(C^3)$, then the mixed moment $E(X_iX_j) = 1/4 = 0.25$ for $i \neq j$ and $1 \leq i,j \leq 3$. The values in the table are the summation of absolute values of difference between the sample moments of $E(X_iX_j)$ and 0.25 for the seven methods. We can see from Table 1.2 that the *glp*-set is not always the best one from the sample moment point of view.

Table 1.2 *Comparisons among various methods*

n	35	135	266	597	1626	2440
gp-1	0.0062	0.0156	0.0077	0.0020	0.0014	0.0015
gp-2	0.0386	0.0255	0.0262	0.0178	0.0013	0.0037
H-W	0.0619	0.0119	0.0073	0.0024	0.0015	0.0013
Haber	0.0706	0.0318	0.0171	0.0120	0.0073	0.0084
Halton	0.0815	0.0254	0.0168	0.0086	0.0034	0.0024
Hammersley	0.0305	0.0116	0.0074	0.0053	0.0034	0.0012
glp	0.0094	0.0183	0.0126	0.0054	0.0011	0.0011

1.5 The NT-nets on a ball, a sphere and a simplex

We have introduced the methods for finding the NT-nets on C^s in the previous sections. Now we want to find NT-nets on D where D is the ball, sphere, simplex, etc. This is a more difficult problem. For example, if we want to obtain a set of n points that is uniformly

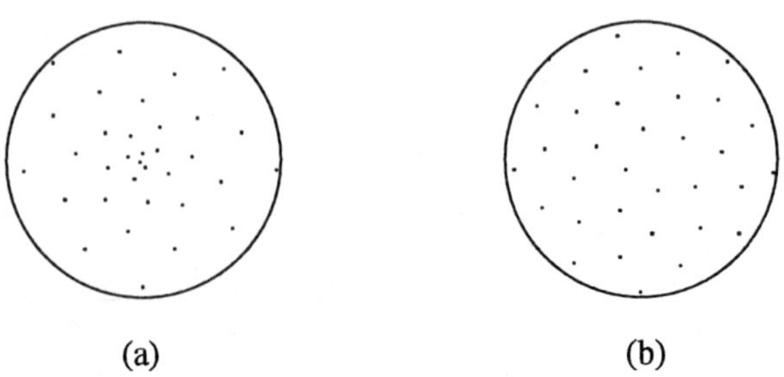

Figure 1.9 Two sets of points on B_2

scattered on a unit disk:

$$B_2 = \{(x, y) : x^2 + y^2 \leq 1\}, \tag{1.5.1}$$

we might use the polar coordinates

$$\begin{cases} x = r\cos(2\pi\theta), \\ y = r\sin(2\pi\theta), \end{cases} \tag{1.5.2}$$

where $(r, \theta) \in C^2$. Take an NT-net $\mathcal{P}_n = \{(r_k, \theta_k), k = 1, \cdots, n\}$ on C^2, and then we obtain n points

$$\begin{aligned} x_k &= r_k \cos(2\pi\theta_k), \\ y_k &= r_k \sin(2\pi\theta_k), \qquad i = 1, \cdots, n, \end{aligned} \tag{1.5.3}$$

on B_2 by substituting $\mathcal{P}_n$ into (1.5.2). It is obvious that these points are not scattered uniformly on B_2 (Figure 1.9 (a)) and that most of these points are concentrated around the center of the B_2.

Now we explain the reason for this phenomenon. Suppose that (X, Y) is a random vector which is uniformly distributed on B_2. Then the (r, θ) defined by (1.5.2) is also a random vector. It can be easily shown that r and θ are independent with respective p.d.f.'s

$$\begin{aligned} p_r(x) &= 2x, \ \ 0 \leq x \leq 1, \\ p_\theta(x) &= 1, \quad 0 \leq x \leq 1, \end{aligned}$$

and c.d.f.'s

$$\begin{aligned} F_r(x) &= x^2, \ 0 \le x \le 1, \\ F_\theta(x) &= x, \quad 0 \le x \le 1. \end{aligned}$$

The (r, θ) is is not the uniform distribution on C^2, and so the set of points (1.5.3) is not scattered uniformly on B_2. Inspired by the Monte Carlo method for generating random variables of continuous distribution, we substitute $\{(F_r^{-1}(r_k), F_\theta^{-1}(\theta_k))\} = \{(\sqrt{r_k}, \theta_k)\}$, instead of $\{(r_k, \theta_k)\}$, into (1.5.3), and obtain

$$\begin{aligned} x_k &= \sqrt{r}_k \cos(2\pi\theta_k), \\ y_k &= \sqrt{r}_k \sin(2\pi\theta_k), \end{aligned} \qquad k = 1, \cdots, n. \tag{1.5.4}$$

We see from the Figure 1.9 (b) that the points (1.5.4) are scattered quite uniformly.

Why is the set $\mathcal{P}_F = \{(x_k, y_k)\}$ uniformly scattered on B_2? Suppose that $\mathcal{P} = \{(r_k, \theta_k)\}$ has a low discrepancy d on C^2, then it is easy to show that $\{(\sqrt{r_k}, \theta_k)\}$ has a F-discrepancy d with respect to $F(x, y) = F_r(x)F_\theta(y)$ (Example 1.5).

Given $(x, y) \in B_2$, there is a $(q, \phi) \in C^2$ such that

$$x = \sqrt{q}\cos(2\pi\phi), \quad y = \sqrt{q}\sin(2\pi\phi).$$

We use $N((r, \theta), \mathcal{P})$ to denote the number of points in $\mathcal{P}$ falling inside the rectangle $[0, r] \times [0, \theta]$ which is equal to the number of points in $\mathcal{P}_F$ belonging to the fan-shaped region

$$\begin{aligned} G = \{(x, y) : x = \tilde{q}\cos(2\pi\phi), \ y = \tilde{q}\sin(2\pi\phi), \\ 0 \le \tilde{q} \le \sqrt{r}, \ 0 \le \phi \le \theta\}. \end{aligned} \tag{1.5.5}$$

The areas of the rectangle $[0, r] \times [0, \theta]$ and the fan-shaped region G are $r\theta$ and $\pi r\theta$ respectively (Fig. 1.10). Therefore, we have

$$\begin{aligned} \left| \frac{N((r, \theta), \mathcal{P})}{n} - r\theta \right| &= \left| \frac{N(G, \mathcal{P}_F)}{n} - \frac{v(G)}{u(B_2)} \right| \\ &= \left| \frac{N(G, \mathcal{P}_F)}{n} - \frac{\pi r\theta}{\pi} \right|, \end{aligned} \tag{1.5.6}$$

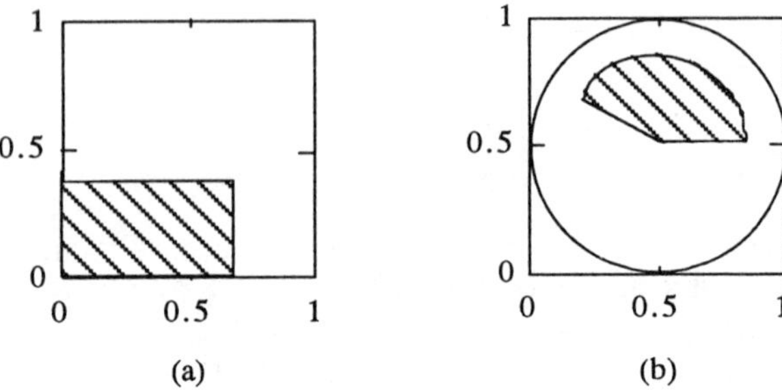

Figure 1.10 Illustration of transformation from C^2 to B_2

where $N(G, \mathcal{P}_F)$ denotes the number of points of $\mathcal{P}_F$ falling in G. If we use the fan-shaped region to replace the rectangle in the definition of F-discrepancy, the formula (1.5.6) gives a measure for uniformity of $\mathcal{P}_F$ on B_2. So we need to extend the concept of F-discrepancy.

Definition 1.9

Let $\boldsymbol{x} \in R^s$ be a random vector with a c.d.f. $F(\boldsymbol{x})$ and have a stochastic representation $\boldsymbol{x} = \boldsymbol{h}(\boldsymbol{z})$, where $\boldsymbol{z} \sim U(C^t), t \leq s$. Let $\{\boldsymbol{c}_k, k = 1, \cdots, n\}$ be an NT-net on C^t with discrepancy d. Then we call that the set $\mathcal{P}_F = \{\boldsymbol{h}(\boldsymbol{c}_k), k = 1, \cdots, n\}$ has **quasi F-discrepancy** d with respect to $F(\boldsymbol{x})$.

For practical use the following statement is useful. Let $\boldsymbol{x}$ be an s-dimensional random vector and have a stochastic representation

$$\boldsymbol{x} = \boldsymbol{x}(\boldsymbol{\phi}), \tag{1.5.7}$$

where $\boldsymbol{\phi} = (\phi_1, \cdots, \phi_t)$ is a t-dimensional random vector with independent components. Denote by $F(\boldsymbol{x})$ and $G(\boldsymbol{\phi})$ the c.d.f.'s of $\boldsymbol{x}$ and $\boldsymbol{\phi}$ respectively. Let $\{\boldsymbol{y}_k, k = 1, \cdots, n\}$ be a set of points on R^t with F-discrepancy d with respect to G. Then the set

$$\mathcal{P}_F = \{\boldsymbol{x}(\boldsymbol{y}_k), k = 1, \cdots, n\} \tag{1.5.8}$$

has quasi F-discrepancy d with respect to F.

When $\boldsymbol{x} \sim U(D)$, where D is a closed and bounded domain in R^s and $U(D)$ the uniform distribution on D we shall argue whether the quasi F-discrepancy gives a reasonable measure of uniformity on D. The p.d.f. of $\boldsymbol{x}$ is given by

$$p(\boldsymbol{x}) = \begin{cases} 1/v(D), & \text{if } \boldsymbol{x} \in D, \\ 0, & \text{if } \boldsymbol{x} \notin D. \end{cases}$$

From Appendix B, the joint density of the corresponding $\phi_1, \cdots, \phi_t$ is

$$\begin{cases} \frac{1}{v(D)} J(\boldsymbol{\phi}), & \text{if } \boldsymbol{\phi} \in C^t, \\ 0, & \text{otherwise,} \end{cases}$$

where

$$J(\boldsymbol{\phi}) = \det(\boldsymbol{T}\boldsymbol{T}')^{1/2} \quad \text{and} \quad \boldsymbol{T} = \left(\frac{\partial x_j}{\partial \phi_i}\right).$$

From the assumption we have

$$\frac{1}{v(D)} J(\boldsymbol{\phi}) = \prod_1^t p_i(\phi_i)$$

where $p_i(\phi_i)$ is the p.d.f. of ϕ_i, or

$$\frac{1}{v(D)} \int_{\boldsymbol{\phi} \leq \boldsymbol{r}} J(\boldsymbol{\phi}) d\boldsymbol{\phi} = \prod_1^t F_i(r_i),$$

where F_i is the c.d.f. of ϕ_i and $\boldsymbol{r} = (r_1, \cdots, r_t)$.

Given a set $\mathcal{P} = \{\boldsymbol{c}_k,\ k = 1, \cdots, n\}$ on C^t with discrepancy d, then the set $\mathcal{P}_F = \{\boldsymbol{x}_k,\ k = 1, \cdots, n\}$, where

$$\boldsymbol{x}_k = \boldsymbol{x}(F_1^{-1}(c_{k1}), \cdots, F_t^{-1}(c_{kt})), \quad k = 1, \cdots, n, \tag{1.5.9}$$

has a quasi F-discrepancy d by Definition 1.9. For each $\boldsymbol{r} \in C^t$, let

$$G_{\boldsymbol{r}} = \{\boldsymbol{x} : \boldsymbol{x} = \boldsymbol{x}(\boldsymbol{\phi}),\ \boldsymbol{\phi} \leq \boldsymbol{r}\} \tag{1.5.10}$$

be a domain. It is clear that

$$P(\boldsymbol{x} \in G_{\boldsymbol{r}}) = v(G_{\boldsymbol{r}})/v(D),$$

as $\boldsymbol{x} \sim U(D)$. Let $N(\mathcal{P}_F, G_{\boldsymbol{r}})$ denote the number of points $\boldsymbol{x}_k$ falling in $G_{\boldsymbol{r}}$. Then

$$GD(\mathcal{P}_F, D) = \sup_{\boldsymbol{r}\in C^t} \left| \frac{N(\mathcal{P}_F, G_{\boldsymbol{r}})}{n} - \frac{v(G_{\boldsymbol{r}})}{v(D)} \right| \tag{1.5.11}$$

gives a measure for uniformity of $\mathcal{P}_F$ on D. We shall prove that $GD(\mathcal{P}_F, D)$ is just the quasi F-discrepancy. It is easy to see that if $\boldsymbol{c}_k$ falls in the rectangle $R_{\boldsymbol{r}} = \{\boldsymbol{\phi} : \boldsymbol{0} \leq \boldsymbol{\phi} \leq \boldsymbol{r}\}$, then $\boldsymbol{x}_k = \boldsymbol{x}(\boldsymbol{c}_k)$ falls in $G_{\boldsymbol{r}}$. Therefore, $N(\mathcal{P}, \boldsymbol{r}) = N(\mathcal{P}_F, G_{\boldsymbol{r}})$, where $N(\mathcal{P}, \boldsymbol{r})$ denote the number of points $\{\boldsymbol{c}_k\}$ falling in $R_{\boldsymbol{r}}$, and

$$\left|\frac{N(\mathcal{P}_F, G_{\boldsymbol{r}})}{n} - \frac{v(G_{\boldsymbol{r}})}{v(D)}\right| = \left|\frac{N(\mathcal{P}, \boldsymbol{r})}{n} - \frac{v(G_{\boldsymbol{r}})}{v(D)}\right| = \left|\frac{N(\mathcal{P}, \boldsymbol{r})}{n} - \prod_1^t F_i(r_i)\right|,$$

where the last step follows by

$$v(G_{\boldsymbol{r}}) = \int_{\boldsymbol{\phi}\leq\boldsymbol{r}} J(\boldsymbol{\phi}) d\boldsymbol{\phi} = v(D) \prod_{i=1}^t F_i(r_i).$$

Thus

$$\begin{aligned} GD(\mathcal{P}_F, D) &= \sup_{\boldsymbol{r}\in C^t} \left| \frac{N(\mathcal{P}_F, G_{\boldsymbol{r}})}{n} - \prod_{i=1}^t F_i(r_i) \right| \\ &= \sup_{\boldsymbol{r}\in C^t} \left| \frac{N(\mathcal{P}_F, G_{\boldsymbol{r}})}{n} - F(\boldsymbol{r}) \right|, \end{aligned}$$

where $F(\boldsymbol{r}) = \prod_{i=1}^t F_i(r_i)$. The right hand side is just the F-discrepancy with respect to $F(\boldsymbol{r})$ and equals to the discrepancy of $\mathcal{P} = \{\boldsymbol{c}_k\}$. We summarize the above argument with the following theorem.

Theorem 1.6

Let D be a closed and bounded domain in R^s and $\boldsymbol{x} \sim U(D)$. Suppose $\boldsymbol{x}$ has a stochastic representation (1.5.7) where $\boldsymbol{\phi}$ is a t-dimensional random vector with independent marginal p.d.f.

$p_i(\phi_i)$ and c.d.f. $F_i(\phi_i)$. Let $\mathcal{P} = \{\boldsymbol{c}_k\}$ be a set on C^t with discrepancy d. Then the set $\mathcal{P}_F$ defined by (1.5.9) has a quasi F-discrepancy d which equals $GD(\mathcal{P}_D, D)$ as defined in (1.5.11).

The method of generating the NT-nets on D mentioned in Theorem 1.6 is called the **inverse transform method**. By applying the method to some specific domains D we can find sets of points uniformly scattered on D, i.e. **NT-nets on D**. In particular, we consider in this section the following domains:

$$A_s = \{(x_1, \cdots, x_s) : 0 \le x_1 \le \cdots \le x_s \le 1\} \tag{1.5.12}$$
$$B_s = \{(x_1, \cdots, x_s) : x_1^2 + \cdots + x_s^2 \le 1\} \tag{1.5.13}$$
$$U_s = \{(x_1, \cdots, x_s) : x_1^2 + \cdots + x_s^2 = 1\} \tag{1.5.14}$$
$$V_s = \{(x_1, \cdots, x_s) \in R_+^s : x_1 + \cdots + x_s \le 1\} \tag{1.5.15}$$
$$T_s = \{(x_1, \cdots, x_s) \in R_+^s : x_1 + \cdots + x_s = 1\} \tag{1.5.16}$$

where

$$R_+^s = \{(x_1, \cdots, x_s) : x_i \ge 0, i = 1, \cdots, s\} \tag{1.5.17}$$

is the nonnegative part of R^s.

1.5.1 NT-net on A_s

Let $\boldsymbol{x} = (X_1, \cdots, X_s)$ be a random vector which is uniformly distributed on the simplex A_s. Let

$$\begin{cases} X_1 & = \phi_1\phi_2\cdots\phi_s \\ X_2 & = \phi_2\phi_3\cdots\phi_s \\ & \cdots \\ X_{s-1} & = \phi_{s-1}\phi_s \\ X_s & = \phi_s, \end{cases} \tag{1.5.18}$$

where $\boldsymbol{\phi} = (\phi_1, \cdots, \phi_s) \in C^s$. This transformation maps C^s into A_s. We shall show in Appendix B.1 that

(a) $\phi_1, \cdots, \phi_s$ are mutually independent, and
(b) ϕ_j has p.d.f. $j\phi^{j-1}$ and c.d.f.

$$F_j(\phi) = \phi^j, \tag{1.5.19}$$

for $j = 1, \cdots, s$, where $0 \le \phi \le 1$. The inverse function of $F_j(\phi)$ is

$$F_j^{-1}(\phi) = \phi^{1/j}.$$

Suppose that $\{\boldsymbol{c}_k\}$ is an NT-net on C^s, then $\{\boldsymbol{x}_k\}$ is NT-net on A_s, where $\boldsymbol{x}_k = (x_{k1}, \cdots, x_{ks})$ and

$$\begin{cases} x_{k1} = c_{k1} c_{k2}^{1/2} \cdots c_{ks}^{1/s}, \\ x_{k2} = c_{k2}^{1/2} \cdots c_{ks}^{1/s}, \\ \cdots \\ x_{ks} = c_{ks}^{1/s}, \end{cases} \qquad k = 1, 2, \cdots, n. \tag{1.5.20}$$

We suggest that the reader tries Exercise 1.12 of this chapter to see the distributions of sets (1.5.20).

1.5.2 NT-net on B_s

Let $\boldsymbol{x}$ be a random vector which is uniformly distributed on the unit ball B_s. We use the **spherical coordinate transformation (SCT)**

$$\begin{aligned} X_j &= \phi_1 S_2 \cdots S_j C_{j+1}, \ j = 1, \cdots s-1 \\ X_s &= \phi_1 S_2 \cdots S_{s-1} S_s, \end{aligned} \tag{1.5.21}$$

where

$$\begin{aligned} S_i &= \sin(\pi\phi_i), \quad C_i = \cos(\pi\phi_i), \ i = 1, \cdots, s-1, \\ S_s &= \sin(2\pi\phi_s), \ C_s = \cos(2\pi\phi_s), \end{aligned}$$

in which $\boldsymbol{\phi} = (\phi_1, \cdots, \phi_s) \in C^s$. This maps C^s into B_s. It follows from Appendix B.2 that

(a) $\phi_1, \cdots, \phi_s$ are mutually independent, and
(b) ϕ_j has p.d.f.

$$p_j(\phi) = \begin{cases} s\phi^{s-1}, \ \text{if } j = 1, \\ \dfrac{\pi(\sin(\pi\phi))^{s-j}}{B(\frac{1}{2}, \frac{s-j+1}{2})}, \qquad \text{if } j = 2, \cdots, s, \end{cases}$$

where $B(a,b)$ denotes the beta function. In particular we have $p_s(\phi) = 1$ which shows that ϕ_s is uniformly distributed over [0, 1]. The c.d.f. of ϕ_j is

$$F_j(\phi) = \begin{cases} \phi^s, & \text{if } j = 1 \\ \dfrac{\pi}{B(\frac{1}{2}, \frac{s-j+1}{2})} \int_0^\phi (\sin \pi x)^{s-j} dx, & \text{if } j = 2, \cdots, s. \end{cases} \tag{1.5.22}$$

Since the analytic expression of the integral in the above formula has many terms if s is large and j is small, it is difficult to find a simple expression of the inverse function of $F_j(\phi)$.

By the inverse transform method, we may obtain the uniformly scattered set on B_s: Let $\{\boldsymbol{c}_k\}$ be an NT-net on C^s. Compute

$$\begin{aligned} b_{k1} &= c_{k1}^{1/s}, \\ b_{ki} &= F_i^{-1}(c_{ki}), \qquad k = 1, \cdots, n, i = 2, \cdots, s, \end{aligned} \tag{1.5.23}$$

where $\{b_{ki}\}$ can be determined by some approximation formula (Yamauti (1972)). Let

$$\begin{aligned} x_{kj} &= b_{k1} \prod_{i=2}^{j} S_{ki} C_{k,j+1}, j = 1, \cdots, s-1, \\ x_{ks} &= b_{k1} \prod_{i=2}^{s} S_{ki}, \end{aligned} \tag{1.5.24}$$

where

$$\begin{aligned} S_{ki} &= \sin(\pi b_{ki}),\ C_{ki} = \cos(\pi b_{ki}), \qquad i = 2, \cdots, s-1, \\ S_{ks} &= \sin(2\pi b_{ks}),\ C_{ks} = \cos(2\pi b_{ks}), \qquad k = 1, \cdots, n. \end{aligned}$$

The set $\{\boldsymbol{x}_k = (x_{k1}, \cdots, x_{ks}),\ k = 1, \cdots, n\}$ is an NT-net on B_s.

Figure 1.11 gives the distributions of two such sets $\{\boldsymbol{x}_k\}$ obtained by the *glp* sets generated by (31; 1, 12) and (233; 1, 144). These two sets appear to be scattered uniformly on B_2.

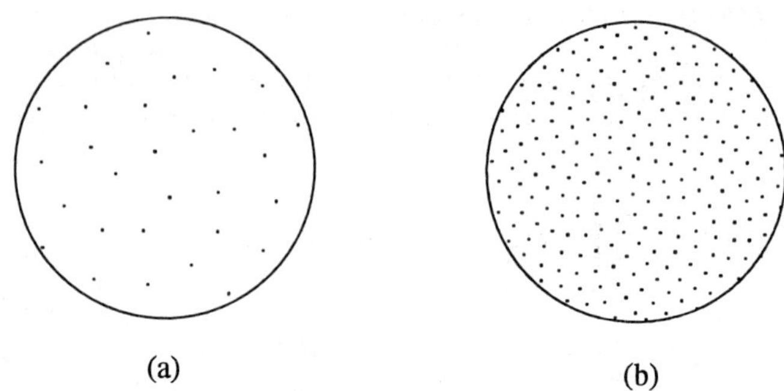

Figure 1.11 Two NT-nets on B_2

Example 1.9

We often meet the problems which need us to provide an NT-net on B_3. By (1.5.22) we know that the c.d.f.'s of $\phi_i (i = 1, 2, 3)$ are

$$\begin{aligned} F_1(\phi) &= \phi^3, \\ F_2(\phi) &= \frac{1}{2}(1 - \cos(\pi\phi)), \\ F_3(\phi) &= \phi. \end{aligned} \tag{1.5.25}$$

If $\{\boldsymbol{c}_k = (c_{k1}, c_{k2}, c_{k3}),\ k = 1, \cdots, n\}$ is an NT-net on C^3, then by (1.5.23) and (1.5.25) we obtain

$$b_{k1} = c_{k1}^{1/3},\ b_{k2} = \frac{1}{\pi} \arccos(1 - 2c_{k2}),\ b_{k3} = c_{k3}$$

and consequently

$$\begin{aligned} x_{k1} &= c_{k1}^{1/3}(1 - 2c_{k2}), \\ x_{k2} &= b_{k1} \sin(\pi b_{k2}) \cos(2\pi b_{k3}) \\ &= 2\sqrt{c_{k2}(1 - c_{k2})}\, c_{k1}^{1/3} \cos(2\pi c_{k3}), \\ x_{k3} &= b_{k1} \sin(\pi b_{k2}) \sin(2\pi b_{k3}) \\ &= 2\sqrt{c_{k2}(1 - c_{k2})}\, c_{k1}^{1/3} \sin(2\pi c_{k3}) \end{aligned} \tag{1.5.26}$$

by (1.5.24). The set $\{\boldsymbol{x}_k = (x_{k1}, x_{k2}, x_{k3}),\ k = 1, \cdots, n\}$ is an NT-net on B_3.

1.5.3 NT-net on U_s

Let $\boldsymbol{x}$ be a random vector which is uniformly distributed on the unit sphere U_s. We use the spherical coordinate transformation

$$\begin{aligned} X_j &= \prod_{i=1}^{j-1} S_i C_j, \qquad j = 1, \cdots, s-1, \\ X_s &= \prod_{i=1}^{s-1} S_i, \end{aligned} \tag{1.5.27}$$

where

$$\begin{aligned} S_i &= \sin(\pi\phi_i), \qquad C_i = \cos(\pi\phi_i), \qquad i = 1, \cdots, s-2, \\ S_{s-1} &= \sin(2\pi\phi_{s-1}), \qquad C_{s-1} = \cos(2\pi\phi_{s-1}). \end{aligned}$$

This transformation maps C^{s-1} into U_s. It follows from Appendix B.2 that

(a) $\phi_1, \cdots, \phi_{s-1}$ are mutually independent, and
(b) The p.d.f. and c.d.f. of ϕ_j are

$$p_j(\phi) = \frac{\pi}{B(\frac{1}{2}, \frac{s-j}{2})} (\sin(\pi\phi))^{s-j-1}$$

and

$$F_j(\phi) = \int_0^\phi p_j(t)dt,$$

respectively for $j = 1, \cdots, s-1$. Then using a similar algorithm to that for finding the NT-net on B_s, we obtain an NT-net on U_s.

When $s = 2$, there is only one random variable ϕ_1 that is uniformly distributed on [0, 1] with c.d.f. $F_1(\phi) = \phi$. By Example 1.1, we know that $Q = \{(2k-1)/2n : k = 1, \cdots, n\}$ is the set on [0, 1] with the lowest discrepancy $1/2n$, and so the set

$$\begin{aligned} x_{k1} &= \cos(2\pi(2k-1)/(2n)), \\ x_{k2} &= \sin(2\pi(2k-1)/(2n)), \qquad k = 1, \cdots, n \end{aligned}$$

is uniformly scattered on the unit circle which coincides with our intuition.

When $s = 3$, there are two independent random variables ϕ_1 and ϕ_2 with c.d.f.'s

$$F_1(\phi) = \frac{1}{2}(1 - \cos \pi\phi) \text{ and } F_2(\phi) = \phi$$

respectively. If $\{(c_{k1}, c_{k2}), k = 1, \cdots, n\}$ is an NT-net on C^2, then by a simple calculation the set $\{\boldsymbol{x}_k = (x_{k1}, x_{k2}, x_{k3}), k = 1, \cdots, n\}$ is an NT-net on U_3, where

$$\begin{aligned} x_{k1} &= 1 - 2c_{k1}, \\ x_{k2} &= 2\sqrt{c_{k1}(1 - c_{k1})}\cos(2\pi c_{k2}), \\ x_{k3} &= 2\sqrt{c_{k1}(1 - c_{k1})}\sin(2\pi c_{k2}), \;\; k = 1, \cdots, n, \end{aligned} \tag{1.5.28}$$

which is different from the Monte Carlo method of Marsaglia (1972). In section 4.3 we shall introduce an alternative method for generating an NT-net on B_s as well as on U_s which is more efficient.

1.5.4 NT-net on V_s

Let $\boldsymbol{x}$ be a random variable which is uniformly distributed on the unit l_1-ball V_s. Let

$$\begin{aligned} X_i &= \phi_1 \prod_{j=2}^{i} S_j^2 C_{i+1}^2, \qquad i = 1, \cdots, s-1, \\ X_s &= \phi_1 \prod_{j=2}^{s} S_j^2, \end{aligned} \tag{1.5.29}$$

where

$$S_i = \sin(\pi\phi_i/2), \;\; C_i = \cos(\pi\phi_i/2), \;\; i = 2, \cdots, s. \tag{1.5.30}$$

This transformation maps C^s into V_s. By Appendix B.3, we know that

(a) $\phi_1, \cdots, \phi_s$ are mutually independent, and
(b) the p.d.f. and c.d.f. of ϕ_j are

$$p_j(\phi) = \begin{cases} s\phi^{s-1}, & \text{if } j = 1, \\ \pi(s-j+1)\Big(\sin\left(\frac{\pi\phi}{2}\right)\Big)^{2s-2j+1} \cos\left(\frac{\pi\phi}{2}\right), & \text{if } j = 2, \cdots, s, \end{cases}$$

and

$$F_j(\phi) = \begin{cases} \phi^s, & \text{if } j = 1, \\ (\sin(\frac{\pi\phi}{2}))^{2(s-j+1)}, & \text{if } j = 2, \cdots, s. \end{cases}$$

Let $\{\boldsymbol{c}_k = (c_{k1}, \cdots, c_{ks}), k = 1, \cdots, n\}$ be an NT-net on C^s. Let

$$b_{kj} = F_j^{-1}(c_{kj}), \qquad j = 1, \cdots, s, \; k = 1, \cdots, n.$$

Then $\{\boldsymbol{x}_k = (x_{k1}, \cdots, x_{ks}), \; k = 1, \cdots, n\}$ is an NT-net on V_s, where

$$x_{ki} = F_1^{-1}(c_{k1}) \prod_{j=2}^{i-1} \Big(\sin\left(\frac{\pi b_{kj}}{2}\right)\Big)^2 \Big(\cos\left(\frac{\pi b_{k,i+1}}{2}\right)\Big)^2,$$
$$i = 1, \cdots, s-1,$$
$$x_{ks} = F_1^{-1}(c_{k1}) \prod_{j=2}^{s} \Big(\sin\left(\frac{\pi b_{kj}}{2}\right)\Big)^2.$$

Since

$$b_{kj} = \frac{2}{\pi} \arcsin(c_{kj}^{\frac{1}{2(s-j+1)}}), \; j = 2, \cdots, s,$$

we have

$$\sin^2\left(\frac{\pi b_{kj}}{2}\right) = c_{kj}^{\frac{1}{s-j+1}},$$
$$\cos^2\left(\frac{\pi b_{kj}}{2}\right) = 1 - c_{kj}^{\frac{1}{s-j+1}}, \; j = 2, \cdots, s,$$

and therefore we have the following simplifications for the x_{kj}:

$$
\begin{aligned}
x_{ki} &= c_{k1}^{1/s} \prod_{j=2}^{i} c_{kj}^{\frac{1}{s-j+1}} (1 - c_{k,i+1}^{\frac{1}{s-i}}), \qquad i = 1, \cdots, s-1. \\
x_{ks} &= c_{k1}^{1/s} \prod_{j=2}^{s} c_{kj}^{\frac{1}{s-j+1}}, \qquad k = 1, \cdots, n.
\end{aligned}
\tag{1.5.31}
$$

When $s = 2$, we have

$$
\begin{aligned}
x_{k1} &= \sqrt{c_{k1}}(1 - c_{k2}), \\
x_{k2} &= \sqrt{c_{k1}} c_{k2}, \qquad k = 1, \cdots, n.
\end{aligned}
$$

1.5.5 NT-net on T_s

Let $\boldsymbol{x}$ be a random vector which is uniformly distributed on the simplex T_s. Let

$$
\begin{aligned}
X_i &= \prod_{j=1}^{i-1} S_j^2 C_i^2, \qquad i = 1, \cdots, s-1, \\
X_s &= \prod_{j=1}^{s-1} S_j^2.
\end{aligned}
\tag{1.5.32}
$$

where

$$
S_i = \sin\left(\frac{\pi \phi_i}{2}\right), \; C_i = \cos\left(\frac{\pi \phi_i}{2}\right), \; i = 1, \cdots, s-1.
$$

This transformation maps C^{s-1} into T_s. It follows from Appendix B.3 that

(a) $\phi_1, \cdots, \phi_{s-1}$ are mutually independent, and
(b) the c.d.f. of ϕ_j is

$$
F_j(\phi) = \left(\sin\left(\frac{\pi \phi}{2}\right)\right)^{2(s-j)}
$$

for $j = 1, 2, \cdots, s-1$.

Let $\{\boldsymbol{c}_k,\ k = 1, \cdots, n\}$ be an NT-net on C^{s-1}. Then similar to the set on V_s, we know that $\{\boldsymbol{x}_k, k = 1, \cdots, n\}$ is an NT-net on T_s, where

$$\begin{aligned} x_{ki} &= \prod_{j=1}^{i-1} c_{kj}^{\frac{1}{s-j}} (1 - c_{ki}^{\frac{1}{s-i}}), \quad i = 1, 2, \cdots, s-1, \\ x_{ks} &= \prod_{j=1}^{s-1} c_{kj}^{\frac{1}{s-j}}, \quad k = 1, \cdots, n. \end{aligned} \tag{1.5.33}$$

1.5.6 NT-net on a closed and bounded domain D

Let D be a closed and bounded domain in R^s with dimension s. If D has a parametric representation (1.5.7) and satisfies the conditions mentioned above, then we may use the inverse transform method to find the set of points that is uniformly scattered on D. If D does not satisfy the above requirements, then we propose the following algorithm: Let $D \subset [\boldsymbol{a}, \boldsymbol{b}]$. Suppose that $\{\boldsymbol{c}_k : k = 1, \cdots, n\}$ is uniformly scattered on C^s. Define

$$b_{kj} = a_j + (b_j - a_j)c_{kj},\ j = 1, \cdots, s,\ k = 1, \cdots, n. \tag{1.5.34}$$

Then $\{\boldsymbol{b}_k = (b_{k1}, \cdots, b_{ks}),\ k = 1, \cdots, n\}$ is uniformly scattered on $[\boldsymbol{a}, \boldsymbol{b}]$. Let $\mathcal{P}_m = \{\boldsymbol{x}_k : k = 1, \cdots, m\}$ be the set of points of $\{\boldsymbol{b}_k\}$ that lie in the domain D. Then $\mathcal{P}_m$ is approximately scattered uniformly on D (Figure 1.12). This algorithm is simple, efficient and in many cases practical, but it needs some further modifications when the dimension of D is less than s. When D has a parametric representation (1.5.7), as in the cases of A_s, B_s and U_s, the inverse transform method is usually better than this algorithm.

1.6 Other methods

Let $\boldsymbol{x} \sim U(D)$ where D is an s-dimensional domain, $F_1(x)$ be the c.d.f. of X_1, and $F(x_i|x_1, \cdots, x_{i-1})$ the conditional c.d.f. of X_i given $X_j = x_j$, $j = 1, \cdots, i-1$; $i = 2, \cdots, s$. Set

$$\begin{cases} Y_1 = F_1(X_1), \\ Y_i = F_i(X_i|X_1, \cdots, X_{i-1}), \quad i = 2, \cdots, s. \end{cases} \tag{1.6.1}$$

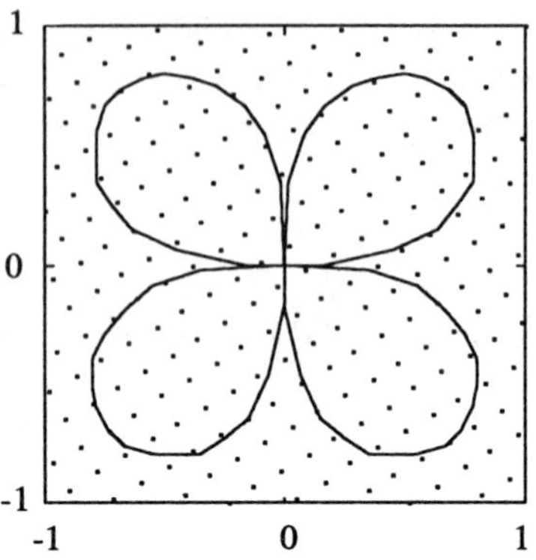

Figure 1.12 An NT-net on D

This is the so-called **Rosenblatt transformation** (Rosenblatt (1952)). By an argument similar to the proof of Theorem 1.1 we can show that each Y_i has a uniform distribution $U(0,1)$ which does not depend on $Y_1, \cdots, Y_{i-1}$, and hence $Y_1, \cdots, Y_s$ are i.i.d. $U(0,1)$. Clearly, the transformation $(X_1, \cdots, X_s) \to (Y_1, \cdots, Y_s)$ is one-to-one and maps D into C^s. The Jacobian of $(X_1, \cdots, X_s) \to (Y_1, \cdots, Y_s)$ is the p.d.f. of $\boldsymbol{x}$.

Zhang (1991) uses the Rosenblatt transformation to find the corresponding transformations for A_s, B_s, U_s, V_s, and T_s, which are essentially the same as given in section 1.5 and suggested by Wang and Fang (1990a).

Note that the representation of the inverse transform method is not unique. In the inverse transform method we assume that $\boldsymbol{x}$ is uniformly distributed on D and can be represented by

$$\boldsymbol{x} = \boldsymbol{f}(\boldsymbol{z}), \qquad \boldsymbol{z} \in C^t, \qquad t \leq s,$$

where $\boldsymbol{z} = (Z_1, \cdots, Z_t)$ has mutually independent components with respective c.d.f. F_j, $j = 1, \cdots, t$. Let

$$\boldsymbol{F}^{-1} = (F_1^{-1}, \cdots, F_t^{-1})$$

Then by Theorem 1.1, it follows that $\boldsymbol{y} = \boldsymbol{F}(\boldsymbol{z})$ is uniformly distributed on C^t, and

$$\boldsymbol{x} = \boldsymbol{f}(\boldsymbol{F}^{-1}(\boldsymbol{y})).$$

Since the stochastic representation of $\boldsymbol{x}$ is not unique we may have many different sets from different representations. Let us examine the following two examples.

Example 1.10

Let m be the integer satisfying $s \le 2m \le s+1$, and suppose that the random vector $\boldsymbol{y}$ is distributed uniformly on C^{2m}. Set

$$\begin{aligned} Z_{2i-1} &= \sqrt{-2\log Y_{2i-1}}\cos(2\pi Y_{2i}), \\ Z_{2i} &= \sqrt{-2\log Y_{2i-1}}\sin(2\pi Y_{2i}), \qquad i = 1, \cdots, m. \end{aligned} \tag{1.6.2}$$

This is so-called **Box-Muller transformation** (**BMT**) (Rubinstein (1981)) and we can prove that $\boldsymbol{z} = (Z_1, \cdots, Z_s)' \sim N_s(\boldsymbol{0}, \boldsymbol{I}_s)$, the s-dimensional standard normal distribution. It then follows from (2.2.23) below that

$$\boldsymbol{x} = \frac{\boldsymbol{z}}{\sqrt{\boldsymbol{z}'\boldsymbol{z}}} \tag{1.6.3}$$

is uniformly distributed on U_s. By (1.6.2) and (1.6.3), we know that $\boldsymbol{x}$ can be expressed as a function of $\boldsymbol{y}$. Write $\boldsymbol{x} = \boldsymbol{g}(\boldsymbol{y})$. Then we can use the inverse transform method to find an NT-net on U_s. This algorithm (called Method II in this section) was proposed by Sibuya (1962) and Shaw (1988) and is good only for large n. Roughly speaking the spherical coordinate transformation method (called Method I in this section) introduced in section 1.5.3 needs only an $(s-1)$-dimensional set of points while Method II requires an s-dimensional set of points when s is even and an $(s+1)$-dimensional set of points when s is odd. Since U_s has dimension $s-1$ it is better to produce an NT-net on U_s from an $(s-1)$-dimensional set of points.

Let us consider the comparison between the spherical coordinate transformation method (called Method I) and the BMT method (called Method II) in terms of the sample moments (section 1.4). Tables 1.3 and 1.4 compare the sample moments of the sets obtained by these two methods, where $n^{(l)}$ denotes the number of points of the set in Method l and $M_{ijk}^{(l)}$ the sample moment of $E(X_iX_jX_k)$ of the set obtained by Method l, for $l = 1, 2,$. For example, $M_{22}^{(1)}$ is the sample moment of $E(X_2X_2) = E(X_2^2)$ obtained by Method I. These two tables show that Method I is superior than Method II for most cases, in particular when s is an odd number. Some further discussion will be given in section 4.3.

Table 1.3 *Comparison between two methods* ($s = 3$, $n^{(1)} = 377$, $n^{(2)} = 440$)

$M_1^{(1)} = -0.00001$	$M_2^{(1)} = 0.00002$	$M_3^{(1)} = 0.00004$
$M_1^{(2)} = -0.00181$	$M_2^{(2)} = 0.00363$	$M_3^{(2)} = -0.00720$
	$EX_i = 0$	
$M_{11}^{(1)} = 0.33332$	$M_{22}^{(1)} = 0.33332$	$M_{33}^{(1)} = 0.33334$
$M_{11}^{(2)} = 0.33718$	$M_{22}^{(2)} = 0.33956$	$M_{33}^{(2)} = 0.32324$
	$EX_i^2 = 1/3$	
$M_{111}^{(1)} = -0.00001$	$M_{222}^{(1)} = -0.00000$	$M_{333}^{(1)} = -0.00000$
$M_{111}^{(2)} = -0.00118$	$M_{222}^{(2)} = 0.00054$	$M_{333}^{(2)} = -0.00649$
	$EX_i^3 = 0$	
$M_{1111}^{(1)} = 0.19999$	$M_{2222}^{(1)} = 0.20000$	$M_{3333}^{(1)} = 0.20000$
$M_{1111}^{(2)} = 0.20176$	$M_{2222}^{(2)} = 0.20595$	$M_{3333}^{(2)} = 0.19088$
	$EX_i^4 = 1/5$	

Example 1.11

Suppose that $\boldsymbol{y} = (Y_1, \cdots, Y_s)$ is uniformly distributed on C^s, i.e. $Y_1, \cdots, Y_s$ i.i.d. and $Y_i \sim U(0,1)$. We know that $F(x) = 1 - e^{-x}$ is the c.d.f. of the standard exponential distribution. Hence $F^{-1}(Y_1), \cdots, F^{-1}(Y_s)$ are i.i.d. and each has the **exponential distribution**. Let

$$\boldsymbol{z} = (Z_1, \cdots, Z_s) = (F^{-1}(Y_1), \cdots, F^{-1}(Y_s)).$$

Then

$$\boldsymbol{x} = (X_1, \cdots, X_s) = \frac{\boldsymbol{z}}{||\boldsymbol{z}||} \sim U(T_s), \tag{1.6.4}$$

where $||\boldsymbol{z}|| = Z_1 + \cdots + Z_s$ is the ℓ_1-norm of $\boldsymbol{z}$ (Fang, Kotz and Ng (1990), p.113). Therefore using the fact that Y_i and $1 - Y_i$ have the same distribution $U(0,1)$ and the formula

$$X_i \stackrel{d}{=} \frac{\log Y_i}{\sum_{j=1}^{s} \log Y_i}, \qquad i = 1, \cdots, s, \tag{1.6.5}$$

where $\stackrel{d}{=}$ means of the two sides of equality have the same distribution, we may obtain the set of points that is uniformly scattered on T_s. Let $\{\boldsymbol{c}_k = (c_{k1}, \cdots, c_{ks}), k = 1, \cdots, n\}$ be an NT-net on C^s and

Table 1.4 *Comparison between two methods* $(s=4,\ n^{(1)}=n^{(2)}=307)$

$M_1^{(1)} = -0.00001$	$M_2^{(1)} = 0.00223$	$M_3^{(1)} = 0.00026$	$M_4^{(1)} = 0.00030$
$M_1^{(2)} = -0.00104$	$M_2^{(2)} = 0.00405$	$M_3^{(2)} = 0.00109$	$M_4^{(2)} = 0.00296$
	$EX_i = 0$		
$M_{11}^{(1)} = 0.24997$	$M_{22}^{(1)} = 0.24977$	$M_{33}^{(1)} = 0.25015$	$M_{44}^{(1)} = 0.25008$
$M_{11}^{(2)} = 0.25024$	$M_{22}^{(2)} = 0.24993$	$M_{33}^{(2)} = 0.24994$	$M_{44}^{(2)} = 0.24987$
	$EX_i^2 = 1/4$		
$M_{111}^{(1)} = -0.00001$	$M_{222}^{(1)} = 0.00224$	$M_{333}^{(1)} = 0.00002$	$M_{444}^{(1)} = 0.00003$
$M_{111}^{(2)} = -0.00026$	$M_{222}^{(2)} = 0.000473$	$M_{333}^{(2)} = 0.00142$	$M_{444}^{(2)} = 0.00612$
	$EX_i^3 = 0$		
$M_{1111}^{(1)} = 0.12495$	$M_{2222}^{(1)} = 0.12481$	$M_{3333}^{(1)} = 0.12507$	$M_{4444}^{(1)} = 0.12507$
$M_{1111}^{(2)} = 0.12356$	$M_{2222}^{(2)} = 0.12673$	$M_{3333}^{(2)} = 0.12458$	$M_{4444}^{(2)} = 0.12576$
	$EX_i^4 = 1/8$		

$$x_{ki} = \frac{\log c_{ki}}{\sum_{j=1}^{s} \log c_{kj}}, \qquad i = 1, \cdots, s, \ \ k = 1, \cdots, n. \tag{1.6.6}$$

Then $\{\boldsymbol{x}_k = (x_{k1}, \cdots, x_{ks}), k = 1, \cdots, n\}$ is an NT-net on T_s.

Since T_s has dimension $s-1$, the $(s-1)$-dimensional set of points obtained by this method is induced from a set in s-dimensional space. The uniformity of the present method should be worse than that of the method stated in section 1.5.

Exercises

1.1 Find the discrepancy of the following 5 points: (0.12, 0.05), (0.25, 0.34), (0.44, 0.86), (0.67, 0.12), (0.89, 0.95).

1.2 Find the *glp* set generated by the generating vector (31; 1, 12) and a set of 31 random numbers in C^2. Then compare the uniformity of these two sets by examining plots.

1.3 Find the discrepancy of the *glp* set given in Example 1.6.

1.4 Show that the *glp* sets generated by (5; 1, 2) and (5; 1, 3) have the same discrepancies. (Hint: $2 \cdot 3 \equiv 1(\bmod 5)$).

1.5 Let $\beta_1 = 1.5$, $\beta_2 = 1.75$ and $\beta_3 = 2.0$. Give the first $n = 100$ points of the *gp* set defined by the method of Example 1.7.

1.6 Give the *gp* set with $s = 20$ and $n = 15$ by the cyclotomic field method [(1.3.9)].

1.7 Give the H-set for $p_1 = 11, p_2 = 13$, and $n = 10$, and plot these ten points.

1.8 Let $\boldsymbol{\alpha} = (\alpha_1, \cdots, \alpha_s)$, where α_i are distinct irrational numbers. Let $\boldsymbol{x}_k = k\boldsymbol{\alpha}(\bmod \mathbf{1}), k = 1, 2, \cdots$, where $\mathbf{1} = (1, \cdots, 1)$. Then $\{\boldsymbol{x}_k\}$ is an infinite set in C^s which is uniformly scattered or even well uniformly scattered on C^s if $\boldsymbol{\alpha}$ is suitably chosen. We introduce the following two methods for the choices of $\boldsymbol{\alpha}$:
   (a) set $\boldsymbol{\alpha} = (\sqrt{p_1}, \cdots, \sqrt{p_s})$, where p_i are distinct prime numbers;
   (b) set $\boldsymbol{\alpha} = (p^{\frac{1}{s+1}}, \cdots, p^{\frac{s}{s+1}})$, where p is a prime.
   Generate the sets given by the above two methods for $s = 4$.

1.9 Let $\boldsymbol{x}_k = \frac{k(k+1)}{2}(\sqrt{p_1}, \cdots, \sqrt{p_s})(\bmod \mathbf{1})$, $k = 1, 2, \cdots$, where the p_i's are distinct primes. Then $\{\boldsymbol{x}_k\}$ is called a **Haber** set (section 1.3.4). Show that the Haber set is a subset of the set given by Exercise 1.8, (a), and give Haber sets for $s = 2, n = 50$ and $s = 4, n = 50$.

1.10 Let F_n be defined by $F_0 = 0, F_1 = 1$ and $F_{n+1} = F_n + F_{n-1} (n \geq 1)$. Then $\{F_n\}$ is called the **Fibonacci sequence**. We may prove that the *glp* set generated by $(F_m; 1, F_{m-1})$ is a well uniformly scattered set with discrepancy $O(F_m^{-1} \log F_m)$ $(m \geq 3)$. Generate the *glp* set given by Fibonacci sequence for $n = 34$ (Hint: $F_9 = 34$).

1.11 The Fibonacci sequence can be generalized as follows: let $a_1, \cdots, a_s$ be s positive integers, and let $F_i = a_i (i = 1, \cdots, s)$, and $F_m = F_{m-1} + F_{m-2} + \cdots + F_{m-s} (m = s+1, s+2, \cdots)$. Then $\{F_m\}$ is called a **generalized Fibonacci sequence**. Generate an s-dimensional *glp* set $(F_m; 1, F_{m-1}, F_{m-2} \cdots, F_{m-s+1})$ for $s = 3$.

1.12 Generate two NT-nets on A_2 by *glp* sets with generating vectors $(31; 1, 14)$ and $(144; 1, 89)$, and then plot them.

1.13 Generate four NT-nets on B_2 and U_3 with generating vectors $(31; 1, 14)$ and $(144; 1, 89)$.

1.14 Generate two NT-nets on T_3 with the same generating vectors given by Exercise 1.13.

1.15 Let $\boldsymbol{A}$ be an $s \times s$ positive definite matrix with real entries, and let $E_s = \{\boldsymbol{x} : \boldsymbol{x}'\boldsymbol{A}\boldsymbol{x} \leq 1\}$ be the ellipsoid in R^s and let $ED_s = \{\boldsymbol{x} : \boldsymbol{x}'\boldsymbol{A}\boldsymbol{x} = 1\}$ be the surface of E_s. Find $v(E_s), v(ED_s)$ and NT-nets on E_s and ED_s (Hint: There exists a matrix $\boldsymbol{T}$ such that $\boldsymbol{T}'\boldsymbol{T} = \boldsymbol{A}$.)

1.16 Let $\boldsymbol{X}$ be uniformly distributed on $D = \{(x, y) : y \leq \sin x, 0 \leq x \leq \pi\}$. Find a way to generate an NT-net on D.

1.17 Give a recurrence formula for computing $\{x_{ki}\}$ in (1.5.33) (section 4.3).

CHAPTER 2

Numerical evaluation of multiple integrals in statistics

Let D be a closed and bounded domain in R^s and $f(\boldsymbol{x})$ be a continuous function defined on D. Suppose we want to calculate the definite integral

$$I(f) = \int_D f(\boldsymbol{x})dv,$$

where dv is the volume element of D. In this chapter we shall give some quadrature formulae for $I(f)$ for frequently used domains D such as R^s, R^s_+, a rectangle, ball, sphere or simplex. Then we shall use these formulae to evaluate the probabilities and moments of various multivariate continuous distributions. For the convenience of the reader we also give a brief introduction to the various multivariate distributions, such as the elliptically symmetric distributions, the multivariate Liouville distributions, and the additive logistic elliptical distributions.

2.1 NTM for numerical integration in a rectangle

First, let D be C^s and $f(\boldsymbol{x})$ be defined on C^s. Consider the integral

$$I(f) = \int_{C^s} f(\boldsymbol{x})d\boldsymbol{x}. \tag{2.1.1}$$

Note that the integrand here should be a member of a suitable class of functions, for instances, the continuous functions, the functions with bounded variation or the differentiable functions. The multivariate normal distribution often appears in multivariate statistics. Consequently there are several effective methods for evaluating its probabilities and moments, and these methods may be more efficient than the methods introduced in this section. The latter, however, can be applied to most multivariate continuous distributions whilst the former works only for the multivariate normal distribution. There are several ways below to evaluate numerically the integral (2.1.1).

2.1.1 Classical method

We reduce the multiple integral (2.1.1) into the repeated simple integrals over $[0, 1]$, namely

$$I(f) = \int_0^1 dx_1 \int_0^1 dx_2 \cdots \int_0^1 f(x_1, \cdots, x_s) dx_s. \tag{2.1.2}$$

Then we can apply a quadrature formula such as Simpson's rule to each integral in (2.1.2). If m points are used in each integral, the total number of points for numerical evaluating (2.1.2) is $n \equiv m^s$. When s is large, n increases rapidly, and the rate of convergence is only $O(n^{-1/s})$ if $\partial^s f/\partial x_1 \cdots \partial x_s$ is continuous in C^s or $O(n^{-2/s})$ if $\partial^2 f/\partial^2 x_i$ is continuous. (Hua and Wang (1981)). Therefore the method is nearly useless.

2.1.2 Monte Carlo method

Another approach to multiple quadrature arose in the 1940s as a part of a Monte Carlo method developed by S. Ulam and J. Von Neumann. The basic idea of the Monte Carlo method is to replace an analytic problem by a probabilistic problem with the same solution, and then investigate the latter problem by statistical simulation. A well-known example is the **sample-mean method**.

Suppose that $\boldsymbol{x} = (X_1, \cdots, X_s)$ is a random vector which is uniformly distributed on C^s. Then $f(\boldsymbol{x})$ is a random variable with expectation and standard deviation

$$E(f(\boldsymbol{x})) = \int_{C^s} f(\boldsymbol{x}) d\boldsymbol{x} = I(f)$$

and

$$\sigma(f) \equiv \sigma(f(\boldsymbol{x})) = \Big(\int_{C^s} f(\boldsymbol{x})^2 d\boldsymbol{x} - E(f(\boldsymbol{x}))^2 \Big)^{1/2}$$

respectively, if they exist. If we take n independent samples $\boldsymbol{x}_1, \cdots, \boldsymbol{x}_n$ of $\boldsymbol{x}$, an unbiased estimator of $I(f)$ is its sample mean

$$I(f,n) = \frac{1}{n}\sum_{i=1}^{n} f(\boldsymbol{x}_i). \tag{2.1.3}$$

By the strong law of large numbers $I(f,n)$ converges to $I(f)$ with probability one as $n \to \infty$. Moreover $I(f,n)$ is approximately normally distributed when n is large by the central limit theorem. Therefore we may conclude that for example, the probability of

$$|I(f,n) - I(f)| \leq 1.96\sigma(f)n^{-1/2} \tag{2.1.4}$$

is about 0.95. Although the error term $O(n^{-1/2})$ in (2.1.4) is better than $O(n^{-2/s})$ of the first method when $s > 4$, it is in the sense of probability and not the usual meaning of absolute error. The law of the iterated logarithm shows that with probability one

$$\lim_{n\to\infty} \sup \sqrt{\frac{n}{2\ln(\ln n)}} \left|\frac{1}{n}\sum_{1}^{n} f(\boldsymbol{x}_i) - I(f)\right| = \sigma^2(f).$$

Therefore the rate of convergence of the Monte Carlo method is in no case worse than $O(\sqrt{\ln(\ln(n))/n})$.

2.1.3 Number-theoretic method

The number-theoretic method for numerical evaluation of the multiple integral is based on the theory of the uniform distribution in the sense of Weyl (sections 1.2-1.5).

Let $\mathcal{P}_n = \{\boldsymbol{c}_k^{(n)}, k = 1, \cdots, n\}$ be an NT-net on C^s with low discrepancy. Then we may use

$$I(f, \mathcal{P}_n) = \frac{1}{n}\sum_{k=1}^{n} f(\boldsymbol{c}_k^{(n)}) \tag{2.1.5}$$

as an approximation for $I(f)$ which we call the **NT-mean method**. What is the error in estimation when using the NTM? To answer this question we shall introduce the concept of **bounded variation** in the sense of Hardy and Krause (Hua and Wang (1981)). For simplicity, we treat the case $s = 2$. For any positive integers l, m and a partition of C^2 given by

$$\sigma : \begin{cases} 0 = x_0 < x_1 < \cdots < x_l = 1 \\ 0 = y_0 < y_1 < \cdots < y_m = 1 \end{cases}.$$

Let

$$\begin{aligned} \Delta_{10} f(x_i, y) &= f(x_{i+1}, y) - f(x_i, y), \\ \Delta_{01} f(x, y_j) &= f(x, y_{j+1}) - f(x, y_j), \end{aligned}$$

and

$$\begin{aligned} \Delta_{11} f(x_i, y_j) =& f(x_{i+1}, y_{j+1}) - f(x_{i+1}, y_j) \\ & - f(x_i, y_{j+1}) + f(x_i, y_j) \\ =& \Delta_{10} \Delta_{01} f(x_i, y_j), \end{aligned}$$

where $f(x, y)$ is a function defined on C^2. The sum

$$V_\sigma = \sum_{i=0}^{l-1} \sum_{j=0}^{m-1} |\Delta_{11} f(x_i, y_j)| + \sum_{i=0}^{l-1} |\Delta_{10} f(x_i, 1)| + \sum_{j=0}^{m-1} |\Delta_{01} f(1, y_j)|$$

is called the variation with respect to σ. If $V(f) \equiv \sup_\sigma V_\sigma < \infty$, then $f(x, y)$ is called a function of **bounded variation** in the sense of Hardy and Krause, and $V(f)$ the **total variation** of f on C^2. This concept is easily generalized to any dimension $s (> 2)$ (Hua and Wang (1981)). This class of functions turns out to be rather technical to define and not easy to handle, especially for large values of s. A more pleasant, and almost identical class, which is made up by the functions with finite variation in the measure sense, was given by Bouleau (1990) and developed by Pagès and Xiao (1991).

Theorem 2.1
Suppose that $\mathcal{P}_n$ is a set of points on C^s with discrepancy $D(n)$ and that $f(\boldsymbol{x})$ is a function of bounded variation with total variation $V(f)$ over C^s in the sense of Hardy and Krause. Then we have

$$|I(f) - I(f, \mathcal{P}_n)| \leq V(f)D(n). \tag{2.1.6}$$

The case of $s = 1$ of this theorem was established by Koksma, and then it was generalized by Hlawka to $s > 1$ (Hua and Wang (1981), Chap. 5).

Suppose that the derivative

$$\frac{\partial^s f}{\partial x_1 \cdots \partial x_s} \tag{2.1.7}$$

exists with all its lower derivatives bounded by L over C^s. Then we can derive that

$$V(f) \leq 2^s L,$$

and therefore we have

$$|I(f) - I(f, \mathcal{P}_n)| \leq 2^s L D(n), \tag{2.1.8}$$

which is applicable in practice since the value L is comparatively easy to be estimated.

It follows from Theorem 2.1 that $I(f, \mathcal{P}_n) \to I(f)$ as $n \to \infty$, and the rate of convergence is $O(D(n))$. We can find a $\mathcal{P}_n$ such that $D(n) = O(n^{-1+\varepsilon})$ or even $D(n) = O(n^{-1}(\log n)^s)$ which is much better than the error $O(n^{-1/2})$ by the sample mean Monte Carlo method for numerical integration when $s > 2$. Moreover the expressions of the points of $\mathcal{P}_n$ are not complicated and easy to produce on a computer. Therefore the NTM for numerical integration has more advantages compared with the classical and Monte Carlo methods.

Actually, the rate of convergence $O(n^{-1}(\log n)^s)$ can still be improved when the integrand is smooth enough. For example, the rate of convergence for quadrature formulas such as $O(n^{-\alpha}(\log n)^{\alpha s})$ $(\alpha > 1)$ or $O(n^{-1}(\log n)^{s/2})$ can be reached (Hua and Wang (1981), Zhu (1990) and Pagès and Xiao (1991)).

Let us see a simple example. Suppose that the random vector $\boldsymbol{x} \sim N_3(\boldsymbol{0}, \boldsymbol{I}_3)$, the standard normal distribution in R^3, and we want to evaluate the probability of $\boldsymbol{x}$ falling in C^3, i.e.,

$$\begin{aligned} p &= \int_0^1 \int_0^1 \int_0^1 (2\pi)^{-3/2} \exp\{-\frac{1}{2}(x_1^2 + x_2^2 + x_3^2)\} dx_1 dx_3 dx_3 \\ &= \int_{C^3} f(\boldsymbol{x}) d\boldsymbol{x}, \end{aligned}$$

say. In fact, the probability $p = [\Phi(1) - \Phi(0)]^3 \cong 0.039772181953$, where $\Phi(x)$ is the c.d.f. of $N(0,1)$. Suppose that $\{\boldsymbol{x}_k, k = 1, \cdots, n\}$ is a set of points in C^3 and we estimate p by

$$\hat{p} = \frac{1}{n} \sum_{k=1}^{n} f(\boldsymbol{x}_k).$$

Let us choose the following three sets of points: 1) the equi-lattice points: $\left\{\frac{2i-1}{2m}, \frac{2j-1}{2m}, \frac{2k-1}{2m} : i, j, k = 1, \cdots, m\right\}$; 2) random numbers in C^3 generated by the Monte Carlo method; 3) a *glp* set (section 1.3). Since the number of points in 1) should be of the form $n = m^3$, we choose the minimum m such that $m^3 > n$. Table 2.1 gives comparisons among three sets of points. Obviously, the NTM is the best.

Table 2.1 *Errors $p - \hat{p}$ by the three sets of points*

n	Equi-Lattice Points	n	Random Numbers	n	Good Lattice Points
64	-2.215e-04	35	-2.121e-03	35	-1.162e-04
125	-1.415e-04	101	1.937e-03	101	5.380e-05
729	-4.358e-05	597	6.349e-05	597	-3.979e-06
1728	-2.451e-05	1626	-6.560e-05	1626	6.688e-06
5832	-1.090e-05	5037	-3.248e-05	5037	2.632e-07
39304	-3.072e-06	39029	1.342e-05	39029	2.438e-09

If $D = [\boldsymbol{a}, \boldsymbol{b}] = [a_1, b_1] \times \cdots \times [a_s, b_s]$, then the integral

$$I(f) = \int_{[\boldsymbol{a},\boldsymbol{b}]} f(\boldsymbol{x}) d\boldsymbol{x} \tag{2.1.9}$$

can be approximated by the sum

$$I(f, \mathcal{P}_n) = \frac{1}{n}\sum_{k=1}^{n} f(a_1 + (b_1 - a_1)c_{k1}, \cdots, a_s + (b_s - a_s)c_{ks}), \quad (2.1.10)$$

where $\mathcal{P}_n = \{\boldsymbol{c}_k = (c_{k1}, \cdots, c_{ks}),\ k = 1, \cdots, n\}$ is an NT-net on C^s, and thus the set

$$\{(a_1 + (b_1 - a_1)c_{k1}, \cdots, a_s + (b_s - a_s)c_{ks}),\ k = 1, \cdots, n\} \quad (2.1.11)$$

is an NT-net on $[\boldsymbol{a}, \boldsymbol{b}]$.

To improve the efficiency of the Monte Carlo method for numerical integration, a number of techniques for variance reduction have been developed, such as stratified sampling, importance sampling, correlated sampling, and the method of antithetic variates (Rubinstein (1981), Ch.4). We explain the last method in the one-dimensional case

$$I(f) = \int_0^1 f(x)dx.$$

It is easy to see

$$I(f) = \int_0^1 \frac{1}{2}[f(x) + f(1-x)]dx = \int_0^1 g(x)dx,$$

say. Let $\{x_1, \cdots, x_n\}$ be a set of random numbers. Then

$$I(g,n) = \frac{1}{n}\sum_{i=1}^{n} g(x_i) = \frac{1}{2n}\sum_{i=1}^{n}[f(x_i) + f(1-x_i)]$$

is an estimate of $I(f)$. The variance of $I(f,n)$ in (2.1.3) is $\sigma^2(f)/n$. Since there are $2n$ function values of f involved in $I(g,n)$, the variance of $I(g,n), \sigma^2(g)/n$, should be compared with the quantity $\sigma^2(f)/(2n)$. If f is a continuous monotone function on $[0,1]$, we can prove that

$$\sigma^2(g) \le \frac{1}{2}\sigma^2(f) \quad (2.1.12)$$

which shows the variance reduction (Rubinstein (1981)). This method can be extended to high dimensional integration. For simplicity we illustrate the method only for the case $s = 2$:

$$\begin{aligned} I(f) &= \int_0^1 \int_0^1 f(x, y) dx dy \\ &= \frac{1}{4} \int_0^1 \int_0^1 [f(x, y) + f(1 - x, y) + f(x, 1 - y) \\ &\qquad + f(1 - x, 1 - y)] dx dy \\ &= \int_0^1 \int_0^1 g(x, y) dx dy, \end{aligned} \tag{2.1.13}$$

where

$$g(x, y) = \frac{1}{4}[f(x, y) + f(1 - x, y) + f(x, 1 - y) + f(1 - x, 1 - y)].$$

We then apply the formula (2.1.3) to the function g. Fang and Zhu (1993) show that a similar conclusion to (2.1.12) holds for the s-dimensional case. This method can be applied in NT-integration and is called the **symmetrization method**. There are also many other methods to symmetrize the integrand (Hua and Wang (1981), Chap. 6).

We introduced several methods to produce NT-nets in section 1.3. It follows from Theorem 2.1 that if the discrepancy of an NT-net $\mathcal{P}_n$ is small then the absolute difference between $I(f, \mathcal{P}_n)$ and $I(f)$ is also small. The number of points n is often large in (2.1.5) and the discrepancy of $\mathcal{P}_n$ is not easily estimated, the discrepancy is not regarded as the only measurement for the error estimation of quadrature formula. Many authors use the following method for comparison: Choose several methods for producing the NT-nets and several integrands that can be defined for any dimension s such that the values of their integral are known, and then the results given by different NT-nets and integrands are compared for distinct s. This is only a heuristic method, but it is very useful in practice. Now we will cite some data from Pagès and Xiao (1991). The numerical methods they used were:

(a) gp–1: the gp–set with $\boldsymbol{\gamma}$ (1.3.7);
(b) H–set;
(c) SH–set: the scrambled Halton set;
(d) HL–set: the Halton-like set;

(e) F–set: the Faure set;

(f) MC–set: the Monte Carlo method with a classical congruential generator:

$$x_{n+1} = ax_n \ (mod\ N) \text{ with } a = 31415821, \ N = 10^8 \text{ and } x_0 = 10^4.$$

The integrands on C^s are

$$f_1(\boldsymbol{x}) = \sin\left(2\pi \sum_{i=1}^{s} x_i\right);$$

$$f_2(\boldsymbol{x}) = \left(\frac{3}{2}\right)^s \prod_{i=1}^{s} \left[(2x_i - 1)\exp\left(1 \Big/ \sum_{i=1}^{s}(2x_i - 1)^s\right)\right];$$

$$f_3(\boldsymbol{x}) = \prod_{i=1}^{s} |4x_i - 1|;$$

and

$$f_4(\boldsymbol{x}) = \prod_{i=1}^{s} \exp(-2(x_i - 0.5)^2),$$

for $s = 2, 3, 5, 10, 15, 20, 25, 30, 40$.

By their numerical results they conclude that the Monte Carlo method is clearly the worst performer. The gp-1 sequence turns out to be the best performer in most cases and the Halton-like sequence also has a good efficiency as well. But, Braaten and Weller (1979) pointed out in their work that the scrambling Halton sequence has a significant improvement in rate of convergence over the Halton sequence.

There are many other useful techniques for numerical integration (Haselgrove (1961), Hsu and Zhou (1980), and Davis and Rabinowitz (1984)) which are out of the scope of this book.

2.2 Spherically and elliptically symmetric distributions

We have mentioned that many problems related to the multivariate normal distribution have been thoroughly treated. But for most non-normal multivariate distributions the situation is quite different. To illustrate the use of the NTM it is better to work on non-normal distributions. In this section, classes of spherically and elliptically symmetric distributions are defined that are extensions of multivariate normal distributions, and a brief discussion of these classes will be given.

Definition 2.1

An $s \times 1$ random vector $\boldsymbol{x}$ is said to have a **spherically symmetric distribution** (or simply **spherical distribution**) if for every $s \times s$ orthogonal matrix $\boldsymbol{P}$

$$\boldsymbol{Px} \stackrel{d}{=} \boldsymbol{x}. \tag{2.2.1}$$

The following theorem gives the important properties of spherical distributions and is adapted from Theorem 2.5 of Fang, Kotz and Ng's book (1990).

Theorem 2.2

Let $\boldsymbol{x} = (X_1, \cdots, X_s)'$ be an $s \times 1$ random vector. The following statements are equivalent:

(a) $\boldsymbol{x} \stackrel{d}{=} \boldsymbol{Px}$ for every orthogonal matrix $\boldsymbol{P}$;
(b) The characteristic function of $\boldsymbol{x}$ is of the form $\phi(\boldsymbol{t}'\boldsymbol{t})$ for some scalar function $\phi : R_+ \to R_+$, where $R_+ = \{x : x \geq 0\}$;
(c) $\boldsymbol{x}$ has a stochastic representation

$$\boldsymbol{x} \stackrel{d}{=} R\boldsymbol{u}^{(s)}, \tag{2.2.2}$$

where $R \in R_+$ is a random variable and is independent of $\boldsymbol{u}^{(s)} \sim U(U_s)$;
(d) For any $\boldsymbol{a} \in R^s$ the formula

$$\boldsymbol{a}'\boldsymbol{x} \stackrel{d}{=} \|\boldsymbol{a}\| X_1, \tag{2.2.3}$$

holds, where $\|\boldsymbol{a}\|^2 = \boldsymbol{a}'\boldsymbol{a}$ and X_1 is the first component of $\boldsymbol{x}$.

The stochastic representation (2.2.2) indicates that the class of spherical distributions is directly related to the class of all nonnegative random variables (R). Therefore we shall call R the **generating variate** of $\boldsymbol{x}$. A spherical distribution, in general, does not necessarily possesses a density. The density of a spherical distribution, if it exists, must be of the form $g(\boldsymbol{x}'\boldsymbol{x})$ for some function $g : R_+ \to R_+$, and in this case we will write $\boldsymbol{x} \sim S_s(g)$ and call g the **density generator**. Furthermore we have the following fact.

Theorem 2.3
If $\boldsymbol{x} \stackrel{d}{=} R\boldsymbol{u}^{(s)}$ has a spherical distribution, then $\boldsymbol{x}$ possesses a density generator $g(\,\cdot\,)$ if and only if R has a density $f(\,\cdot\,)$, and the relationship between $g(\,\cdot\,)$ and $f(\,\cdot\,)$ is as follows:

$$f(r) = \frac{2\pi^{s/2}}{\Gamma(s/2)} r^{s-1} g(r^2). \tag{2.2.4}$$

The reader can refer p.35 of Fang, Kotz and Ng (1990) for a proof. The moments of $\boldsymbol{x} = R\boldsymbol{u}^{(s)}$, provided they exist, can be expressed in terms of moments of R. For even integers $2t_i \geq 0$, $i = 1, \cdots, s$, we have

$$E\Big(\prod_{i=1}^{s} x_i^{2t_i}\Big) = E(R^{2t})\pi^{-s/2} \frac{\Gamma(s/2)}{\Gamma(s/2+t)} \prod_{i=1}^{s} \Gamma\Big(\frac{1}{2} + t_i\Big), \tag{2.2.5}$$

where $t = t_1 + \cdots + t_s$. In particular, we have the mean vector and covariance matrix

$$E(\boldsymbol{x}) = \boldsymbol{0} \qquad \text{and} \qquad \text{Cov}(\boldsymbol{x}) = \frac{E(R^2)\boldsymbol{I}_s}{n}, \tag{2.2.6}$$

respectively.

Theorem 2.4
Suppose that $\boldsymbol{x} = (x_1, \cdots, x_s)' \sim S_s(g)$, $s \geq 2$. Consider the transformation to spherical coordinates for $\boldsymbol{x}$,

$$\begin{cases} x_j = R(\prod_{k=1}^{j-1} \sin\theta_k)\cos\theta_j, \quad 1 \leq j \leq s-1 \\ x_s = R(\prod_{k=1}^{s-1} \sin\theta_k), \end{cases} \tag{2.2.7}$$

where $R \geq 0$, $\theta_k \in [0, \pi)$, $k = 1, \cdots, s-2$, $\theta_{s-1} \in [0, 2\pi)$. Then
(a) R, $\theta_1, \cdots, \theta_{s-1}$ are independent;

(b) The p.d.f. of R is given by (2.2.4);
(c) The p.d.f. of $\theta_k, k = 1, \cdots, s-2$, is

$$\frac{1}{B(\frac{1}{2},(s-k)/2)} \sin^{s-k-1}\theta_k, 0 \le \theta_k < \pi; \tag{2.2.8}$$

(d) $\theta_{s-1} \sim U(0, 2\pi)$.

For details of proof and further discussion, we refer to section 2.2 of Fang, Kotz and Ng (1990).

Definition 2.2
An $s \times 1$ random vector $\boldsymbol{x}$ is said to have an **elliptically symmetric distribution** (or simply **elliptical distribution**) with parameters $\boldsymbol{\mu} : s \times 1$ and $\boldsymbol{\Sigma} : s \times s$, $\boldsymbol{\Sigma} \ge 0$ if

$$\boldsymbol{x} \stackrel{d}{=} \boldsymbol{\mu} + \boldsymbol{\Sigma}^{1/2}\boldsymbol{y} \stackrel{d}{=} \boldsymbol{\mu} + R\boldsymbol{\Sigma}^{1/2}\boldsymbol{\mu}^{(s)} \qquad (2.2.9))$$

where $\boldsymbol{\Sigma}^{1/2}$ is the square root of $\boldsymbol{\Sigma}$ and $\boldsymbol{y}$ has an s–dimensional spherical distribution. We will write $\boldsymbol{x} \sim EC_s(\boldsymbol{\mu}, \boldsymbol{\Sigma}, g)$ if $\boldsymbol{y} \sim S_s(g)$ and $\boldsymbol{\Sigma} > 0$. The variable R is still called the generating variate of $\boldsymbol{x}$.

When $\boldsymbol{x} \sim EC_s(\boldsymbol{\mu}, \boldsymbol{\Sigma}, g)$, the p.d.f. of $\boldsymbol{x}$ is

$$|\det \boldsymbol{\Sigma}|^{-1/2} g((\boldsymbol{x}-\boldsymbol{\mu})'\boldsymbol{\Sigma}^{-1}(\boldsymbol{x}-\boldsymbol{\mu})). \tag{2.2.10}$$

The following examples are important in our book as well as in practice.

Example 2.1 The multivariate normal distributions.
If $\boldsymbol{x} \sim EC_s(\boldsymbol{\mu}, \boldsymbol{\Sigma}, g)$ with

$$g(u) = (2\pi)^{-s/2} \exp\Big(-\frac{1}{2}u\Big), \tag{2.2.11}$$

$\boldsymbol{x}$ is said to have **multivariate normal** (or **multinormal**) **distribution** $N_s(\boldsymbol{\mu}, \boldsymbol{\Sigma})$. From (2.2.10) and (2.2.11) the p.d.f. of

$N_s(\boldsymbol{\mu}, \boldsymbol{\Sigma})$ is

$$(2\pi)^{-s/2}|\det \boldsymbol{\Sigma}|^{-1/2} \exp\Big\{-\frac{1}{2}(\boldsymbol{x}-\boldsymbol{\mu})'\boldsymbol{\Sigma}^{-1}(\boldsymbol{x}-\boldsymbol{\mu})\Big\}. \quad (2.2.12)$$

with

$$E(\boldsymbol{x}) = \boldsymbol{\mu} \quad \text{and} \quad \mathrm{Cov}(\boldsymbol{x}) = \boldsymbol{\Sigma}. \quad (2.2.13)$$

Example 2.2 Symmetric Kotz type distributions
If $\boldsymbol{x} \sim EC_s(\boldsymbol{\mu}, \boldsymbol{\Sigma}, g)$ with

$$g(u) = C_s u^{N-1} \exp(-bu^t), \; b, t > 0, \; 2N + s > 2, \quad (2.2.14)$$

where C_s is the normalizing constant, we say that $\boldsymbol{x}$ possesses a **symmetric Kotz type distribution**. When $b = 1/2$, $t = 1$ and $N = 1$, the distribution reduces to a multivariate normal distribution.

More detailed discussion on the symmetric Kotz type distributions can be found in section 3.2 of Fang, Kotz and Ng (1990).

Example 2.3 Symmetric multivariate Pearson Type VII distributions
If $\boldsymbol{x} \sim EC_s(\boldsymbol{\mu}, \boldsymbol{\Sigma}, g)$ with

$$g(u) = C_s(1 + t/m)^{-N}, \quad N > s/2, \; m > 0, \quad (2.2.15)$$

we say $\boldsymbol{x}$ to have a **symmetric multivariate Pearson Type VII distribution**, where C_s is the normalizing constant

$$C_s = (\pi m)^{-s/2}\Gamma(N)/\Gamma(N - s/2). \quad (2.2.16)$$

We shall denote this distribution by $\boldsymbol{x} \sim \mathrm{MPVII}_s(\boldsymbol{\mu}, \boldsymbol{\Sigma}, g)$.

Symmetric Multivariate Pearson Type VII distributions include some useful distributions such as the **multivariate t-distribution** (for $N = \frac{1}{2}(m+s)$) and the **multivariate Cauchy distribution** (for $m = 1$ and $N = \frac{1}{2}(s+1)$). We shall denote the multivariate

t–distribution by $\boldsymbol{x} \sim Mt_s(m, \boldsymbol{\mu}, \boldsymbol{\Sigma})$ which can be defined for the case of m being a positive integer by

$$\boldsymbol{x} = \boldsymbol{z}/W, \tag{2.2.17}$$

where $\boldsymbol{z} \sim N_s(\boldsymbol{\mu}, \boldsymbol{\Sigma})$ is independent of $W \sim \chi_m$, the **chi–distribution** with degrees of freedom m.

Let $\boldsymbol{x} \sim \mathrm{MPVII}_s(\boldsymbol{\mu}, \boldsymbol{\Sigma}, g)$, where g is given by (2.2.15), and let R be the generating variate of $\boldsymbol{x}$. Then $T = R^2$ has a p.d.f.

$$\frac{1}{B(s/2, N - s/2)} m^{-s/2} t^{s/2-1} (1 + t/m)^{-N}, \quad t > 0. \tag{2.2.18}$$

(Fang, Kotz and Ng (1990), p.82), and

$$E(\boldsymbol{x}) = \boldsymbol{\mu}, \quad \mathrm{Cov}(\boldsymbol{x}) = \frac{m}{2N - s - 2} \boldsymbol{\Sigma}. \tag{2.2.19}$$

Example 2.4 Symmetric multivariate Pearson Type II distributions

If $\boldsymbol{x} \sim EC_s(\boldsymbol{\mu}, \boldsymbol{\Sigma}, g)$ with

$$g(u) = \frac{\Gamma(s/2 + m + 1)}{\Gamma(m + 1)\pi^{s/2}} (1 - u)^m, \quad 0 \le u \le 1, m > -1, \tag{2.2.20}$$

then $\boldsymbol{x}$ is said to have a **symmetric multivariate Pearson Type II distribution**, and is denoted by $\boldsymbol{x} \sim \mathrm{MPII}_s(\boldsymbol{\mu}, \boldsymbol{\Sigma}, g)$. The p.d.f. of the generating variate R of $\boldsymbol{x}$ is

$$\frac{2\Gamma(s/2 + m + 1)}{\Gamma(s/2)\Gamma(m + 1)} r^{s-1} (1 - r^2)^m, \quad 0 \le r \le 1, \tag{2.2.21}$$

and therefore $R^2 \sim Be(s/2, m + 1)$, the beta distribution with parameters $(s/2, m + 1)$. The mean vector and covariance matrix of $\boldsymbol{x}$ are

$$E(\boldsymbol{x}) = \boldsymbol{\mu} \quad \text{and} \quad \mathrm{Cov}(\boldsymbol{x}) = (s + 2m + 2)^{-1} \boldsymbol{\Sigma}. \tag{2.2.22}$$

Example 2.5 Multiuniform distributions
Let $\boldsymbol{u}^{(s)}$ be uniformly distributed on U_s, i.e. $\boldsymbol{u}^{(s)} \sim U(U_s)$. Then

$$\boldsymbol{u}^{(s)} \stackrel{d}{=} \boldsymbol{x}/\|\boldsymbol{x}\| \tag{2.2.23}$$

for any $\boldsymbol{x} \sim EC_s(\boldsymbol{0}, \boldsymbol{I}_s, g)$. In particular, we can take $\boldsymbol{x} \sim N_s(\boldsymbol{0}, \boldsymbol{I}_s)$ as has been used in Example 1.11.

Let $\boldsymbol{x} \sim U(B_s)$, the uniform distribution on B_s. Then $\boldsymbol{x}$ has a spherical distribution and its generating variate R has a p.d.f.

$$f(r) = \begin{cases} sr^{s-1}, & \text{if } 0 \le r \le 1, \\ 0, & \text{otherwise.} \end{cases} \tag{2.2.24}$$

The reader can find more detailed discussion on all the examples mentioned in this section in Fang, Kotz and Ng (1990), Chap. 3.

2.3 Probability calculation of a multivariate distribution

In this section we shall give some examples to illustrate how to apply the NTM to evaluate the probabilities of a continuous multivariate distribution. We shall consider only the class of elliptical distributions as an example.

Suppose $\boldsymbol{x} \sim N_s(\boldsymbol{\mu}, \boldsymbol{\Sigma})$. Then $\boldsymbol{x}$ has p.d.f. (2.2.12). In general, the probability of $\boldsymbol{x}$ falling in a given rectangle $[\boldsymbol{a}, \boldsymbol{b}]$ has no analytic expression, and it can be found only by numerical calculation. There are a number of ways to approach this problem (Tong (1990)) . These methods usually give us a good approximation, but they only apply to the multinormal distribution. However, the NTM can be applied to any continuous multivariate distribution.

Example 2.6 Orthant probability of a multinormal distribution
If the expectation of the random vector $\boldsymbol{x}$ is $\boldsymbol{0}$, then the probability that $\boldsymbol{x}$ falls in the first orthant is called the **orthant probability**. There is much literature on this problem for the case $\boldsymbol{x} \sim N_s(\boldsymbol{0}, \boldsymbol{R})$. For example, Gupta (1963) provided a comprehensive review. Steck (1962) gave a substantial review of the results pertaining to orthant probabilities in the equicorrelated case, and Johnson and Kotz (1972) presented some additional results and

references. In order to show the precision of the NTM on this problem, we give the following example.

Denote by $\boldsymbol{R} = (\rho_{ij})$ the covariance matrix of $\boldsymbol{x}$ in R^3. It is known that the orthant probability is equal to

$$\begin{aligned} p &= \int_0^\infty \int_0^\infty \int_0^\infty (2\pi)^{-3/2} |\det \boldsymbol{R}|^{-\frac{1}{2}} \exp\left\{ -\frac{1}{2}\boldsymbol{x}'\boldsymbol{R}^{-1}\boldsymbol{x}\right\} d\boldsymbol{x} \\ &= \frac{1}{8} + \frac{1}{4\pi}(\arcsin \rho_{12} + \arcsin \rho_{13} + \arcsin \rho_{23}). \end{aligned} \tag{2.3.1}$$

Now we use the NTM to calculate p. The normal distribution is essentially zero outside of $[0, 5]$ so we approximate p by

$$p \approx \int_0^5 \cdots \int_0^5 (2\pi)^{-3/2} |\det \boldsymbol{R}|^{-1/2} \exp\left\{ -\frac{1}{2}\boldsymbol{x}'\boldsymbol{R}^{-1}\boldsymbol{x}\right\} d\boldsymbol{x}. \tag{2.3.2}$$

Using the *glp* sets and (2.1.10) we obtain the approximations for the values on the right-hand side of (2.3.2). The results are listed in Table 2.2 for the case $\rho_{12} = \rho_{13} = \rho_{23} = 0.5$. From the table we see that the result generally improves as n increases. But this is not always the case, for instance, the result corresponding to $n = 2,440$ is better than that of $n = 10,007$.

Table 2.2 *Orthant Probability* ($\rho_{12} = \rho_{13} = \rho_{23} = 0.5$)

n	Approximate value of p	Error
135	0.2599914	1.00×10^{-2}
1010	0.2505930	5.93×10^{-4}
2440	0.2502970	2.97×10^{-4}
4040	0.2503471	3.47×10^{-4}
5037	0.2498223	-1.77×10^{-4}
10,007	0.2512536	1.25×10^{-3}
39,024	0.2500323	3.23×10^{-5}
∞	0.25	

Although the integrand in (2.3.2) is not sharp, the precision of calculation is possibly to be enhanced if the method of symmetrization for integrand is applied. Note that the points for evaluating the integral are increased in the symmetrization method, i.e. the points used for a s-fold integral is a 2^s multiple of the original one.

In Table 2.3, the real values in the column of n are the values in the brackets (which equal $2^3 \times n = 8n$). Comparing Tables 2.3 and 2.2, we see that the precision of calculation is enhanced if the symmetrization method is applied.

Table 2.3 *Orthant probability (Symmetrization method)*

n	Approximate value of p	Error
135(1080)	0.2508592	8.59×10^{-4}
597(4776)	0.2501168	1.17×10^{-4}
1010(8080)	0.2500185	1.85×10^{-5}
2440(19520)	0.2500008	8.00×10^{-7}
∞	0.25	

Example 2.7 Orthant probability of a multivariate t–distribution
The multivariate t–distribution was defined in Example 2.3. Let $\boldsymbol{x} \sim Mt_s(m, \boldsymbol{0}, \boldsymbol{R})$, where $\boldsymbol{R}$ is the correlation matrix of $\boldsymbol{x}$. It was pointed out in section 2.7 of Fang, Kotz and Ng (1990) that $Mt_s(m, \boldsymbol{0}, \boldsymbol{R})$ and $N_s(\boldsymbol{0}, \boldsymbol{R})$ have the same orthant probability. The p.d.f. of $Mt_s(m, \boldsymbol{\mu}, \boldsymbol{\Sigma})$ which can be obtained by Example 2.3 is

$$\frac{\Gamma((s+m)/2)}{(\pi m)^{s/2}\Gamma(m/2)}|\boldsymbol{\Sigma}|^{-1/2}(1 + m^{-1}(\boldsymbol{x} - \boldsymbol{\mu})'\boldsymbol{\Sigma}^{-1}(\boldsymbol{x} - \boldsymbol{\mu}))^{-(s+m)/2}. \tag{2.3.3}$$

When $\boldsymbol{x} \sim Mt_3(2, \boldsymbol{0}, \boldsymbol{R})$ with $\rho_{12} = \rho_{13} = \rho_{23} = 0.5$, we know by Example 2.6 that the orthant probability is 0.25. We now use the approximation of

$$\begin{aligned} p &= \frac{\Gamma(2.5)}{(2\pi)^{1.5}\Gamma(1)}|R|^{-1/2}\int_0^\infty\int_0^\infty\int_0^\infty \left(1 + \frac{1}{2}\boldsymbol{x}'R^{-1}\boldsymbol{x}\right)^{-2.5} d\boldsymbol{x} \\ &\cong \frac{\Gamma(2.5)}{(2\pi)^{1.5}}|R|^{-1/2}\int_0^A\int_0^A\int_0^A \left(1 + \frac{1}{2}\boldsymbol{x}'R^{-1}\boldsymbol{x}\right)^{-2.5} d\boldsymbol{x} \end{aligned} \tag{2.3.4}$$

and then use the NTM for numerical integration to evaluate p. Since the convergence rate of the function $(1 + \frac{1}{2}x^2)^{-2.5}$ is lower than that of $\exp(-\frac{1}{2}x^2)$ when $x \to \infty$, then value A should be taken greater than 5 in (2.3.4). How do we choose A? In general this is a difficult problem for the application of the NTM to numerical integration.

We should take A such that the value of the integral over the eliminated domain is less than the pre-assigned precision ε. If A is taken too small, then the approximate value of the integral becomes much smaller than its real value. Otherwise we use too many points in ensuring that the approximate value of integral has a higher precision because the integrand is nearly close to zero in a large region. Therefore we often use the **testing method** to choose a proper value of A. In this example, we try the values $A = 8, 9, 10, 11, 12$ for the integrals which are calculated by the symmetrization method for the integrand, and the results are listed in Table 2.4. We see that $A = 10$ is the best one. However, there is no universal approach to treating the problem of reducing an infinite integral to an integral over a bounded domain. Some further discussion will be given in section 4.7

Table 2.4 *Orthant probability of a multivariate t-distribution*

n	$A=8$	$A=9$	$A=10$	$A=11$	$A=12$
101	0.2680093	0.2878872	0.3150274	0.3509449	0.3969012
135	0.2631638	0.2788493	0.2992300	0.3254924	0.3587046
307	0.2470927	0.2531881	0.2605061	0.2696726	0.2812716
597	0.2428794	0.2465284	0.2504974	0.2551893	0.2609337
701	0.2426810	0.2460799	0.2495917	0.2535489	0.2582208

2.4 Numerical integration over a bounded domain

Let D be a closed and bounded domain in R^s and $f(\boldsymbol{x})$ be a continuous function defined on D. Consider the integral

$$I(f) = \int_D f(\boldsymbol{x}) dv, \tag{2.4.1}$$

where dv is the volume element of D. We have discussed the numerical integration of $I(f)$ when $D = [\boldsymbol{a}, \boldsymbol{b}]$ in previous sections. Now we introduce three methods for numerical calculations of $I(f)$ as follows:

2.4.1 Indicator function method

Suppose that D has dimension s and $[\boldsymbol{a}, \boldsymbol{b}]$ is the circumscribed rectangle of D. We have

$$I(f) = \int_{[\boldsymbol{a},\boldsymbol{b}]} I_D(\boldsymbol{x}) f(\boldsymbol{x}) d\boldsymbol{x}, \tag{2.4.2}$$

where $I_D(\boldsymbol{x})$ is the indicator function of D, i.e.

$$I_D(\boldsymbol{x}) = \begin{cases} 1, & \text{if } \boldsymbol{x} \in D, \\ 0, & \text{otherwise.} \end{cases} \tag{2.4.3}$$

Then we can apply the method of numerical integration on $[\boldsymbol{a}, \boldsymbol{b}]$ to the integral (2.4.2). Suppose that $\{\boldsymbol{x}_k; k = 1, \cdots, n\}$ is an NT-net on $[\boldsymbol{a}, \boldsymbol{b}]$, and that there are m points $\{\boldsymbol{x}_{i_k}, k = 1, \cdots, m\}$ of $\{\boldsymbol{x}_k\}$ falling on D. Then the integral $I(f)$ can be approximated by

$$\tilde{I}(f, n) = \frac{1}{n} v([\boldsymbol{a}, \boldsymbol{b}]) \sum_{k=1}^{m} f(\boldsymbol{x}_{i_k}), \tag{2.4.4}$$

where $v([\boldsymbol{a}, \boldsymbol{b}])$ is the volume of $[\boldsymbol{a}, \boldsymbol{b}]$. We may rewrite (2.4.4) as

$$\begin{aligned} \tilde{I}(f, n) &= \frac{1}{m} \Big(\frac{m}{n} v([\boldsymbol{a}, \boldsymbol{b}]) \Big) \sum_{k=1}^{m} f(\boldsymbol{x}_{i_k}) \\ &\cong \frac{1}{m} v(D) \sum_{k=1}^{m} f(\boldsymbol{x}_{i_k}), \end{aligned} \tag{2.4.5}$$

where $\frac{m}{n} v([\boldsymbol{a}, \boldsymbol{b}])$ can be regarded as an approximation of the volume of D, if the points are uniformly scattered on D (section 1.5). Hence (2.4.5) is reasonable, and the indicator function method is effective in many cases. However the points on the boundary of D are usually the discontinuities of the function $I_D(\boldsymbol{x}) f(\boldsymbol{x})$ on $[\boldsymbol{a}, \boldsymbol{b}]$ and the distribution of the set $\{\boldsymbol{x}_{i_k}\}$ on D may be not very uniform, so that the precision of this method is not high. If $v(D)$ can be obtained exactly, then the formula

$$I^*(f,n) = \frac{1}{m} v(D) \sum_{k=1}^{m} f(\boldsymbol{x}_{i_k}) \tag{2.4.6}$$

is recommended to approximate $I(f)$. Some examples show that $I^*(f,n)$ is slightly better than $I(f,n)$.

Example 2.8
Find the approximate value of the integral

$$I = \int\int_{x^2+y^2\leq 1} (1+x+y)e^{-\frac{1}{2}(x^2+y^2)}dxdy \tag{2.4.7}$$

by the indicator function method.
We know that

$$I = \int\int_{x^2+y^2\leq 1} e^{-\frac{1}{2}(x^2+y^2)}dxdy = 2\pi(1-e^{-\frac{1}{2}}) \cong 2.4722408.$$

It follows from (2.4.4) and (2.4.6) that

$$\tilde{I}(f,n) = \frac{4}{n} \sum_{k=1}^{m} f(\boldsymbol{x}_{i_k})$$

and

$$I^*(f,n) = \frac{\pi}{m} \sum_{k=1}^{m} f(\boldsymbol{x}_{i_k}),$$

here $v(D) = v(B_2) = \pi$. The results of computation are listed in Table 2.5, from which we see that $I^*(f,n)$ usually gives a better result than $\tilde{I}(f,n)$.

Table 2.5 *The indicator function method*

n	m	Error for $\tilde{I}(f,n)$	Error for $I^*(f,n)$
144	109	9.422×10^{-2}	4.862×10^{-3}
233	183	1.744×10^{-2}	1.737×10^{-2}
377	298	2.241×10^{-3}	1.360×10^{-2}
610	483	2.000×10^{-2}	1.900×10^{-4}
987	772	1.065×10^{-3}	1.125×10^{-2}
1587	1248	1.876×10^{-3}	1.276×10^{-3}
2584	2029	1.896×10^{-3}	1.359×10^{-3}
4181	3276	5.357×10^{-3}	4.446×10^{-4}
6765	5306	2.266×10^{-3}	1.060×10^{-3}
10,946	8585	2.969×10^{-3}	4.403×10^{-4}

2.4.2 Transformation method

Before we introduce the so-called transformation method, we return to Example 2.9. Let

$$x = r\cos(2\pi\theta), \quad y = r\sin(2\pi\theta).$$

Then I defined by (2.4.7) is reduced to an integral of the type (2.1.1)

$$I = \int_0^1 \int_0^1 (1 + r\cos(2\pi\theta) + r\sin(2\pi\theta))2\pi r e^{-\frac{1}{2}r^2} d\theta dr.$$

We apply the formula (2.1.3) to calculate $I(f,n)$, and the results are listed in the second column of Table 2.6. Comparing with Table 2.5, we see that the transformation method is in general better than the indicator function method, and the error tends to zero quickly in the former method when n increases. The last column of Table 2.6 will be explained later.

Table 2.6 *Transformation method and direct method*

n	Error in transformation method	Error in direct method
144	1.820×10^{-2}	3.087×10^{-1}
233	4.860×10^{-3}	5.436×10^{-3}
377	6.994×10^{-3}	5.395×10^{-3}
610	1.890×10^{-3}	2.115×10^{-3}
987	2.679×10^{-3}	1.347×10^{-3}
1587	2.403×10^{-3}	8.310×10^{-4}
2584	1.022×10^{-3}	5.060×10^{-4}
4181	2.849×10^{-4}	3.170×10^{-4}
6765	3.939×10^{-4}	1.970×10^{-4}
10,946	1.027×10^{-4}	1.220×10^{-4}

Now we introduce the transformation method. Suppose that D has representation

$$x_j = x_j(\phi_1, \cdots, \phi_t) = x_j(\boldsymbol{\phi}),\ j = 1, \cdots, s(\geq t), \tag{2.4.8}$$

where $\boldsymbol{\phi} = (\phi_1, \cdots, \phi_t) \in C^t$, and that $x_j(\boldsymbol{\phi}), j = 1, \cdots, s$, have continuous derivatives with respect to each $\phi_i, i = 1, \cdots, t$, over [0, 1]. Let

$$\boldsymbol{T} = \left(\frac{\partial x_j}{\partial \phi_i}\right), \qquad i = 1, \cdots, t,\ j = 1, \cdots, s$$

be a $t \times s$ matrix and let

$$J(\boldsymbol{\phi}) = \det(\boldsymbol{T}\boldsymbol{T}')^{1/2}. \tag{2.4.9}$$

When $t = s$, $J(\boldsymbol{\phi})$ is just the Jacobian of transformation (2.4.8). Then we have

$$I(f) = \int_D f(\boldsymbol{x})dv = \int_{C^t} f(\boldsymbol{x}(\boldsymbol{\phi}))J(\boldsymbol{\phi})d\boldsymbol{\phi}, \tag{2.4.10}$$

where $\boldsymbol{x}(\boldsymbol{\phi}) = (x_1(\boldsymbol{\phi}), \cdots, x_s(\boldsymbol{\phi}))$ and $d\boldsymbol{\phi} = \prod_{j=1}^t d\phi_j$ (Hua (1984)). A quadrature formula over C^t induces a quadrature formula over D. In particular, when $f \equiv 1$, we have the integral representation of the volume of D

$$v(D) = \int_{C^t} J(\boldsymbol{\phi})d\boldsymbol{\phi}. \tag{2.4.11}$$

Let $\{\boldsymbol{c}_k, k = 1, \cdots, n\}$ be an NT-net on C^t. Now (2.1.5) and (2.4.10) suggest estimating $I(f)$ by

$$I(f,n) = \frac{1}{n}\sum_{k=1}^{n} f(\boldsymbol{x}(\boldsymbol{c}_k))J(\boldsymbol{c}_k). \tag{2.4.12}$$

For convenience of application, we give the corresponding formula (2.4.10) and the respective quadrature formulas for the domains A_s, B_s, U_s, T_s and V_s (section 1.5). In what follows we always denote an NT-net on C^t by $\{\boldsymbol{c}_k = (c_{k1}, \cdots, c_{kt}) : k = 1, \cdots, n\}$ and the set of points of $\{\boldsymbol{c}_k\}$ falling in D by $\{\boldsymbol{c}_{i_k} : k = 1, \cdots, m\}$.

(a) The integration on A_s

The domain A_s is defined by (1.5.12), and it has representation (1.5.18). From Appendix B.1 we have

$$J(\boldsymbol{\phi}) = \prod_{i=2}^{s} \phi_i^{i-1} \tag{2.4.13}$$

and

$$v(A_s) = \int_{C^s} \Big(\prod_{i=2}^{s} \phi_i^{i-1}\Big) d\boldsymbol{\phi} = \frac{1}{s!}. \tag{2.4.14}$$

Hence

$$I(f) = \int_{A_s} f(\boldsymbol{x})dv = \int_{C^s} f(\boldsymbol{x}(\boldsymbol{\phi}))\Big(\prod_{i=2}^{s} \phi_i^{i-1}\Big) d\boldsymbol{\phi}. \tag{2.4.15}$$

By (2.4.12) and (2.4.6) we have

$$I(f,n) = \frac{1}{n}\sum_{k=1}^{n} f(\boldsymbol{x}(\boldsymbol{c}_k))\Big(\prod_{i=2}^{s} c_{ki}^{i-1}\Big) \tag{2.4.16}$$

and

$$I^*(f,n) = \frac{1}{m \cdot s!}\sum_{k=1}^{m} f(\boldsymbol{x}(\boldsymbol{c}_{i_k})). \tag{2.4.17}$$

In next section, we shall use these two formulas to evaluate the moments of order statistics.

(b) The integration on B_s

The domain B_s is defined by (1.5.13), and it has representation (1.5.21). By Appendix B.2, we obtain

$$J(\boldsymbol{\phi}) = 2\pi^{s-1}\phi_1^{s-1}\prod_{i=2}^{s} S_i^{s-i},$$

where $S_i = \sin(\pi\phi_i), i = 2, \cdots, s$. By (2.4.11) we have

$$\begin{aligned} v(B_s) &= \int_{C^s}\Big[2\pi^{s-1}\phi_1^{s-1}\prod_{i=2}^{s}(\sin(\pi\phi_i))^{s-i}\Big]d\boldsymbol{\phi} \\ &= 2\pi^{s-1}\int_0^1 \phi_1^{s-1}d\phi_1 \prod_{i=2}^{s}\int_0^1 (\sin(\pi\phi_i))^{s-i}d\phi_i \\ &= \frac{2}{s}\prod_{i=2}^{s} B\Big(\frac{1}{2}, \frac{s-i+1}{2}\Big). \end{aligned}$$

Since for positive integer m we have

$$\int_0^1 (\sin(\pi x))^m dx = \frac{1}{\pi}B\Big(\frac{1}{2}, \frac{m+1}{2}\Big).$$

Using $B(a,b) = \Gamma(a)\Gamma(b)/\Gamma(a+b)$, we obtain

$$v(B_s) = \frac{2\pi^{s/2}}{s\Gamma(s/2)}. \tag{2.4.18}$$

Therefore

$$\begin{aligned} I(f) &= \int_{B_s} f(\boldsymbol{x})d\boldsymbol{x} \\ &= \int_{C^s} \left[f(\boldsymbol{x}(\boldsymbol{\phi}))2\pi^{s-1}\phi_1^{s-1}\prod_{i=2}^{s}(\sin(\pi\phi_i))^{s-i}\right] d\boldsymbol{\phi}, \end{aligned} \tag{2.4.19}$$

$$I(f,n) = \frac{2\pi^{s-1}}{n}\sum_{k=1}^{n} f(\boldsymbol{x}(\boldsymbol{c}_k))c_{k_1}^{s-1}\prod_{i=2}^{s}(\sin(\pi c_{ki}))^{s-i}, \tag{2.4.20}$$

and

$$I^*(f,n) = \frac{2\pi^{s/2}}{sm\Gamma(s/2)}\sum_{k=1}^{m} f(\boldsymbol{x}(\boldsymbol{c}_{i_k})). \tag{2.4.21}$$

Obviously, Example 2.8 is a special case of $s = 2$ and $f(x,y) = (1+x+y)e^{-\frac{1}{2}(x^2+y^2)}$.

(c) The integration on U_s

The domain U_s is defined by (1.5.14), and it has representation (1.5.27). By Appendix B.2 and (2.4.11) we obtain

$$J(\boldsymbol{\phi}) = 2\pi^{s-1}\prod_{i=1}^{s-2}(\sin(\pi\phi_i))^{s-i-1} \tag{2.4.22}$$

and

$$v(U_s) = 2\pi\prod_{i=1}^{s-2} B(\frac{1}{2},\frac{s-i}{2}) = \frac{2\pi^{s/2}}{\Gamma(s/2)}. \tag{2.4.23}$$

Therefore

$$\begin{aligned} I(f) &= \int_{U_s} f(\boldsymbol{x})dv \\ &= \int_{C^{s-1}} \left[f(\boldsymbol{x}(\boldsymbol{\phi}))2\pi^{s-1}\prod_{i=1}^{s-2}(\sin(\pi\phi_i))^{s-i-1}\right] d\boldsymbol{\phi} \end{aligned} \tag{2.4.24}$$

and

$$I(f,n) = \frac{2\pi^{s-1}}{n} \sum_{k=1}^{n} f(\boldsymbol{x}(\boldsymbol{c}_k)) \prod_{i=1}^{s-2} (\sin \pi c_{ki})^{s-i-1}. \tag{2.4.25}$$

Note that there exists no respective quadrature formula $I^*(f,n)$ because the dimension of U_s is $t = s-1 < s$. Stroud (1967) obtained some formulas for $I(f)$ when the integrand $f(\boldsymbol{x})$ is a polynomial of degree ≤ 7,

(d) The integration on V_s

The domain V_s is defined by (1.5.15), and it has representation (1.5.29). By Appendix B.3 and (2.4.11) we have

$$J(\boldsymbol{\phi}) = (\pi\phi_1)^{s-1} \prod_{j=2}^{s} \left(\sin\left(\frac{\pi\phi_j}{2}\right)\right)^{2s-2j+1} \cos\left(\frac{\pi\phi_j}{2}\right) \tag{2.4.26}$$

and

$$\begin{aligned} v(V_s) =& \pi^{s-1} \int_0^1 \phi_1^{s-1} d\phi_1 \prod_{j=2}^{s} \int_0^1 \left(\sin \frac{\pi\phi_j}{2}\right)^{2s-2j+1} \cos\left(\frac{\pi\phi_j}{2}\right) d\phi_j \\ =& \frac{1}{s!}. \end{aligned} \tag{2.4.27}$$

Therefore

$$\begin{aligned} I(f) =& \int_{V_s} f(\boldsymbol{x}) dv \\ =& \pi^{s-1} \int_{C^s} \phi_1^{s-1} f(\boldsymbol{x}(\boldsymbol{\phi})) \\ & \left[\prod_{j=2}^{s} \left(\sin\left(\frac{\pi\phi_j}{2}\right)\right)^{2s-2j+1} \cos\left(\frac{\pi\phi_j}{2}\right)\right] d\boldsymbol{\phi}, \end{aligned} \tag{2.4.28}$$

$$I(f,n) = \frac{\pi^{s-1}}{n} \sum_{k=1}^{n} f(\boldsymbol{x}(\boldsymbol{c}_k)) c_{k_1}^{s-1}$$
$$\prod_{j=2}^{s} \left(\sin\left(\frac{\pi c_{kj}}{2}\right)\right)^{2s-2j+1} \cos\left(\frac{\pi c_{kj}}{2}\right), \tag{2.4.29}$$

and

$$I^*(f,n) = \frac{1}{s!m} \sum_{k=1}^{m} f(\boldsymbol{x}(\boldsymbol{c}_{i_k})). \tag{2.4.30}$$

(e) The integration on T_s

The domain T_s is defined by (1.5.16), and it has representation (1.5.32). By Appendix B.3 and (2.4.11) we obtain

$$J(\boldsymbol{\phi}) = \pi^{s-1}\sqrt{s} \prod_{j=1}^{s-1} \left(\sin\left(\frac{\pi \phi_j}{2}\right)\right)^{2s-2j-1} \cos\left(\frac{\pi \phi_j}{2}\right) \tag{2.4.31}$$

and

$$v(T_s) = \int_{C^{s-1}} J(\boldsymbol{\phi}) d\boldsymbol{\phi} = \sqrt{s}/(s-1)!. \tag{2.4.32}$$

Therefore we have

$$I(f,n) = \frac{\pi^{s-1}\sqrt{s}}{n} \sum_{k=1}^{n} f(\boldsymbol{x}(\boldsymbol{c}_k))$$
$$\prod_{j=1}^{s-1} \left(\sin\left(\frac{\pi c_{kj}}{2}\right)\right)^{2s-2j-1} \cos\left(\frac{\pi c_{kj}}{2}\right). \tag{2.4.33}$$

Since $t = s - 1$, there is no formula for $I^*(f,n)$.

By some well known transformations we may find more quadrature formulas, for instance, the quadrature formulas for ellipsoidal D or for D being the surface of an ellipsoid (Exercise 2.10).

2.4.3 Direct method

Let $\{\boldsymbol{x}_k, k = 1, \cdots, n\}$ be an NT-net on D. Then we have

$$I(f) = v(D) \int_D f(\boldsymbol{x}) \frac{1}{v(D)} d\boldsymbol{x} = v(D) \int_D f(\boldsymbol{x}) p(\boldsymbol{x}) d\boldsymbol{x},$$

where $p(\boldsymbol{x}) = 1/v(D)$ is the p.d.f. of $v(D)$. By using the NT-mean method we have

$$I(f) \cong I(f,n) = \frac{1}{n} v(D) \sum_{k=1}^{n} f(\boldsymbol{x}_k). \qquad (2.4.34)$$

This is the **direct method** for numerical integration. For example, let $D = B_2$. Then we obtain NT-net on B_2 by the method stated in section 1.5, and apply it to Example 2.9, and get the results in (2.4.34) which have almost the same precision as those obtained by the transformation method (the third column of Table 2.6). More precisely, the direct method often leads to better results than transformation method (Example 2.10 in section 2.5). Now we investigate the error properties of the formula (2.4.34).

Suppose that the random vector $\boldsymbol{x}$ is uniformly distributed on D and that $\boldsymbol{x}$ has a stochastic representation

$$X_j = x_j(\phi_1, \cdots, \phi_t) = x_j(\boldsymbol{\phi}), \; j = 1, \cdots, t,$$

where $t \le s$ and $\boldsymbol{\phi} \in C^t$. We suppose also that $\phi_1, \cdots, \phi_t$ are mutually independent with respective c.d.f. $G_1(x_1), \cdots, G_t(x_t)$. If $\{\boldsymbol{c}_k, k = 1, \cdots, n\}$ is a set of points on C^t with discrepancy d. Then it follows that $\{\boldsymbol{b}_k = (G_1^{-1}(c_{k1}), \cdots, G_t^{-1}(c_{kt})) : k = 1, \cdots, n\}$ has F-discrepancy d (Example 1.5). Consequently the set $\mathcal{P}_n = \{\boldsymbol{x}(\boldsymbol{b}_k), k = 1, \cdots, n\}$ has quasi F-discrepancy d with respect to $G(\boldsymbol{x}) = \prod_{i=1}^{t} G_i(x_i)$. Set

$$h(\boldsymbol{\psi}) = f(\boldsymbol{x}(\boldsymbol{G}^{-1}(\boldsymbol{\psi}))),$$

where $\boldsymbol{G}^{-1}(\boldsymbol{\psi}) = (G_1^{-1}(\psi_1), \cdots, G_t^{-1}(\psi_t))$.

Theorem 2.5
Suppose that $h(\boldsymbol{\psi})$ has continuous derivative $\frac{\partial^t h}{\partial\psi_1\cdots\partial\psi_t}$ and that

$$\left|\frac{\partial^l h(\boldsymbol{\psi})}{\partial\psi_{i_1}\cdots\partial\psi_{i_l}}\right| \leq C,\ 1 \leq i_1 < \cdots < i_l \leq t, 1 \leq l \leq t$$

for some constant C. Then

$$\left|\int_D f(\boldsymbol{x})dv - \frac{v(D)}{n}\sum_{k=1}^{n} f(\boldsymbol{x}_k)\right| \leq 2^s C v(D) d.$$

PROOF Since

$$\int_D f(\boldsymbol{x})dv = \int_{C^t} f(\boldsymbol{x}(\boldsymbol{\phi}))J(\boldsymbol{\phi})d\boldsymbol{\phi}$$

and by (2.4.12),

$$\int_{C^t} \frac{1}{v(D)} J(\boldsymbol{\phi})d\boldsymbol{\phi} = \int_{C^t} dG_1(\phi_1)\cdots dG_t(\phi_t) = \int_{C^t} d\psi_1\cdots d\psi_t = 1,$$

where we set

$$G_i^{-1}(\psi_i) = \phi_i,\ i = 1,\cdots,t,$$

we have

$$\int_D f(\boldsymbol{x})dv = v(D)\int_{C^t} f(\boldsymbol{x}(\boldsymbol{G}^{-1}(\boldsymbol{\psi})))d\boldsymbol{\psi} = v(D)\int_{C^t} h(\boldsymbol{\psi})d\boldsymbol{\psi}.$$

Then by (2.1.8)

$$\left|\int_D f(\boldsymbol{x})d\boldsymbol{x} - \frac{v(D)}{n}\sum_{k=1}^{n} f(\boldsymbol{x}_k)\right| = v(D)\left|\int_{C^t} h(\boldsymbol{\psi})d\boldsymbol{\psi} - \frac{1}{n}\sum_{k=1}^{n} h(\boldsymbol{c}_k)\right|$$
$$\leq 2^s C v(D) d.$$

The theorem is proved. □

Remarks 2.1

(a) The indicator function method can be applied only when D has dimension s, where D can be any closed and bounded domain.
(b) If there is a differentiable mapping which maps C^t to D, and the derivatives of f are continuous, then we can use the transformation method.
(c) If we can find an NT-net on D, then the direct method is recommended.
(d) The direct method is often the best one among these three methods, and transformation method is usually better than the indicator function method. In many cases, it will be beneficial if the exact value of $v(D)$ appears in the quadrature formula. We shall give some applications of the NTM for numerical integration to the problems in statistics in the following sections.
(e) In section 4.7 we shall introduce some alternative methods for numerical integration in statistics.

2.5 Moments of order statistics

Order statistics have wide applications in statistics. David (1981) gave a systematic exposition on the theory and applications of order statistics. The moments of order statistics usually have no analytic expression, and therefore we can only use numerical methods to find their approximate values. Arnold and Balakrishnan (1989) gave a detailed exposition on the numerical evaluation of the moments of order statistics. In this section we give only the numerical calculation of the moments of order statistics (Zhang (1990)) and do not involve the other aspects of order statistics. Let $X_1, \cdots, X_N$ be a sample of the population distribution $F(x)$, and $X_{(1)} \leq X_{(2)} \leq \cdots \leq X_{(N)}$ be its order statistics. Write $Y_i = X_{(i)}, i = 1, \cdots, N$.

(a) The joint density of $Y_1, \cdots, Y_N$ is

$$p(y_1, \cdots, y_N) = N! \prod_{i=1}^{N} p(y_i), \tag{2.5.1}$$

where $p(y)$ is the p.d.f. of $F(y)$;

(b) The p.d.f. of $Y_r = X_{(r)}$ is

$$p_r(y) = \frac{N!}{(r-1)!(N-r)!} F^{r-1}(y)(1-F(y))^{N-r} p(y); \quad (2.5.2)$$

(c) The p.d.f. of any order statistics $Y_{i_1}, \cdots, Y_{i_s}$ is

$$p_{i_1,\cdots,i_s}(y_1,\cdots,y_s) = N! \prod_{j=0}^{s} \frac{(F(y_{j+1}) - F(y_j))^{i_{j+1}-i_j-1}}{(i_{j+1}-i_j-1)!} \prod_{k=1}^{s} p(y_k), \quad (2.5.3)$$

where $0 < s \le N, 0 = i_0 < i_1 < \cdots < i_s < i_{s+1} = N+1$ and $-\infty = y_0 < y_1 < \cdots < y_s < y_{s+1} = \infty$ (David (1981)). Therefore the mixed moment of order $r_1, \cdots, r_s$ of $Y_{i_1}, \cdots, Y_{i_s}$ is

$$M_{i_1,\cdots,i_s}(r_1,\cdots,r_s) = E(Y_{i_1}^{r_1} \cdots Y_{i_s}^{r_s})$$
$$= N! \int_D \prod_{k=1}^{s} (y_k^{r_k} p(y_k)) \prod_{j=0}^{s} \frac{(F(y_{j+1}) - F(y_j))^{i_{j+1}-i_j-1}}{(i_{j+1}-i_j-1)!} dv, \quad (2.5.4)$$

where

$$D = \{\boldsymbol{x} : \boldsymbol{x} \in R^s, x_1 < \cdots < x_s\} \quad (2.5.5)$$

and dv is the volume element of D. We have not discussed this domain in previous section. Note that the value of (2.5.4) is unchanged if some of "$<$" are changed by "$\le$" in the definition (2.5.5), so it makes no difference if some of "$<$" are replaced by "$\le$" in the definition of D.

If the values of $p(x)$ outside [0,1] are zero, then the respective D is $A_s = \{0 \le X_1 \le \cdots \le X_s \le 1\}$. The formulas (2.4.16) and (2.4.17) give the methods for evaluating (2.5.4) by the transformation method and by direct method, respectively. The latter gives

$$M_{i_1,\cdots,i_s}(r_1,\cdots,r_s) \cong \frac{N!}{s!n} \sum_{k=1}^{s} \left[\prod_{l=1}^{s} x_{kl}^{r_l} p(x_{kl}) \right]$$
$$\prod_{j=0}^{s} \frac{(F(x_{k,j+1}) - F(x_{kj}))^{i_{j+1}-i_j-1}}{(i_{j+1} - i_j - 1)!}. \tag{2.5.6}$$

In particular when $F(x)$ is the uniform distribution on [0,1], the mixed moments of order statistics have analytic expression

$$M_{i_1,\cdots,i_s}(r_1,\cdots,r_s) = \prod_{j=1}^{s} \prod_{k=i_j}^{i_{j+1}-1} \frac{k}{\sum_{l=1}^{j} r_l + k}$$
$$= \frac{i_1^{[r_1]}(r_1+i_2)^{[r_2]}\cdots(r_1+\cdots+r_{s-1}+i_s)^{[r_s]}}{(N+1)^{[r_1+\cdots+r_s]}}, \tag{2.5.7}$$

where $[x]^r = x(x+1)\cdots(x+r-1)$. From (2.5.7) we immediately obtain that

$$E(Y_i^r) = \frac{(r+i-1)!N!}{(i-1)!(N+r)!}, \quad r = 1, 2, \cdots$$

and

$$E(Y_iY_j) = \frac{i(j+1)}{(N+1)(N+2)}, \quad 1 \le i < j \le N.$$

Since the moments of order statistics of the uniform distribution have an analytic expression, we may use them to check the errors of estimation of the NTM when the latter is applied to calculate the moments of order statistics.

Example 2.9

Take $s = 3, N = 7, i_1 = 1, i_2 = 3$ and $i_3 = 5$. Table 2.7 lists the values of the moments of the order statistics, and those values obtained by transformation and direct methods of numerical integration, where we have taken $n = 10\,007$. We can see that the direct method is usually better than the transformation method.

If the support of the p.d.f. $p(x)$ is in an infinite interval, we should evaluate the integral with domain (2.5.5). Now D must be reduced to a finite domain such that the NTM can be applied. It means that we should find a finite interval (A,B) satisfying $-\infty < A < B < \infty$, and use

Table 2.7 *Moments of order statistics of $U(0,1)$*

(r_1, r_2, r_3)	real value	Error for transformation method	Error for direct method
(1,0,0)	0.125	2.0×10^{-7}	1.0×10^{-7}
(2,0,0)	0.02777778	8.0×10^{-8}	1.0×10^{-8}
(3,0,0)	8.33333333	6.2×10^{-8}	1.6×10^{-8}
(1,1,0)	0.05555556	1.0×10^{-7}	1.0×10^{-8}
(1,2,0)	0.02777778	1.0×10^{-7}	1.0×10^{-8}
(1,3,0)	0.01515152	1.6×10^{-7}	1.0×10^{-8}
(1,2,3)	0.01165501	1.4×10^{-7}	1.0×10^{-8}

$$D(A,B) = \{\boldsymbol{x} : A \le X_1 \le \cdots < X_s \le B\} \tag{2.5.8}$$

instead of D such that the error between the integrals over D and $D(A,B)$ is less than the pre-assigned positive number. Using a linear transformation

$$y_i = (x_i - A)/(B - A), \; i = 1, \cdots, s,$$

we have

$$\int_{D(A,B)} f(\boldsymbol{x}) d\boldsymbol{x}$$

$$= (B-A)^s \int_{A_s} f(A + (B-A)y_1, \cdots, A + (B-A)y_s) dv,$$

and then we can obtain the approximate value of the above integral by the methods stated in previous sections. Hence the rest of the problem is how to find two suitable numbers A and B. This is similar to Example 2.7, and there is no universal method for solving this problem. We may use also the testing method, i.e. we try several intervals $(A_i, B_i), i = 1, \cdots, m$, and then find the best one, say (A_{i_0}, B_{i_0}) among them. We often choose an interval (A, B) for an integral which we have the known value of the integral, and then use this interval for the calculation of other integrals in which we do not know their values. We now illustrate this method by the following example.

Example 2.10

Consider the order statistics of standard normal distribution. When $N = 7, i_1 = 6$, and $i_2 = 7$, it is known that

$$M_{6,7}(1,1) = 1.2203041356$$

(Teichroew (1956)). By the symmetry of normal distribution about the origin, we take $B = -A = 4,\ 4.5,\ 5,\ 5.5$, and we use the transformation and direct methods for numerical integration with $n = 28\,657,\ 46\,368,\ 75\,025,\ 121\,393$, and different intervals (A, B) respectively. The errors are listed in Table 2.8, where the first value in each cell denotes the absolute error occurred by the transformation method and the second one by the direct method. We may draw two conclusions from Table 2.8: (a) $(-4.5, 4.5)$ is the best one chosen by two methods; (b) The direct method is often better than the transformation method. This coincides with the conclusion of Example 2.9.

In general we do not know how to choose (A, B), in particular, for the moments of high order. So how do we determine (A, B)? We could use the following identities,

$$M_{i_1,\cdots,i_s}(1,\cdots,1) = (-1)^s M_{N-i_s+1,\cdots,N-i_1+1}(1,\cdots,1) \tag{2.5.9}$$

$$M_{i_1,\cdots,i_s}(1,\cdots,1) = 0, \text{if } N \text{ and } s \text{ are odd numbers,}$$
$$\text{and } i_j = s+1-i_{s+1-j},\ j = 1,\cdots,\tfrac{s+1}{2}.$$

Table 2.8 *Comparison of integral domains*

(A, B) / n	$(-4, 4)$	$(-4.5, 4.5)$	$(-5, 5)$	$(-5.5, 5.5)$
28,657	9.45×10^{-4}	2.69×10^{-4}	5.78×10^{-4}	7.56×10^{-4}
	1.12×10^{-3}	9.50×10^{-5}	3.30×10^{-4}	6.28×10^{-4}
46,368	1.03×10^{-3}	1.19×10^{-4}	3.60×10^{-4}	4.75×10^{-4}
	1.13×10^{-3}	1.50×10^{-5}	2.72×10^{-4}	4.01×10^{-4}
75,025	1.08×10^{-3}	2.90×10^{-5}	2.25×10^{-4}	3.04×10^{-4}
	1.15×10^{-3}	4.40×10^{-5}	1.69×10^{-4}	2.55×10^{-4}
121,393	1.23×10^{-3}	4.00×10^{-5}	1.41×10^{-4}	1.89×10^{-4}
	1.15×10^{-3}	8.40×10^{-5}	1.06×10^{-4}	1.65×10^{-4}

We refer the reader to David(1981), Patel, Kapadia and Owen (1976) for more identities. One chooses an interval (A, B) to calculate approximately the moments on each side of (2.5.9), and use their difference as the measurement for an estimate of the error. Then we may choose the best one (A_{i_0}, B_{i_0}) among a number of intervals. Besides we may also use these identities to determine the number N for the numerical evaluation of integrals (Zhang (1990)). In his paper, Zhang gives various mixed moments of order statistics of the standard normal distribution, for instance, for $N = 10$

$$\begin{aligned} E(X_{(1)}X_{(2)}X_{(3)}) &\cong -1.468, \\ E(X_{(1)}X_{(2)}X_{(6)}) &\cong 0.041, \\ E(X_{(7)}X_{(8)}X_{(9)}) &\cong 0.509. \end{aligned}$$

2.6 Distributions on the simplex T_s

Following Aitchison (1986) any vector $\boldsymbol{x} \in T_s$ is called a composition, where T_s is defined in (1.4.8). This type of data appears in many different fields, such as geochemical compositions of rocks, sediments at different depths and industrial production. In order to study compositional data we need various distributions on the simplex T_s which are called **compositional distributions** (Fang, Kotz and Ng (1992)). The most popular and important distribution on T_s is the **Dirichlet distribution**.

2.6.1 The Dirichlet distribution

Recall that a random variable Y has a **gamma distribution** with parameters $\beta > 0$ and $\lambda > 0$ if Y has the p.d.f.

$$\lambda^{\beta}(\Gamma(\beta))^{-1}y^{\beta-1}e^{-\lambda y}, \quad y > 0, \tag{2.6.1}$$

and we write $Y \sim Ga(\beta, \lambda)$.

When $\lambda = 1$, $Ga(\beta, 1)$ is called the standard gamma distribution; when $\beta = n/2$ and $\lambda = 1/2$, $Ga(n/2, 1/2)$ reduces to the chi–square distribution χ^2_n; when $\beta = 1$, $Ga(1, \lambda)$ is just the exponential distribution $\text{Exp}(\lambda)$ with p.d.f.

$$p(x) = \begin{cases} \lambda e^{-\lambda x}, & \text{if } x > 0, \\ 0, & \text{otherwise.} \end{cases} \tag{2.6.2}$$

Definition 2.3
Let $X_1, \cdots, X_s$ be independent random variables and $X_i \sim Ga(\alpha_i, \lambda)$, $\alpha_i > 0$, $\lambda > 0$, $i = 1, \cdots, s$. Set

$$Y_i = X_i \Big/ \sum_{j=1}^{s} X_j, \quad i = 1, \cdots, s. \tag{2.6.3}$$

Then we say $(Y_1, \cdots, Y_s)$ has a **Dirichlet distribution** with parameters $\alpha_1, \cdots, \alpha_s$ and write $(Y_1, \ldots, Y_s) \sim D_s(\alpha_1, \cdots, \alpha_s)$.

The following properties of Dirichlet distribution are useful, and for more details the reader may refer to Johnson and Kotz (1972), and Fang, Kotz and Ng (1990):

(a) Suppose $(Y_1, \cdots, Y_s) \sim D_s(\alpha_1, \cdots, \alpha_s)$. Then $(Y_1, \cdots, Y_s) \in T_s$ and $(Y_1, \cdots, Y_{s-1})$ has a p.d.f.

$$p(y_1, \cdots, y_{s-1}) = \begin{cases} \frac{\Gamma(\alpha_1 + \cdots + \alpha_s)}{\Gamma(\alpha_1) \cdots \Gamma(\alpha_s)} \prod_{i=1}^{s} y_i^{\alpha_i - 1}, & \text{if } (y_1, \cdots, y_s) \in T_s, \\ 0, & \text{otherwise.} \end{cases} \tag{2.6.4}$$

Therefore, the distribution of $(Y_1, \cdots, Y_s)$ is independent of λ.

(b) Moments of $(Y_1, \cdots, Y_s) \sim D_s(\alpha_1, \cdots, \alpha_s)$ are given by

$$E\Big(\prod_{j=1}^{s} Y_j^{r_j}\Big) = \Big(\prod_{j=1}^{s} \alpha_j^{[r_j]}\Big) \Big/ \alpha^{[r_1+\cdots+r_s]}, \tag{2.6.5}$$

where $\alpha = \alpha_1 + \cdots + \alpha_s$ and $x^{[r]} = x(x+1)\cdots(x+r-1)$ are descending factorials. In particular,

$$E(Y_j) = \alpha_j/\alpha, \qquad \mathrm{Var}(Y_j) = \frac{\alpha_j(\alpha-\alpha_j)}{\alpha^2(\alpha+1)}$$

and

$$\mathrm{Cov}(Y_i, Y_j) = \frac{-\alpha_i\alpha_j}{\alpha^2(\alpha+1)}.$$

(c) When $\alpha_1 = \cdots = \alpha_s = 1$, $D_s(1, \cdots, 1)$ reduces to the uniform distribution $U(T_s)$. When $s = 2$, $D_2(\alpha_1, \alpha_2)$ is the beta distribution $Be(\alpha_1, \alpha_2)$.

(d) Suppose $\boldsymbol{y} \equiv (Y_1, \cdots, Y_s) \sim D_s(\alpha_1, \cdots, \alpha_s)$. Then $\boldsymbol{y}$ has the following stochastic representations:

$$\boldsymbol{y} \stackrel{d}{=} \Big(\prod_{i=1}^{s-1} B_i, (1-B_1)\prod_{i=2}^{s-1} B_i, \cdots, 1-B_{s-1}\Big), \tag{2.6.6}$$

where $B_1, \cdots, B_{s-1}$ are independent and $B_i \sim Be(\sum_{j=1}^{i}\alpha_j, \alpha_{i+1})$, $i = 1, \cdots, s-1$, or

$$\boldsymbol{y} \stackrel{d}{=} \Big(B_1, B_2(1-B_1), \cdots, B_{s-1}\prod_{i=1}^{s-2}(1-B_i), \prod_{i=1}^{s-1}(1-B_i)\Big), \tag{2.6.7}$$

where $B_1, \cdots, B_{s-1}$ are independent and $B_i \sim Be(\alpha_i, \sum_{j=i+1}^{s}\alpha_j)$, $i = 1, \cdots, s-1$.

2.6.2 Additive logistic elliptical distributions

Let

$$T_s^+ = \{\boldsymbol{x} = (x_1, \cdots, x_s) : x_i > 0, i = 1, \cdots, s, x_1 + \cdots + x_s = 1\}. \tag{2.6.8}$$

For any $\boldsymbol{x} \in T_s^+$, let

$$\boldsymbol{y}_{\boldsymbol{x}}' = (y_1, \cdots, y_{s-1}) = (\log \frac{x_1}{x_s}, \cdots, \log \frac{x_{s-1}}{x_s}). \tag{2.6.9}$$

This is a one to one mapping which maps T_s^+ onto R^{s-1}, and it has inverse mapping

$$\begin{aligned} x_i &= e^{y_i} \Big/ \Big(1 + \sum_{j=1}^{s-1} e^{y_j}\Big), i = 1, \cdots, s-1, \\ x_s &= 1 \Big/ \Big(1 + \sum_{j=1}^{s-1} e^{y_j}\Big) \end{aligned} \tag{2.6.10}$$

which is the so-called **logistic transformation**.

Definition 2.4

A random vector $\boldsymbol{x} \in T_s^+$ is said to have an **additive logistic elliptical distribution** if its associate $\boldsymbol{y}_{\boldsymbol{x}} \sim EC_{s-1}(\boldsymbol{\mu}, \Sigma, g)$ and is denoted by $\boldsymbol{x} \sim ALE_s(\boldsymbol{\mu}, \Sigma, g)$, where g is called the density generator. In particular, if $\boldsymbol{y}_{\boldsymbol{x}} \sim N_{s-1}(\boldsymbol{\mu}, \Sigma)$, $\boldsymbol{x}$ is said to have an **additive logistic normal distribution** (Aitchison (1986)) and is denoted by $\boldsymbol{x} \sim ALN_s(\boldsymbol{\mu}, \Sigma)$.

Assume $\boldsymbol{x} = (X_1, \cdots, X_s)' \sim ALE_s(\boldsymbol{\mu}, \Sigma, g)$. Then the density of $X_1, \cdots, X_{s-1}$ is given by

$$(\det \boldsymbol{\Sigma})^{-1/2} \prod_{i=1}^{s} x_i^{-1} g[(\boldsymbol{y}_x - \boldsymbol{\mu})' \boldsymbol{\Sigma}^{-1} (\boldsymbol{y}_x - \boldsymbol{\mu})]. \tag{2.6.11}$$

Since the additive logistic elliptical distribution is defined by the ratios of the components of $\boldsymbol{x}$, we can find the formulae for

$E(X_i/X_j)$, $\text{cov}(X_i/X_j,\ X_k/X_l)$, $E(\log \frac{X_i}{X_j})$, $\text{cov}(\log \frac{X_i}{X_j}, \log \frac{X_k}{X_l})$ but there is no analytic expression for $E(X_i)$ and $\text{cov}(X_i,\ X_j)$ which are useful in many practical problems. Using the method stated in section 2.4, we will give in this section a numerical method for evaluating any mixed moments of the components of $\boldsymbol{x}$. For simplicity we consider the ALN distribution only.

Since the density function of $ALN_s(\boldsymbol{\mu}, \Sigma)$ is

$$(2\pi)^{\frac{-(s-1)}{2}} (\det \Sigma)^{-\frac{1}{2}} \prod_{i=1}^{s} x_i^{-1} \exp\Big\{ -\frac{1}{2}(\boldsymbol{y_x} - \boldsymbol{\mu})'\Sigma^{-1}(\boldsymbol{y_x} - \boldsymbol{\mu})\Big\}, \tag{2.6.12}$$

where $\boldsymbol{y_x}$ is defined by (2.6.9), the mixed moment of order $r_1, \cdots, r_s$ of $\boldsymbol{x}$ is

$$\begin{aligned} E(X_1^{r_1} \cdots X_s^{r_s}) &= (2\pi)^{-(s-1)/2} (\det \Sigma)^{-1/2} \\ &\times \int_{T_s^+} \prod_{i=1}^{s} x_i^{r_i - 1} \exp\Big\{ -\frac{1}{2}(\boldsymbol{y_x} - \boldsymbol{\mu})'\Sigma^{-1}(\boldsymbol{y_x} - \boldsymbol{\mu})\Big\} dv \end{aligned} \tag{2.6.13}$$

where dv is the volume element of T_s^+.

In section 2.4 we introduced the numerical methods for evaluating the integrals over T_s, and thus we can obtain the approximate value for the mixed moment (2.6.13) by the transformation method and direct method. (Wang and Fang (1990a)).

2.7 Applications in Bayesian statistics

Numerical integration plays a central role in practical **Bayesian statistics**. The implementation of Bayesian inference procedures can be made to appear deceptively simple. Given a likelihood function $l(\boldsymbol{x};\boldsymbol{\theta})$, where $\boldsymbol{\theta}$ is parameter, and **prior density** $q(\boldsymbol{\theta})$, we apply Bayes' theorem to obtain the **posterior density** of $\boldsymbol{\theta}$

$$p(\boldsymbol{\theta}|\boldsymbol{x}) = \frac{l(\boldsymbol{x};\boldsymbol{\theta})q(\boldsymbol{\theta})}{\int l(\boldsymbol{x};\boldsymbol{\theta})q(\boldsymbol{\theta})d\boldsymbol{\theta}}, \tag{2.7.1}$$

which is the basis of statistical inference in Bayesian statistics. If we are interested in the marginal density of $\boldsymbol{\theta}_{(1)}$, where $\boldsymbol{\theta}_{(1)}$ is a subset of $\boldsymbol{\theta} = (\boldsymbol{\theta}'_{(1)}, \boldsymbol{\theta}'_{(2)})'$, we then simply integrate over $\boldsymbol{\theta}_{(2)}$ to obtain

$$p(\boldsymbol{\theta}_{(1)}|\boldsymbol{x}) = \int p(\boldsymbol{\theta}|\boldsymbol{x})d\boldsymbol{\theta}_{(2)}. \tag{2.7.2}$$

When $l(\boldsymbol{x};\boldsymbol{\theta})$ belongs to an **exponential family** (Patel, Kapadia and Owen (1976)) and $p(\boldsymbol{\theta})$ belongs to the corresponding conjugate family respectively, it is well known that $p(\boldsymbol{\theta}|\boldsymbol{x})$ and $p(\boldsymbol{\theta}_{(1)}|\boldsymbol{x})$ can be performed analytically. However the integrals in (2.7.1) and (2.7.2) must either be performed numerically or analytically. A lot of efficient approaches to this integration problem have been proposed by many authors (Reilly (1976), Naylor and Smith (1982), Smith *et al.* (1985), and Shaw (1988)). In particular Shaw considered the NTM approach.

Consider an integral of the form

$$I_s(q) = \int_{R^s} q(\boldsymbol{\theta})f(\boldsymbol{x};\boldsymbol{\theta})d\boldsymbol{\theta}. \tag{2.7.3}$$

Shaw (1988) suggested that $I_s(q)$ be numerically estimated by a sum of the form

$$\widehat{I}_s(q) = \sum_{i=1}^{n} w_i q(\boldsymbol{\theta}_i)f(\boldsymbol{x};\boldsymbol{\theta}_i), \tag{2.7.4}$$

where w_i and $\{\boldsymbol{\theta}_i\}$ are called **weights** and **nodes**. How are $\{w_i\}$ and $\{\boldsymbol{\theta}_i\}$ chosen? Shaw introduced a flexible class of so-called **importance sampling distributions** defined implicitly by

$$\begin{aligned} &x_j = c_j[A_j f_j(U_j) - (1-A_j)f_j(1-U_j) + b_j], \quad j = 1,\cdots,s \\ &\boldsymbol{x} = (x_1,\cdots,x_s), \end{aligned}$$

where $0 \le A_j \le 1$, f_j is a monotonic increasing function on $(0,1)$ such that $f_j(u) \to -\infty$ as $u \to 0^+$, $\boldsymbol{u} = (U_1,\cdots,U_s) \sim U(C^s)$, and b_j and c_j are constants determined by f_j and A_j. Let $\{\boldsymbol{c}_k\}$ be an NT–net of n points on C^s and $\{v_j\}$ be a discrete probability distribution on $\{\boldsymbol{c}_k\}$. Then the nodes of the corresponding integral are $\{\boldsymbol{\theta}_k\}$, where the jth coordinate of $\boldsymbol{\theta}_k$ is

$$\theta_{kj} = c_j[A_j f_j(c_{kj}) - (1-A_j)f_j(1-c_{kj}) + b_j],$$

with corresponding weight

$$w_k = v_k \prod_{j=1}^{s} [c_j(A_j f_j'(c_{kj}) - (1 - A_j) f_j'(1 - c_{kj}))],$$

where f_j' denotes the derivative of f_j. Typically, $v_i = 1/n$ for all i. Shaw gave some examples and related discussions on his approach, which in fact can be considered to be a mixture of Monte Carlo method and the NTM.

An early work of Cranley and Patterson (1976) had a similar idea: Consider the integral (2.1.1). A quadrature rule $R(f)$ may be randomized when it is possible to construct a family of rules $R(f, q)$ indexed by a parameter q with similar integration properties. When q is sampled from a distribution and there is no reason to suppose that one member of the family will be more accurate than any other on an arbitrary integrand, this set of rules is called a **stochastic family of quadrature rules**. It is not necessary that every rule in a family has the same chance of selection. Suppose that $q \sim G(u)$. Then

$$I = \int R(f, q) dG(q).$$

Cranley and Patterson (1976) considered employing the *glp*–set and the uniform distribution on C^t for q. They suggested using a mixture of the *glp*–set and a fixed but random vector for q. Some numerical examples and detailed discussion were also given. Obviously, their method can be applied to Bayesian statistics.

When the dimension s is large or infinite, Bouleau (1990) proposed a method based on the implementation of the Bernoulli shift operator by pointers. Therefore, his work can evaluate the expectation of a function of stochastic processes.

It is known that Gibbs sampling plays an important role in Bayesian Statistics. This approach provides an algorithm to generate a posterior Monte Carlo sample (Gelfand *et al.* (1990), Müller (1992)). The problem of finding an efficient algorithm based on a combination of the NTM and Gibbs sampling is still open.

Exercises

2.1 Compute the integral

$$\iint_{x^2+y^2\le 1} dxdy$$

by a classical method (repeated Simpson rule or repeated trapezoid rule), a Monte Carlo method and the NTM (*glp* set), and compare the results with the real value of π.

2.2 Compute the integral

$$\iiint_{x^2+y^2+z^2\le 1} dxdydz$$

by classical method, Monte Carlo method and NTM, and compare the results with the real value of $4\pi/3$.

2.3 Use different NT-nets to compute the following integrals
(a) $I_1 = \int_0^1\int_0^1\int_0^1 x_1x_2x_3dx_1dx_2dx_3$,
(b) $I_2 = \int_0^1\int_0^1 e^{-x^2-y^2}dxdy$,
(c) $I_3 = \int_0^1\int_0^1\int_0^1 \exp(x_1x_2x_3)dx_1dx_2dx_3$,
(d) $I_4 = \int_0^1\int_0^1\int_0^1 \exp(\sin x_1 \sin x_2 \sin x_3)dx_1dx_2dx_3$.

2.4 Use different NT-nets to evaluate the integrals

$$I_i = \int_{C^s} f_i(\boldsymbol{x})d\boldsymbol{x}, \quad 1\le i\le 4,$$

where f_i $(i=1,\cdots,4)$ are given in section 2.1. Give comparisons for $s=2,3,4$ and different NTMs.

2.5 Give the p.d.f.s of $\boldsymbol{x}\sim\mathrm{MPVII}_s(\boldsymbol{\mu},\boldsymbol{\Sigma},g)$ and $\boldsymbol{x}\sim\mathrm{MPII}_s(\boldsymbol{\mu},\boldsymbol{\Sigma})$.

2.6 Give the corresponding formula to (2.1.13) for the case $s=3$.

2.7 Let $\boldsymbol{x} \sim N_3(\boldsymbol{0}, \boldsymbol{R})$ with $\boldsymbol{R} = (\rho_{ij})$. It is known that the orthant probability p of $\boldsymbol{x} \in R^3_+$ has the formula (2.3.1). Let

$$\begin{aligned}
D_1 &= \{(x,y,z) : x > 0, y > 0, z > 0\} = R^3_+, \\
D_2 &= \{(x,y,z) : x > 0, y > 0, z < 0\}, \\
D_3 &= \{(x,y,z) : x > 0, y < 0, z > 0\}, \\
D_4 &= \{(x,y,z) : x > 0, y < 0, z < 0\}, \\
D_5 &= \{(x,y,z) : x < 0, y > 0, z > 0\}, \\
D_6 &= \{(x,y,z) : x < 0, y > 0, z < 0\}, \\
D_7 &= \{(x,y,z) : x < 0, y < 0, z > 0\},
\end{aligned}$$

and

$$D_8 = \{(x,y,z) : x < 0, y < 0, z < 0\}.$$

(a) Find the corresponding formulas for the orthant probabilities for $\boldsymbol{x} \in D_i$ $(2 \le i \le 8)$.

(b) Find approximate values for

$$p_i = \int_{D_i} n_3(\boldsymbol{x}; \boldsymbol{0}, \boldsymbol{R}) d\boldsymbol{x}, \quad 1 \le i \le 8,$$

where $n_3(\boldsymbol{x}; \boldsymbol{0}, \boldsymbol{R})$ is the p.d.f. of $N_3(\boldsymbol{0}, \boldsymbol{R})$ by the symmetrization method.

2.8 Let $p(\boldsymbol{x})$ be the p.d.f. of $Mt_3(2, \boldsymbol{0}, \boldsymbol{R})$ with $\rho_{12} = \rho_{13} = \rho_{23} = 1/2$. Find the approximate values of

$$q_i = \int_{D_i} p(\boldsymbol{x}) d\boldsymbol{x}, \quad 1 \le i \le 8,$$

where the D_i are given in exercise 2.7, by the method of symmetrization. Give comparisons between the p_i and q_i, where the p_i are defined in Exercise 2.7.

2.9 Let $\boldsymbol{u} = (U_1, \cdots, U_s)' \sim U(U_s)$, the uniform distribution on U_s. Evaluate $E(\boldsymbol{u})$ and $\mathrm{Cov}(\boldsymbol{u})$ for $s = 3, 4, 5$ by the transformation and direct methods and compare these two methods.

2.10 Let $D = \{\boldsymbol{x} : \boldsymbol{x} \in R^s, \boldsymbol{x}'\boldsymbol{A}\boldsymbol{x} = 1\}$ be the surface of an ellipsoid $\{\boldsymbol{x} : \boldsymbol{x}'\boldsymbol{A}\boldsymbol{x} \leq 1\}$, where $\boldsymbol{A} > 0$. Let $\boldsymbol{u} \sim U(D)$.

(a) Find the formulas of $E(\boldsymbol{u})$ and $\mathrm{Cov}(\boldsymbol{u})$.

(b) Find formulas of $E(h(\boldsymbol{u}))$ for the transformation and direct methods.

(c) Find approximate values for $E(\boldsymbol{u})$ and $\mathrm{Cov}(\boldsymbol{u})$ by the transformation and direct methods.

2.11 Let $F(x)$ be the standard exponential distribution and$X_{(1)} \leq X_{(2)} \leq \cdots \leq X_{(N)}$ be its order statistics. Find approximate values of $E(X_{(i)})$ and $\mathrm{Cov}(X_{(i)}, X_{(j)})$ for $N = 5$, $i, j = 1, \cdots, 5$.

2.12 Let $\boldsymbol{y} = (Y_1, Y_2, Y_3) \sim D_3(2, 4, 6)$.

(a) Give the real values of $E(\boldsymbol{y})$ and $\mathrm{Cov}(\boldsymbol{y})$.

(b) Give approximate values of $E(\boldsymbol{y})$ and $\mathrm{Cov}(\boldsymbol{y})$ by the transformation and direct methods.

2.13 Let $\boldsymbol{x} \sim ALN_3(\boldsymbol{0}, \boldsymbol{I}_3)$. Find the approximate values of $\mathrm{Cov}(\boldsymbol{x})$ by the transformation and direct methods.

2.14 A random variable B is said to have the beta type II distribution with parameters α and β if B has density

$$\frac{1}{B(\alpha, \beta)} b^{\alpha-1}(1+b)^{-(\alpha+\beta)}, \quad b > 0,$$

and we shall denote it by $B \sim BeII(\alpha, \beta)$. Let $B^* \sim Be(\alpha, \beta)$.

(a) Find the relationship between B and B^*.

(b) It is known that $Be(1,1)$ is the uniform distribution $U(0,1)$. What is the p.d.f. of $BeII(1,1)$? Find the p.d.f. of $BeII(1,1)$ directly from the relationship (a).

2.15 Based on the above exercises give your comments on the different techniques (classical, Monte carlo, NTMs), and the different NTMs for numerical evaluation of multiple integrals.

CHAPTER 3

Optimization and its applications in statistics

Let D be a closed and bounded domain in R^s and let $f(\boldsymbol{x})$ be a continuous function on D. We want to find $\boldsymbol{x}^* \in D$ and M such that

$$M = f(\boldsymbol{x}^*) = \max_{\boldsymbol{x} \in D} f(\boldsymbol{x}).$$

For the case where D is a rectangle in R^s, we will introduce the SNTO algorithm which is a sequential algorithm for finding the approximate solution for $\boldsymbol{x}^*$ and M. We will then apply SNTO and its extension RSNTO to robust regression, nonlinear regression, maximum likelihood estimation and likelihood ratio statistics. A version of SNTO for nonrectangular domains is proposed and applied to many problems.

3.1 A number-theoretic method for optimization

Let $D = [\boldsymbol{a}, \boldsymbol{b}]$ be a rectangle in R^s, i.e. $a_i \leq x_i \leq b_i$, $i = 1, \cdots, s$, and let $f(\boldsymbol{x})$ be a continuous function on D. We want to find $\boldsymbol{x}^*$ and M such that

$$M = f(\boldsymbol{x}^*) = \max_{\boldsymbol{x} \in D} f(\boldsymbol{x}). \tag{3.1.1}$$

M is called the **global maximum** of $f(\boldsymbol{x})$ on D, and $\boldsymbol{x}^*$ is called a **maximum point** on D. There are many numerical methods for solving this problem, such as the downhill simplex method, Newton-Gauss method, quasi-Newton methods, steepest descent method, conjugate gradient methods and restricted step methods (Dixon (1972), Wolfe (1969), Dixon and Szegö (1975), Avriel (1976), Dennis and Moré (1977) Nash and Walker-Smith (1987), Deng (1988), Fletcher (1987), Gill, Murray and Wright (1981)).

However most of these methods require that the function $f(\boldsymbol{x})$ is unimodal and/or differentiable to ensure that the global maximum can be attained. Otherwise only a local maximum may be obtained. Besides these methods will have difficulties in finding maxima of functions containing the expressions "max", "min" or $|\boldsymbol{x}|$, or if $f(\boldsymbol{x})$ is defined piecewise such as

$$f(\boldsymbol{x}) = \begin{cases} f_1(\boldsymbol{x}), & \text{if } \boldsymbol{x} \in D_1, \\ \quad \cdots & \\ f_m(\boldsymbol{x}), & \text{if } \boldsymbol{x} \in D_m, \end{cases} \qquad D_1 \cup \cdots \cup D_m = D,$$

where the derivative often does not exist or is not easily computed on the boundary of each D_i. Horst and Tuy (1990) said: "The enormous practical need for solving global optimization problems coupled with a rapidly advancing computer technology has allowed one to consider problems which a years ago would have been considered computationally intractable". The book written by Horst and Tuy (1990) has collected a number of diverse algorithms for solving a wide variety of multiextremal global optimization problems. On the other hand, in past twenty years there has been considerable activity related to Monte Carlo simulation. There is a growing recognition that topics such as Monte Carlo optimization (Rubinstein (1986)), and simulated annealing algorithms (Bertsimas and Tsitsiklis (1993)). The latter is a global optimization algorithm based on stochastic iteration. Ferrari *et al.* (1993) said "Simulated annealing is a very well known algorithm for global optimization. Its convergence properties have been intensively studied,$\cdots$, it has been applied to many hard problems which could not be solved in a satisfactory way otherwise. The most serious drawback lies in its very slow convergence rate."

In this chapter we shall discuss applications of the NTM in global optimization problems, in particular, in statistical problems. The NTM for optimization competes well with other optimization methods.

The idea of the NTM for optimization is as follows: Take an NT-net $\mathcal{P} = \{\boldsymbol{x}_k, k = 1, \cdots, n\}$ on D. Since $f(\boldsymbol{x})$ is continuous and D is closed and bounded, we may expect that there is a point $\boldsymbol{x}_n^*$ among $\{\boldsymbol{x}_k\}$ such that $f(\boldsymbol{x}_n^*)$ is close to M, and $\boldsymbol{x}_n^*$ is close to

$\boldsymbol{x}^*$ if n is large. Now we let $\boldsymbol{x}_n^* \in \{\boldsymbol{x}_k\}$ be a point satisfying

$$M_n = f(\boldsymbol{x}_n^*) = \max_{1 \le k \le n} f(\boldsymbol{x}_k). \tag{3.1.2}$$

In this section we shall give an estimation for the convergence rate of $M_n \to M$, as $n \to \infty$.

By the assumptions on $f(\boldsymbol{x})$ and D, it follows that $f(\boldsymbol{x})$ is uniformly continuous on D. Hence for any given $\varepsilon > 0$, there exists $\delta > 0$ such that $|f(\boldsymbol{x}_1) - f(\boldsymbol{x}_2)| < \varepsilon$ for any $\boldsymbol{x}_1, \boldsymbol{x}_2 \in D$ and $d(\boldsymbol{x}_1, \boldsymbol{x}_2) < \delta$, where $d(\boldsymbol{x}_1, \boldsymbol{x}_2)$ denotes the Euclidean distance between $\boldsymbol{x}_1$ and $\boldsymbol{x}_2$. Recall that the dispersion of the set $\mathcal{P}$ is defined in section 1.4 by

$$DP(\mathcal{P}, D) = \max_{\boldsymbol{x} \in D} \min_{1 \le k \le n} d(\boldsymbol{x}, \boldsymbol{x}_k).$$

Hence if $DP(\mathcal{P}, D) < \delta$, then for some n we have

$$|f(\boldsymbol{x}_n^*) - f(\boldsymbol{x}^*)| < \varepsilon,$$

and consequently

$$M_n \le M \le M_n + \varepsilon. \tag{3.1.3}$$

Furthermore if we can show that there is a sequence of sets $\{\mathcal{P}_{n_i}\}$ $(n_1 < n_2 < \cdots)$ such that $DP \to 0$ as $i \to \infty$, then the sequence $\{M_{n_i}\}$ is convergent. Theorem 1.5 provides an upper estimate for $DP(\mathcal{P}, D)$ which gives $DP(\mathcal{P}_{n_i}, D) \to 0$ $(i \to \infty)$, and more precisely, we shall prove that $DP(\mathcal{P}_{n_i}, D) = O(n_i^{-1/s} \log n_i)(i \to \infty)$ holds for some NT-nets. To prove Theorem 1.5, we shall need the following lemma.

Lemma 3.1

Suppose that $\boldsymbol{y} \in C^s$ and $r \le \sqrt{s}$. Then we can inscribe a cube C^* into $B(\boldsymbol{y}, r) \cap C^s$ with edge-length $r/\sqrt{s}$, where $B(\boldsymbol{y}, r)$ denotes the s-dimensional ball with centre $\boldsymbol{y}$ and radius r.

PROOF We first take an s-dimensional cube C_0 with centre $\boldsymbol{y}$ and edge-length $2r/\sqrt{s}$. Now we consider the rectangle $C_0 \cap C^s$. Each edge of this rectangle is a one-dimensional interval of the form

$I = [t - r/\sqrt{s}, t + r/\sqrt{s}] \cap [0,1]$ for some $t \in [0,1]$. If $t + r/\sqrt{s} \leq 1$, then I contains the interval $[t, t+r/\sqrt{s}]$ of length $r/\sqrt{s}$. Otherwise I contains the interval $[1 - r/\sqrt{s}, 1]$ of length $r/\sqrt{s}$. Altogether, $C_0 \cap C^s$ contains a cube C^* with edge-length $r/\sqrt{s}$, and so $B(\boldsymbol{y}, r) \cap C^s$ contains C^*. The lemma is proved. □

PROOF OF THEOREM 1.5 We may assume that $DP(\mathcal{P}, D) > 0$ and that $\boldsymbol{x} \in D$ satisfies

$$\min_{1 \leq k \leq n} d(\boldsymbol{x}, \boldsymbol{x}_k) = DP(\mathcal{P}, D).$$

Hence the ball $B(\boldsymbol{x}, DP(\mathcal{P}, D) - \varepsilon)$ contains none of the points $\boldsymbol{x}_1, \cdots, \boldsymbol{x}_n$ for any ε with $0 < \varepsilon < DP(\mathcal{P}, D)$. If follows from *Lemma 3.1* that we can inscribe a cube C^* with edge-length $(DP(\mathcal{P}, D) - \varepsilon)/\sqrt{s}$ in $B(\boldsymbol{x}, DP(\mathcal{P}, D - \varepsilon)) \cap C^s$. Therefore by (1.4.3) we have

$$\begin{aligned} 2D(n)^{1/s} \geq D^*(n)^{1/s} &\geq \left| \frac{N(C^*, \mathcal{P})}{n} - v(C^*) \right|^{1/s} \\ &= v(C^*)^{1/s} = \frac{DP(\mathcal{P}, D) - \varepsilon}{\sqrt{s}}, \end{aligned}$$

and the theorem follows. □

Theorem 3.1

Suppose that $f(\boldsymbol{x})$ is a continuous function defined on a closed and bounded domain D, and that $\{\mathcal{P}_{n_i}\} (n_1 < n_2 < \cdots)$ is a sequence of sets on D which have F-discrepancy or quasi F-discrepancy d_{n_i} such that $d_{n_i} = o(1)$ as $i \to \infty$. Let $\boldsymbol{x}^*_{n_i} \in \mathcal{P}_{n_i}$ be a point satisfying

$$M_{n_i} = f(\boldsymbol{x}^*_{n_i}) = \max_{1 \leq k \leq n_i} f(\boldsymbol{x}_k^{(n_i)}).$$

Then $M_{n_i} \to M$ as $i \to \infty$.

It is difficult to give a quantitative relation between ε and δ such that (3.1.3) holds, since δ depends on ε, $f(\boldsymbol{x})$ and D. However we have the following

Theorem 3.2
Suppose that $\nabla f = (\frac{\partial f}{\partial x_1}, \cdots, \frac{\partial f}{\partial x_s})$ is continuous and

$$\|\nabla f\|_2 = \Big(\sum_{i=1}^{s} \Big(\frac{\partial f}{\partial x_i}\Big)^2\Big)^{1/2} < C$$

over D. Then

$$M_n \le M \le M_n + CDP(\mathcal{P}, D). \tag{3.1.4}$$

PROOF By the mean value theorem we obtain

$$f(\boldsymbol{x}^*) - f(\boldsymbol{x}_i) = \nabla f|_{\boldsymbol{z}} \cdot (\boldsymbol{x}^* - \boldsymbol{x}_i)$$

for some $\boldsymbol{z}$ on the line segment between $\boldsymbol{x}^*$ and $\boldsymbol{x}_i$. Hence it follows by the Schwarz inequality that

$$\begin{aligned} |f(\boldsymbol{x}^*) - f(\boldsymbol{x}_i)| &\le \|\nabla f\|_{\boldsymbol{z}} d(\boldsymbol{x}^*, \boldsymbol{x}_i) \\ &\le Cd(\boldsymbol{x}^*, \boldsymbol{x}_i) \\ &\le CDP(\mathcal{P}, D), \end{aligned} \tag{3.1.5}$$

and

$$M_n \le M \le M_n + CDP(\mathcal{P}, D),$$

and the theorem is proved. □

By Theorems 3.2 and 1.5, we have

$$M_n \le M \le M_n + 2s^{1/2}CD(n, \mathcal{P})^{1/s}. \tag{3.1.6}$$

Concerning the dispersion and its properties, we refer Niederreiter (1983, 1992).

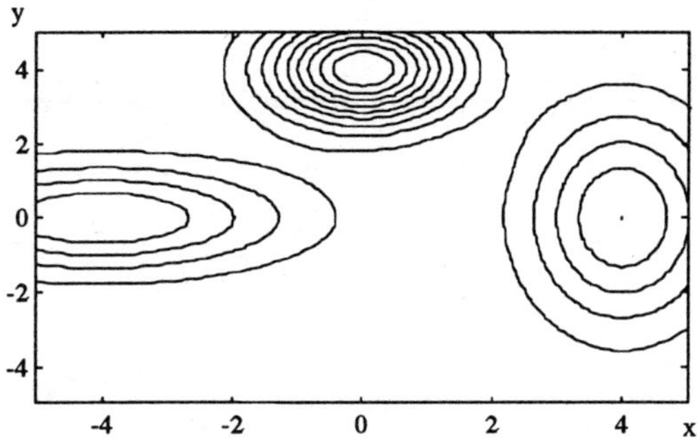

Figure 3.1 Contours of $f(x,y)$

Example 3.1
Consider the function

$$f(x,y) = 2e^{-\frac{1}{2}(x^2+(y-4)^2)} + e^{-\frac{1}{2}((x-4)^2+\frac{y^2}{4})} + e^{-\frac{1}{2}(\frac{(x+4)^2}{4}+y^2)}, \tag{3.1.7}$$

where $(x,y) \in R^2$. Find the global maximum and a maximum point of $f(x,y)$ in R^2.

Using software, for example MATLAB, we know from the contours of f (Figure 3.1) that this function has three extremes which are located at (0,4), (4, 0) and (-4, 0), and that (0,4) is the global maximum point, i.e. $\boldsymbol{x}^* = (0,4)$ and $M = f(\boldsymbol{x}^*) = f(0,4) \cong 2.000091$. If an unsuitable initial point for iteration is chosen in any version of Newton-Gauss method, it is possible that one of the other two extremes $f(4,0) \cong 1.0000003$ or $f(-4,0) \cong 1.0000001$ is attained.

Now we use the NTM to find the approximate solutions of M and $\boldsymbol{x}^*$. We know that the values of the function $f(x,y)$ outside the rectangle $D = [-10,7] \times [-6,7]$ is close to zero, and thus we restrict the domain to be D. We now use Fibonacci sequence to produce the *glp* sets (Exercise 1.10), and the results are listed in Table 3.1. From Table 3.1, we would draw the following conclusions:

(a) All $\boldsymbol{x}_n^* = (x_n^*, y_n^*)$ are close to the global maximum point $\boldsymbol{x}^* = (0,4)$. This is an advantage of NTM for finding the global maximum and maximum point of a function on a domain D.

Table 3.1 *Optimization by NTM*

n	M_n	x_n^*	y_n^*
233	1.876053	0.3240347	3.847639
377	1.848486	0.3938990	3.948276
610	1.947677	−0.1483603	4.176230
987	1.944570	0.1018238	4.214286
1,597	1.988365	−0.1055107	4.024734
2,584	1.982186	0.1282892	4.039280
4,181	1.991075	−0.0890932	3.966874
6,765	1.993981	0.0222206	4.078123
10,946	1.996928	0.0554085	4.010095
17,711	1.999915	0.0040932	3.987268
28,657	1.999322	−0.0276194	4.001658
46,368	1.998314	0.0382052	4.017889
75,025	1.999495	−0.0065374	4.023386

(b) $|M_n - M|$ usually decreases as n increases. The relative error

$$|M_n - M|/M$$

is 2.62% when $n = 610$, and it decreases to 0.586% when $n =$ 1597.

(c) If $n_1 > n_2$, it is possible that $M_{n_1} < M_{n_2}$, because the *glp* sets corresponding to different n's are distinct at all. But if $n_1 >> n_2$, then $M_{n_1} > M_{n_2}$ holds in general.

(d) The rate of convergence $M_n \to M$ is not fast, and those of $(x_n^*, y_n^*) \to \boldsymbol{x}^* = (0, 4)$ is still lower.

In order to avoid the shortcoming incurred by the finite NT-net as stated in comment (c), we suggest the use of infinite NT-net. Suppose that $\{\boldsymbol{x}_1, \boldsymbol{x}_2, \cdots\}$ is an infinite NT-net on D, and let

$$\begin{aligned} M_1 &= f(\boldsymbol{x}_1), \\ M_{k+1} &= \max(M_k, f(\boldsymbol{x}_{k+1})) \quad k = 1, 2, \cdots. \end{aligned} \tag{3.1.8}$$

After a large number n of steps, we may reasonably expect that M_n is close to the maximum M, and additionally M_n is always nondecreasing.

The above two methods based on the NTM are known as **NTM searches**. We shall compare the NTM search with the next example, where the results for optimization are obtained by the NTM search with finite and infinite NT-nets.

Example 3.2
Let

$$f(x, y, z, u) = \exp(xyzu) \sin(x + y + z + u),$$

where $(x, y, z, u) \in C^4$. We already know $M = 1.0261986$.

Table 3.2 gives the comparion of the results obtained by the NTM search with *glp* sets and *H*-sets respectively, where Niederreiter and McCurley (1979) take $p_1 = 2, p_2 = 3, p_3 = 5$ and $p_4 = 11$ in the latter method (section 1.3). The results show that these two methods have almost the same precision. M_n increases monotonically, but the computational effort required to compute the *H*-set is much greater than that required by the *glp* sets. More precisely, the computing time needed for the latter method is more than twice that of the former method, so we recommend the use of *glp* sets.

3.2 A Sequential Algorithm (SNTO)

In the previous section we introduced the NTM for solving the optimization problem (3.1.1), and pointed out that, under certain conditions, the rate of convergence of $M_n \to M$ is $O(n^{-1/s} \log n)$.

Table 3.2 *Comparison between glp set and H-set*

n	$Mn(glp)$	$Mn(H)$
60	1.022111	1.017213
118	1.015450	1.017213
180	1.019175	1.018451
932	1.024526	1.023790
2129	1.025099	1.024064
3001	1.025076	1.024064
5003	1.025139	1.024889
10,007	1.025569	1.025550
28,117	1.025725	1.026118

If the precision requirement is not high, then the approximate solutions M_n and $\boldsymbol{x}_n^*$ may satisfy the requirement when n is suitably large. Otherwise n should be very large. For example, let $a_n = n^{-1/4}\log n$ for the dimension $s = 4$. We show some values of a_n in Table 3.3.

Table 3.3 *Values of* $a_n = n^{-1/4}\log n$

n	10	10^5	10^{10}
a_n	1.4563	0.6474	0.0728
n	10^{20}	10^{30}	10^{40}
a_n	4.605×10^{-4}	2.184×10^{-6}	9.210×10^{-9}

If we require $|M_n - M| < 10^{-5}$, then n must be of an order greater than 10^{20} which is too large, and therefore we need a method to decrease the number n.

Suppose that $[\boldsymbol{a}, \boldsymbol{b}]$ is a rectangle in $\boldsymbol{R}^s$ with volume $v([\boldsymbol{a}.\boldsymbol{b}]) = \prod_{i=1}^{s}(b_i - a_i)$, and that $\{\boldsymbol{y}_k,\ k = 1, \cdots, n\}$ is an NT-net on C^s. Set

$$x_{ki} = a_i + (b_i - a_i)y_{ki},\ i = 1, \cdots, s,$$
$$\boldsymbol{x}_k = (x_{k1}, \cdots, x_{ks}),\ k = 1, \cdots, n.$$

Then $\{\boldsymbol{x}_k,\ k = 1, \cdots, n\}$ is an NT-net on $[\boldsymbol{a}, \boldsymbol{b}]$, and we have

$$\begin{aligned} d(\boldsymbol{x}_i, \boldsymbol{x}_j) &= \Big(\sum_{l=1}^{s}(b_l - a_l)^2(y_{il} - y_{jl})^2\Big)^{1/2} \\ &\leq \max_{1\leq l\leq s}|b_l - a_l| d(\boldsymbol{y}_i, \boldsymbol{y}_j) \end{aligned} \tag{3.2.1}$$

and the dispersion: $DP(\{\boldsymbol{x}_i\}, [\boldsymbol{a}, \boldsymbol{b}]) \leq \max_{1\leq l\leq s}(b_l - a_l) DP(\{\boldsymbol{y}_k\}, C^s)$.

Let $l(D)$ denote the largest edge-length of the rectangle D. If we can find a domain D^* such that $\boldsymbol{x}^* \in D^* \subset D$ and $l(D^*) << l(D)$, then the optimization problem (3.1.1) is reduced to an optimization problem for the region D^*, and thus the same sized NT-net would normally lead to a much more precise approximation to $\boldsymbol{x}^*$. For example, if we take $s = 4$, $D = [\boldsymbol{a}, \boldsymbol{b}]$ and $D^* = [\boldsymbol{a}^*, \boldsymbol{b}^*] \subset D$, where each edge-length of D^* is equal to 10^{-1} times the corresponding edge of D. Roughly speaking, the precision of results obtained in D^* by n points is equivalent to that obtained in D by $10^4 n$ points. This idea was initiated and developed by Niederreiter and Peart (1986), and by Fang and Wang (1990). More precisely, Fang and Wang (1990) suggested a **sequential algorithm for optimization** with NT-nets which we abbreviate by **SNTO**. Now we illustrate SNTO for D being a rectangle $[\boldsymbol{a}, \boldsymbol{b}]$. In what follows we use $\max \boldsymbol{a}$ ($\min \boldsymbol{a}$) to represent $\max_{1 \le i \le n} a_i$ ($\min_{1 \le i \le n} a_i$) where $\boldsymbol{a} = (a_1, \cdots, a_n)$; and n is the number of components of $\boldsymbol{a}$.

SNTO algorithm:

Step 0 **Initialization.** Set $t = 0$, $D^{(0)} = D$, $\boldsymbol{a}^{(0)} = \boldsymbol{a}$ and $\boldsymbol{b}^{(0)} = \boldsymbol{b}$.

Step 1 **Generate an NT-net**. Use a number-theoretic method to generate n_t points $\mathcal{P}^{(t)}$ uniformly scattered on $D^{(t)} = [\boldsymbol{a}^{(t)}, \boldsymbol{b}^{(t)}]$.

Step 2 **Compute a new approximation**. Find $\boldsymbol{x}^{(t)} \in \mathcal{P}^{(t)} \cup \{\boldsymbol{x}^{(t-1)}\}$ and $M^{(t)}$ such that

$$M^{(t)} = f(\boldsymbol{x}^{(t)}) \ge f(\boldsymbol{y}), \quad \forall \boldsymbol{y} \in \mathcal{P}^{(t)} \cup \{\boldsymbol{x}^{(t-1)}\}, \qquad (3.2.2)$$

where $\boldsymbol{x}^{(-1)}$ is the empty set, $\boldsymbol{x}^{(t)}$ and $M^{(t)}$ are the best approximations to $\boldsymbol{x}^*$ and M so far.

Step 3 **Termination criterion**. Let $\boldsymbol{c}^{(t)} = (\boldsymbol{b}^{(t)} - \boldsymbol{a}^{(t)})/2$. If $\max \boldsymbol{c}^{(t)} < \delta$, a pre-assigned small number, then $D^{(t)}$ is small enough; $\boldsymbol{x}^{(t)}$ and $M^{(t)}$ are acceptable; terminate algorithm. Otherwise, proceed to next step.

Step 4 **Contract domain**. Form new domain $D^{(t+1)} = [\boldsymbol{a}^{(t+1)}, \boldsymbol{b}^{(t+1)}]$ as follows:

$$a_i^{(t+1)} = \max(x_i^{(t)} - \gamma c_i^{(t)}, a_i)$$

and

$$b_i^{(t+1)} = \min(x_i^{(t)} + \gamma c_i^{(t)}, b_i)$$

where γ is a predefined **contraction ratio**. Set $t = t + 1$. Go to Step 1.

According to our experience, we suggest to take $n_1 > n_2 = n_3 = \cdots$. The contraction ratio γ is often taken to be 0.5. Niederreiter and Peart (1986) suggested using $\gamma_i = \gamma^i$ as a contraction ratio at the ith stage, where $\gamma > 0$ is a constant.

The steps of SNTO are shown in Figure 3.2, where each edge-length of $D^{(t)}$ is $\frac{1}{2}$ times those of $D^{(t-1)}$. But if $\boldsymbol{x}^{(t)}$ is near the boundary of D as $\boldsymbol{x}^{(1)}$ in Figure 3.2, then $D^{(t)}$ should be required

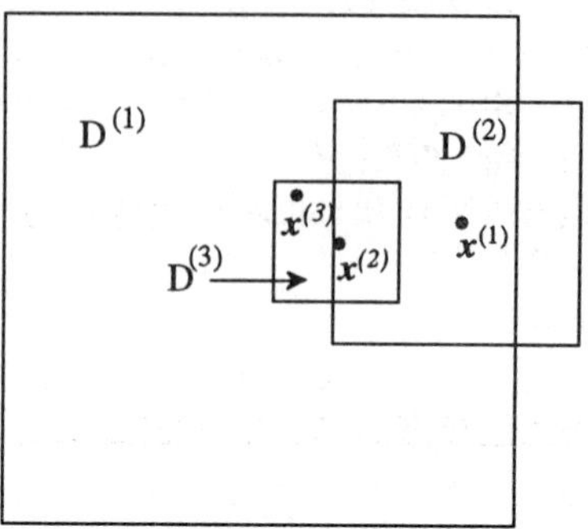

Figure 3.2 SNTO

to fall in D, and so the edge-length may contract to less than half that of $D^{(t-1)}$. Note that it is not necessary to always have $D^{(t+1)} \subset D^{(t)} (t > 1)$. For example, $D^{(3)} \subset D$ but $D^{(3)} \not\subset D^{(2)}$.

An alternative way to contract the domain is to choose $D^{(t+1)} = [\boldsymbol{a}^{(t+1)}, \boldsymbol{b}^{(t+1)}]$ to be the smallest box containing the ρn_t points in $D^{(t)}$ with the maximum function values for some predefined proportion ρ, e.g. $\rho = 0.2$.

Example 3.2 (Continuation)

We now use the SNTO to find the maximum and maximum point $\boldsymbol{x}^*$ of $f(\boldsymbol{x})$ on C^4. Take $n_1 = 1142$, $n_2 = n_3 = \cdots = 118$. Table 3.4 shows $M^{(t)}$ and $\boldsymbol{x}^{(t)}$ when $1 \le t \le 10$. The precision of $M^{(10)}$ is only to 7 d.p. by using only 2204 points, and the precision is less than 7 d.p. if we use the usual method stated in section 3.1 even with 10^5 points.

Remark 3.1

Table 3.4 shows that $M^{(1)} = M^{(2)}$, and this happens often in many other examples. This is the reason that we cannot use $M^{(t+1)} - M^{(t)} \le \varepsilon$ to control the process of termination.

Example 3.3

The function

$$f(x, y, z, u) = -\left(x - \frac{3}{11}\right)^2 - \left(y - \frac{6}{13}\right)^2 - \left(z - \frac{12}{23}\right)^4 - \left(u - \frac{8}{37}\right)^6$$

has its maximum $M = 0$ on C^4 at

$$\begin{aligned} \boldsymbol{x}^* &= \left(\frac{3}{11}, \frac{6}{13}, \frac{12}{23}, \frac{8}{37}\right) \\ &\cong (0.272727, 0.4615384, 0.521739, 0.216216). \end{aligned}$$

Table 3.4 *Approximations of M and $\boldsymbol{x}^*$ by SNTO*

n_t	$M^{(t)}$	$x_1^{(t)}$	$x_2^{(t)}$	$x_3^{(t)}$	$x_4^{(t)}$
1142	1.0242500	0.5153240	0.3638354	0.4470228	0.3183012
118	1.0242500	0.5153240	0.3628354	0.4470228	0.3183012
118	1.0256555	0.4274003	0.4242167	0.3463872	0.4252928
118	1.0259550	0.4141587	0.3823739	0.3998829	0.4512461
118	1.0261160	0.4271354	0.4017061	0.3980291	0.4207906
118	1.0261934	0.4127022	0.4097835	0.4039878	0.4113892
118	1.0261961	0.4113117	0.4093200	0.4056430	0.4127795
118	1.0261983	0.4106828	0.4102800	0.4091189	0.4091050
118	1.0261984	0.4109311	0.4091710	0.4099299	0.4091878
118	1.0261986	0.4104594	0.4096097	0.4099382	0.4094940
2004					

Table 3.5 *Comparison between SNTO and NTM search*

n	$Mn(H)$	n_t	$M^{(t)}$
1	−.0788052	1142	-4.133×10^{-4}
2	−.0428139	118	-4.133×10^{-4}
4	−.0301196	118	-2.299×10^{-4}
22	−.0210772	118	-2.863×10^{-5}
28	−.0163983	118	-2.863×10^{-5}
58	−.0078911	118	-5.478×10^{-6}
316	−.0032230	118	-2.202×10^{-6}
1282	−.0027283	118	-1.593×10^{-6}
1666	−.0026824	118	-1.262×10^{-6}
2002	−.0008884	118	-1.033×10^{-6}
5602	−.0006499	118	-9.940×10^{-7}
9346	−.0006242	118	-9.466×10^{-7}
		118	-9.286×10^{-7}
		2558	

Table 3.5 gives the comparison between the SNTO and NTM search stated in section 3.1, and shows that SNTO is much powerful. Note that $M^{(13)}$ is quite close to M, but $\boldsymbol{x}^{(13)} = (0.2727148, 0.4615162, 0.4907019, 0.2213063)$ is not very close to $\boldsymbol{x}^*$. The reason is that at some stage of the iteration the domain is contracted such that $\boldsymbol{x}^*$ falls outside. In this case $\boldsymbol{x}^*$ does not belong to $D^{(13)}$. In order to enhance the precision of the approximation of $\boldsymbol{x}^*$, we propose the following modification of SNTO:

First we use SNTO to obtain a solution $\boldsymbol{x}^{(t)}$. Then consider a rectangle $\bar{D}^{(1)} = [\bar{\boldsymbol{a}}^{(1)}, \bar{\boldsymbol{b}}^{(1)}]$ with centre at $\boldsymbol{x}^{(t)}$ and with each edge-length which is equal to $1/4 \sim 1/16$ times the respective edge-length of $D = [\boldsymbol{a}, \boldsymbol{b}]$. More precisely,

$$\bar{\boldsymbol{a}}^{(1)} = (\bar{a}_1^{(1)}, \cdots, \bar{a}_s^{(1)}), \ \bar{\boldsymbol{b}}^{(1)} = (\bar{b}_1^{(1)}, \cdots, \bar{b}_s^{(1)}),$$

where

$$\begin{aligned} \bar{a}_i^{(1)} &= \max(a_i, x_i^{(t)} - c_i^{(1)}/m), \\ \bar{b}_i^{(1)} &= \min(b_i, x_i^{(t)} + c_i^{(1)}/m), \ i = 1, \cdots, s, \end{aligned}$$

m is a number satisfying $4 \leq m \leq 16$, and $\boldsymbol{c}^{(1)} = (\boldsymbol{b} - \boldsymbol{a})/2$. We use SNTO again in the domain $\bar{D}^{(1)}$, and then we can define the domain $\bar{D}^{(2)}$ similarly by $\bar{D}^{(1)}$ instead of D, and so on until a satisfactory result is obtained. We call this modified version RSNTO. A similar idea to RSNTO can be found in Niederreiter (1992). The reader may use RSNTO to improve the results of Example 3.3.

Remark 3.2

The rate of convergence may be increased, if the contraction ratio $\gamma < \frac{1}{2}$ is reduced. Tables 3.4 and 3.5 show that it is possible that $\frac{1}{4}$ may be used instead of $\frac{1}{2}$ in the first step of contraction, and we can even use $\frac{1}{10}$ instead of $\frac{1}{2}$ if the function is quite smooth. However if the rate of contraction of domains is too fast, then $\boldsymbol{x}^*$ may lie outside of some $D^{(t)}$, resulting in an incorrect answer.

We shall use SNTO and RSNTO in following sections to treat some problems in statistics.

3.3 Maximum likelihood estimation

The method of maximum likelihood is regarded as one of the most effective methods for estimating the parameters of a distribution. Let $\boldsymbol{x}_1, \cdots, \boldsymbol{x}_N$ be a sample from a population with p.d.f. $g(\boldsymbol{x}, \boldsymbol{\theta})$, where $\boldsymbol{\theta} = (\theta_1, \cdots, \theta_s)$ are parameters, $\boldsymbol{\theta} \in D$, the parameter space. The **maximum likelihood estimate** (MLE) $\hat{\boldsymbol{\theta}}$ of $\boldsymbol{\theta}$ maximizes the **likelihood function**

$$L(\boldsymbol{\theta}) = \prod_{i=1}^{N} g(\boldsymbol{x}_i, \boldsymbol{\theta}), \tag{3.3.1}$$

i.e.

$$L(\hat{\boldsymbol{\theta}}) = \max_{\boldsymbol{\theta} \in D} L(\boldsymbol{\theta}); \tag{3.3.2}$$

or equivalently, the MLE maximizes the logarithm of the likelihood function

$$\mathcal{L}(\boldsymbol{\theta}) = \sum_{i=1}^{N} \log g(\boldsymbol{x}_i, \boldsymbol{\theta}), \tag{3.3.3}$$

i.e.

$$\mathcal{L}(\hat{\boldsymbol{\theta}}) = \max_{\boldsymbol{\theta} \in D} \mathcal{L}(\boldsymbol{\theta}). \tag{3.3.4}$$

For only a few distributions, $\hat{\boldsymbol{\theta}}$ has an analytic expression. In most cases we have to use numerical methods to obtain the MLE $\hat{\boldsymbol{\theta}}$. People very often use the Newton-Raphson method and its extensions to find an approximation of $\hat{\boldsymbol{\theta}}$. However the Newton-Raphson method and its extensions do not always work. Fu (1989) investigated the efficiency of the Newton-Raphson method in estimating the parameters by solving the nonlinear equations. He worked with 1000 samples generated from a Cauchy distribution and found that 43 times among 1000, the method fails to converge and that the method yields local maximum 79 times especially when the initial value is not suitable. The situation becomes worse as the number of parameters increases. This is particularly true for the three-parameter Weibull and the beta distributions where the MLEs have been studied by a number of authors: Cohen (1965), Dubey (1967), Gnanadesikan, Pinkham, and Hughes (1967), Harter and Moore (1965), Menon (1963), Sirvanci and Yang (1984) and Wingo (1972).

However, finding the MLE $\hat{\boldsymbol{\theta}}$ satisfying (3.3.2) or (3.3.4) is a special case of the optimization problem (3.1.1). We might apply the SNTO to find the MLE. In most cases the parameter space D is very large possibly even infinite. It is impossible to use such a large D as the initial domain for the SNTO. Therefore Fang and Yuan (1990) suggested the following method to give the initial domain.

First we find a set of moment estimates $\boldsymbol{\theta}^* = (\theta_1^*, \cdots, \theta_s^*)$ which are not very far from the MLEs of $\boldsymbol{\theta}$ and are not too difficult to find for most useful distributions. Secondly choose a set of positive constants $c_1, \cdots, c_s$ which depends on the population distribution and sample size. Generally speaking, $\{c_i\}$ should be smaller with larger sample size, and bigger with smaller sample size. Then the rectangle

$$D = [\theta_1^* - c_1, \theta_1^* + c_1] \times [\theta_2^* - c_2, \theta_2^* + c_2] \times \cdots \times [\theta_s^* - c_s, \theta_s^* + c_s]$$

can be used for the initial domain of SNTO.

Now we shall apply the above method to the MLEs for the Weibull and the beta distributions.

The p.d.f. of a Weibull distribution with location parameter γ, scale parameter β, and shape parameter α is

$$g(x; \gamma, \beta, \alpha) = \frac{\alpha}{\beta^\alpha}(x-\gamma)^{\alpha-1} \exp\{-[(x-\gamma)/\beta]^\alpha\}. \tag{3.3.5}$$

Its likelihood and logarithm of the likelihood functions are

$$\mathcal{L}(\gamma, \beta, \alpha) = \prod_{i=1}^{N} g(x_i; \gamma, \beta, \alpha), \tag{3.3.6}$$

and

$$\begin{aligned} \mathcal{L}(\gamma, \beta, \alpha) =& N(\log\alpha - \alpha\log\beta) + (\alpha-1)\sum_{i=1}^{N}\log(x_i-\gamma) \\ & - \frac{1}{\beta^\alpha}\sum_{i=1}^{N}(x_i-\gamma)^\alpha, \end{aligned}$$

respectively. Let $\boldsymbol{\theta} = (\gamma, \alpha, \beta)$ and D be the parameter space. Clearly we have

$$D = \{(x, y, z) : x \in R^1,\ y > 0,\ z > 0\}.$$

Menon (1963) and Dubey (1967) gave two different kinds of estimates for the Weibull distributions, which are not maximum likelihood estimates. Cohen (1965), Harter and Moore (1965) and Wingo (1972) discussed methods of getting MLEs for the Weibull distribution. Now we use the SNTO for obtaining the MLE for the Weibull distribution.

First we need to find a set of moment estimates γ^*, β^*, α^* of γ, β, α, as the "initial value". There are several ways to obtain the moment estimates. Let $x_1, \cdots, x_N$ be a sample from the population. The following method is often used:

$\gamma^* = \min(x_1, \cdots, x_N) = x_{(1)}$, the smallest order statistic, α^* is the solution of the equation

$$\frac{(\bar{x} - \gamma^*)^2}{\hat{\sigma}^2 + (x - \gamma^*)^2} = \frac{\Gamma^2(1 + 1/\alpha^*)}{\Gamma(1 + 2/\alpha^*)}, \tag{3.3.7}$$

where $\bar{x}$ and $\hat{\sigma}^2$ are the sample mean and the sample variance respectively, and

$$\beta^* = (\bar{x} - \gamma^*)/\Gamma(1 + 1/\alpha^*). \tag{3.3.8}$$

The values of c_1, c_2 and c_3 depend on the sample size. With the initial values $\{c_i\}$ and the initial region D we have to impose the following conditions on $D = [\boldsymbol{a}, \boldsymbol{b}]$ and its contracted area $D^{(t)} = [\boldsymbol{a}^{(t)}, \boldsymbol{b}^{(t)}]$:

$$a_i^{(t)} \geq 0, \quad i = 1, 2, 3;\ b_1^{(1)} < x_{(1)}. \tag{3.3.9}$$

The five sets of data that we shall use for the Weibull distribution are due to Menon (1963, Set 1, two parameters with $\gamma = 0$), Smith and Dubey (1964, Set 2), Harter and Moore (1965, Sets 3 and 4), and Dubey (1967, Set 5). All of these sets of data except the first are from a three parameter Weibull distribution. We take $\delta = 0.000001$ in the SNTO for all sets. Table 3.6 to Table 3.10 give comparisons of our approximations to the MLEs by the SNTO and the approximations of the MLEs obtained by Menon (1963), Harter and Moore (1965), Cohen (1963) and Wingo (1972), and also the moment estimates. In the last line of each table, n stands for the total points used in the SNTO. LLMAX's of each table denote the values of the function $\mathcal{L}$ at the corresponding estimates of parameters. These tables show that among all the results SNTO's are the best.

Table 3.6 *Set 1 (Menon)*

α	β	LLMAX	Methods
1.22039	0.42855	−36.14313	moment estimates
1.40000	0.57000	−36.04471	Menon's estimates
1.36300	0.50600	−35.67247	Cohen's estimates
1.36016	0.50482	−35.67238	SNTO's estimates
$c_1 = c_2 = 5$	$n_1 = 987$	$n_2 = 233$	$n = 4249$

Table 3.7 *Set 2 (Smith and Dubey)*

γ	α	β	LLMAX	Methods
13.00000	1.73433	52.72173		moment estimates
11.48075	1.78250	54.30704	−466.40332	Wingo's estimates
11.41591	1.80314	54.59485	−466.39307	SNTO's estimates
$c_1 = 10$	$c_2 = 5$	$c_3 = 10$	$n_1 = 1459$	$n_2 = 418, \; n = 6893$

Table 3.8 *Set 3 (Harter and Moore)*

γ	α	β	LLMAX	Methods
15.00000	1.86293	87.47449		moment estimates
2.48000	2.19900	101.80000	−206.25839	H-M's estimates
2.47865	2.19942	101.79488	−206.25839	Wingo's estimates
2.48774	2.19872	101.77040	−206.25826	SNTO's estimates
$c_1 = 15$	$c_2 = 5$	$c_3 = 20$	$n_1 = 1459$	$n_2 = 418, \; n = 6475$

Table 3.9 *Set 4 (Harter and Moore)*

γ	α	β	LLMAX	Methods
40.89999	1.97754	72.10414		moment estimates
30.80000	2.33000	83.50000	−196.51839	H-M's estimates
30.79874	2.32999	83.50906	−196.51833	Wingo's estimates
30.78090	2.33005	83.46425	−196.51828	SNTO's estimates
$c_1 = 20$	$c_2 = 5$	$c_3 = 20$	$n_1 = 1459$	$n_2 = 418,\ n = 8565$

Table 3.10 *Set 5 (Dubey)*

γ	α	β	LLMAX	Methods
0.02723	1.15487	1.05928		moment estimates
0.02248	1.20123	1.09158	−98.55104	Wingo's estimates
0.02240	1.20163	1.07676	−98.53603	SNTO's estimates
$c_1 = 5$	$c_2 = 5$	$c_3 = 5$	$n_1 = 1459$	$n_2 = 418,\ n = 8565$

Now we consider parameter estimation for the beta distribution. The p.d.f. of a beta distribution (section 2.6) is

$$f(x;\alpha,\beta) = \frac{1}{B(\alpha,\beta)} x^{\alpha-1}(1-x)^{\beta-1}, 0 < x < 1, \tag{3.3.10}$$

with $\boldsymbol{\theta} = (\alpha,\beta), \alpha > 0,\ \beta > 0$, and the parameter space $D = \{(x,y) : x > 0,\ y > 0\}$. The logarithm of the likelihood function is

$$\begin{aligned}\mathcal{L}(\alpha,\beta) =& N[\log\Gamma(\alpha+\beta) - \log\Gamma(\alpha) - \log\Gamma(\beta)] \\ &+ (\alpha-1)\sum_{i=1}^{N}\log x_i + (\beta-1)\sum_{i=1}^{N}\log(1-x_i).\end{aligned} \tag{3.3.11}$$

Moment estimates of α and β can be obtained by

$$\alpha^* = \bar{x}\left[\frac{\bar{x}(1-\bar{x})}{s^2} - 1\right],$$

and

$$\beta^* = (1-\bar{x})\left[\frac{\bar{x}(1-\bar{x})}{s^2} - 1\right],$$

respectively, in which $\bar{x}$ and s^2 are the respective sample mean and sample variance.

Gnanadesikan, Pinkham and Hughes (1967) gave a detailed discussion on the MLEs for the beta distribution and some numerical examples which form the data of our Set 6 and Set 7. Using SNTO to get the MLEs of α and β, we take $\delta = .000001$, $c_1 = 5$ and $c_2 = 10$. Table 3.11 and Table 3.12 give the comparisons of SNTO's results with Gnanadesikan's approximations of the MLEs.

Table 3.11 *Set 6 (Gnanadesikan)*

α	β	LLMAX	Methods
1.65405	11.81971		moment estimates
1.79300	12.78100	24.31381	Gnanadesikan's estimates
1.79511	12.79744	24.31425	SNTO's estimates
$c_1 = 5$	$c_2 = 10$	$n_1 = 987$	$n_2 = 233$, $n = 4016$

Table 3.12 *Set 7 (Gnanadesikan)*

α	β	LLMAX	Methods
6.18149	10.50741		moment estimates
6.54300	11.05200	15.55757	Gnanadesikan's estimates
6.55102	11.05593	15.55875	SNTO's estimates
$c_1 = 5$	$c_2 = 10$	$n_1 = 987$	$n_2 = 233$, $n = 4249$

Remark 3.3

The present method can be used to find the MLEs of the parameters of a continuous multivariate distribution, even for a censored samples. This will be discussed further in section 6.1.

Remark 3.4

Likelihood ratio statistics for testing hypotheses about $\boldsymbol{\theta}$ have been widely used. Clearly, SNTO can help to calculate the approximation of a likelihood ratio statistic, since both numerator and denominator have the form (3.3.2) with different D's (sections 1.1 and 6.1).

3.4 Nonlinear regression model

Consider a general regression model

$$EY = g(\boldsymbol{x}; \boldsymbol{\theta}), \tag{3.4.1}$$

where $\boldsymbol{x} = (X_1, \cdots, X_p)$ are independent random variables and $\boldsymbol{\theta} = (\theta_1, \cdots, \theta_s)$ are parameters to be determined. The aim of regression analysis is to use a set of observations $\{Y_i, X_{i1}, \cdots, X_{ip}, i = 1, \cdots, N\}$ to estimate $\theta_1, \cdots, \theta_s$. This problem is often treated by the least squares method. Let

$$L(\boldsymbol{\theta}) = \sum_{i=1}^{N} [Y_i - g(\boldsymbol{x}_i; \boldsymbol{\theta})]^2. \tag{3.4.2}$$

Then the least squares solution $\hat{\boldsymbol{\theta}}$ satisfies

$$L(\hat{\boldsymbol{\theta}}) = \min_{\boldsymbol{\theta}} L(\boldsymbol{\theta}). \tag{3.4.3}$$

When g is a nonlinear function of $\boldsymbol{\theta}$, (3.4.1) is a nonlinear regression model. Ratkowsky (1983, 1990), Nash and Walker-Smith (1987), Seber (1989) and Wei (1989) gave a thorough discussion of nonlinear regression models. The parameter $\boldsymbol{\theta}$ is often estimated by the Newton-Gauss iteration method or the improved Newton-Gauss method. In fact, we can use SNTO to estimate the regression coefficients $\boldsymbol{\theta}$. The following expressions are some known nonlinear regression models:

$$\begin{aligned}
&EY = \exp(\theta_1 - \theta_2 \theta_3^X) \text{ (Gompertz model)},\\
&EY = \frac{\theta_1}{1 + \exp(\theta_2 - \theta_3^X)} \text{ (Logistic model)},\\
&EY = \theta_1 - \theta_2 \exp(-\theta_3 X^{\theta_4}) \text{ (Weibull model)},\\
&EY = \theta_1 - \log(1 + \theta_2 \exp(-\theta_3 X)) \text{ (logarithm model)},\\
&EY = (\theta_1 + \theta_2 X + \theta_3 X^2)^{-1} \text{ (Holliday model)},\\
&EY = \theta_1 + \theta_2 \exp(-\theta_3 X) \text{ (Asymptotic regression model)}.
\end{aligned} \tag{3.4.4}$$

Let D be the parameter space of $\boldsymbol{\theta}$. In general, D is a large region. The problem of how to reduce D to a much smaller domain D^* such that $\hat{\boldsymbol{\theta}} \in D^*$ is important when applying SNTO. It is difficult to establish a general method for contracting the domain D, since there are no corresponding moment estimates. Hence there may be more difficulties in applying SNTO to regression analysis than to MLE. We propose the following three methods.

3.4.1 Linearization method

There are some nonlinear regression models which can be transformed to a linear model, for instance

$$\begin{aligned} EY &= a + b\log X, & EY &= \frac{X}{aX+b}, \\ EY &= \frac{1}{a+be^{-cX}}, & EY &= aX^b, \\ EY &= ae^{bX}, & EY &= ae^{b/X}. \end{aligned} \tag{3.4.5}$$

As an example, we may use the transformation $Y^* = \ln Y$, $a^* = \ln a$, $b^* = b$ and $X^* = X^{-1}$ to reduce the last model in (3.4.5) to the typical linear model

$$EY^* = a^* + b^* X^*. \tag{3.4.6}$$

We first find the least squares estimates $\hat{a}^*$ and $\hat{b}^*$ of a^* and b^* by the method of linear regression model and then obtain the "least squares estimates" $\hat{a}$ and $\hat{b}$ of a and b by the inverse transformation between a, b and a^*, b^*. This method is called **the linearization method**. It is recommended by many books since it is simple and can easily be worked out. However the $\hat{a}$, $\hat{b}$ are not the least squares estimates. For example, the least squares estimates of (3.4.5) are the $\hat{a}$, $\hat{b}$ such that

$$\sum_{i=1}^{N}(Y_i - ae^{b/X_i})^2 \tag{3.4.7}$$

attains its minimum at $a = \hat{a}$ and $b = \hat{b}$. In model (3.4.6) we should find $\hat{a}^*$, $\hat{b}^*$ such that

$$\sum_{i=1}^{N} (\ln Y_i - a^* - b^* X_i)^2 \qquad (3.4.8)$$

reaches its minimum at $a^* = \hat{a}^*$ and $b^* = \hat{b}^*$. The two least squares estimates do not in general have the relation $\hat{a}^* = \ln\hat{a}$ and $\hat{b}^* = \hat{b}$. Therefore we should minimize (3.4.7) directly to find the least squares estimates $\hat{a}$, $\hat{b}$.

Since the parameter space D is often large, for example, $D = R^2$ or $D = \{(x, y) : y > 0\}$, and the efficiency of SNTO is low if $\hat{a}$ and $\hat{b}$ are found directly on D, we should shrink D to a considerably small $D^* \subset D$ such that $(\hat{a}, \hat{b}) \in D^*$ and work on D^*. We propose the use of the following so-called **LNL algorithm**. Suppose that we have the nonlinear regression model (3.4.1). We want to find the least squares estimate $\hat{\boldsymbol{\theta}}$ satisfying (3.4.3). If the model can be linearized to give a linear model that has a "least squares estimate" $\hat{\boldsymbol{\theta}}^*$ then the LNL algorithm are as follows:

Step 1 Use the linearization method to find the least squares estimate $\hat{\boldsymbol{\theta}}^*$ of the corresponding linear model.

Step 2 Find a vector $\boldsymbol{c}$ with positive components according to the type of models and the distribution of the data. Set

$$D^* = \{\boldsymbol{\theta} : \hat{\boldsymbol{\theta}}^* - \boldsymbol{c} \le \boldsymbol{\theta} \le \hat{\boldsymbol{\theta}}^* + \boldsymbol{c}\}. \qquad (3.4.9)$$

Step 3 Use SNTO to find a minimum point $\hat{\boldsymbol{\theta}}$ of $L(\boldsymbol{\theta})$ on D^*. This is the least squares estimate that we want to find.

The following example shows how the LNL algorithm is applied and also shows that the difference between the least squares solution $\hat{\boldsymbol{\theta}}$ and the $\hat{\boldsymbol{\theta}}^*$ obtained by the linearization method may be quite big.

Example 3.4

The following geological data is adopted from Wang (1982):

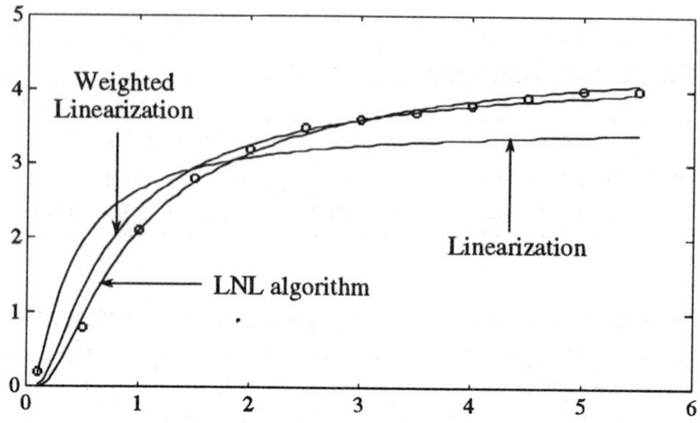

Figure 3.3 *The curves of three methods*

Table 3.13 *Data*

X	0.1	0.5	1.0	1.5	2.0	2.5	3.0	3.5	4.0	4.5	5.0	5.5
Y	0.2	0.8	2.1	2.8	3.2	3.5	3.6	3.7	3.8	3.9	4.0	4.0

Wang suggests to use the last model in (3.4.5) to fit the data by the linearization method and gives the estimates

$$\hat{a}^* = 3.5869 \qquad \text{and} \qquad \hat{b}^* = -0.3064$$

with the correlation $R^2 = 0.8161$ and the residual variance $\hat{\sigma}^2 = 0.3331$.

Now we use the LNL algorithm to find the least squares estimates $\hat{a}$, and $\hat{b}$. Take $c_1 = 2$, $c_2 = 1$ and set $D^* = \{(x, y) : 1.537 \leq x \leq 5.537, -1.306 \leq y \leq 0.694\}$. Applying SNTO for (3.4.7) on D^* we find

$$\hat{a} = 4.7061 \qquad \text{and} \qquad \hat{b} = -0.8119$$

with $R^2 = 0.9955$ and $\hat{\sigma}^2 = 0.008081$. For the comparison, all results are listed in Table 3.14 and shown by Figure 3.3. Both Table 3.14 and Figure 3.3 show that large errors may result if only the linearization method is used, and therefore it is better to use the LNL algorithm.

Table 3.14 *Comparison of three methods*

Methods	$\hat{a}$	$\hat{b}$	$\hat{\sigma}^2$	R^2
Linearization	3.5869	−0.3064	0.3331	0.8161
Weighted linearization	4.3975	−0.6091	0.0413	0.9772
LNL algorithm	4.7061	−0.8119	0.0081	0.9955

As shown by the example above, the parameter estimates by the linearization technique may vary significantly from the nonlinear least squares regression estimates. This difference can be lessened if appropriate weights are used in the linearization technique. In general, we may suppose that the nonlinear model takes the form

$$Y_i = g(\boldsymbol{x}_i, \boldsymbol{\theta}) + \varepsilon_i$$

where $\varepsilon_i \sim N(0, \delta^2)$. Since this model can be linearized, there must be some function h such that $h(g(\boldsymbol{x}, \boldsymbol{\theta})) = \boldsymbol{a}(\boldsymbol{x})'\boldsymbol{\theta}$ where $\boldsymbol{a}$ is some vector function. Linearizing the above model then gives

$$Z_i = h(Y_i) = h(g(\boldsymbol{x}_i, \boldsymbol{\theta}) + \varepsilon_i).$$

If ε_i is assumed to be relatively small, then this model can be approximated using a linear Taylor polymomial.

$$Z_i = h(g(\boldsymbol{x}_i, \boldsymbol{\theta}) + \varepsilon_i) \approx \boldsymbol{a}(\boldsymbol{x}_i)'\boldsymbol{\theta} + h'(g(\boldsymbol{x}_i, \boldsymbol{\theta}))\varepsilon_i \approx \boldsymbol{a}(\boldsymbol{x}_i)'\boldsymbol{\theta} + h'(Y_i)\varepsilon_i. \tag{3.4.10}$$

Since the residuals in this linearization are $h'(Y_i)\varepsilon_i$, it is more appropriate to compute a weighted least squares estimate to $\boldsymbol{\theta}$ according to model (3.4.10) and use this as the initial guess to the nonlinear least squares estimate. The superiority of the weighted least squares estimates are shown in Table 3.14.

3.4.2 Partial linearization method

The above linearization method can be applied to nonlinear regression models with rather simple structures or few parameters. Otherwise we may apply the so-called **partial linearization method** to find an initial estimate $\hat{\boldsymbol{\theta}}_0$ and the least squares estimate $\hat{\boldsymbol{\theta}}$ of $\boldsymbol{\theta}$ by the use of the SNTO. For illustration of the partial linearization method, consider a nonlinear regression model with three parameters θ_1, θ_2 and θ_3 and a domain D. Suppose that we can give several rough estimates $\hat{\theta}_{1i}$ ($i = 1, \cdots, m$, say) for θ_1, the linearization method can be applied to parameters θ_2 and θ_3 and the corresponding estimators are denoted by by $(\hat{\theta}_{2i}, \hat{\theta}_{3i})$ $i = 1, \cdots, m$. An initial estimate $\hat{\boldsymbol{\theta}}_0 = (\hat{\theta}_{10}, \hat{\theta}_{20}, \hat{\theta}_{30})$ can be chosen among the m estimates $\hat{\boldsymbol{\theta}}_i = (\hat{\theta}_{1i}, \hat{\theta}_{2i}, \hat{\theta}_{3i})$, $i = 1, \cdots, m$ such that it has the mimimum sum of squares. Consequently, we can apply SNTO to find the least squares estimate $\hat{\boldsymbol{\theta}}$ with a much smaller domain D^* relative to D, where D^* is constructed by $\hat{\boldsymbol{\theta}}_0$ and a constant vector $\boldsymbol{c}$ (section 3.4.1). Obviously, the partial linearization method is effective for models with small numbers of parameters, in particular with three parameters. Let us examine the following example.

Example 3.5
The following data set

X	0	10	20	30	40
Y	26.2	30.4	36.3	37.8	38.6

is adapted from Set 3 of Appendix 5 of Ratkowsky (1983), for fitting the model

$$EY = \theta_1 + \theta_2 \theta_3^X, \qquad 0 < \theta_3 < 1.$$

Since $EY \to \theta_1$ as $X \to \infty$, we may estimate $\hat{\theta}_{01} = y_{\max} + \Delta$, where $y_{\max} = \max\{Y_1, \cdots, Y_N\}$, $\hat{\boldsymbol{\theta}}_0 = (\hat{\theta}_{10}, \hat{\theta}_{20}, \hat{\theta}_{30})'$ and Δ is a positive number determined below. Then from the model we have

$$\log(\hat{\theta}_{01} - EY) = \log\theta_2 + X\log\theta_3, \tag{3.4.11}$$

and find the linearization least squares estimates of θ_2 and θ_3 and the residual variance $\hat{\sigma}_0^2$. Now $y_{\max} = 38.6$, and take $\Delta = 0.4, 1.4, 2.4, 3.4$, and 4.4. The corresponding estimates of θ_1, θ_2, θ_3 and $\hat{\sigma}_0^2$ are listed in Table 3.15.

Table 3.15 *The initial values of parameters*

$\hat{\theta}_{01}$	$\hat{\theta}_{02}$	$\hat{\theta}_{03}$	$\hat{\sigma}_0^2$	remark
39	15.9951	0.9148	7.1748	
40	14.4961	0.9413	1.8730	
41	14.8135	0.9528	1.5798	*
42	15.4831	0.9599	1.6875	*
43	16.2876	0.9650	1.8658	

From Table 3.15 $\hat{\theta}_{01}$ should be in $[40.5, 42.5]$, because the corresponding $\hat{\sigma}_0^2$ has smaller values, and thus we may take $\hat{\theta}_{01} = 41.5$, and then $\hat{\theta}_{02} = 15.1239$ and $\hat{\theta}_{03} = 0.9567$ from (3.4.11) by the linearization least squares method, i.e.

$$\hat{\boldsymbol{\theta}}_0 = (41.5, 15.1239, 0.9567)'. \tag{3.4.12}$$

Note the possible values of $\hat{\boldsymbol{\theta}}_0$ in Table 3.15. We take

$$\boldsymbol{c} = (1.5, 0.90, 0.01)'$$

and have the initial parameter space

$$\begin{aligned} D^* &= [\hat{\boldsymbol{\theta}}_0 - \boldsymbol{c}, \hat{\boldsymbol{\theta}} + \boldsymbol{c}] \\ &= [40, 43] \times [14.2239, 16.0239] \times [0.9467, 0.9667]. \end{aligned}$$

Using SNTO with $\delta = 10^{-5}$, $n_1 = 1495$, $n_2 = n_3 = \cdots = n_{19} = 597$, we obtain the approximations of $\boldsymbol{\theta}$ and σ^2 as follows

$$\begin{aligned} &\hat{\boldsymbol{\theta}}^{(19)} = (41.7079, 15.8388, 0.95628)' \\ &(\hat{\sigma}^{(19)})^2 = 1.454474 \end{aligned}$$

which coincide with the least squares estimates $\hat{\boldsymbol{\theta}}$ and $\hat{\sigma}^2$ (Ratkowsky (1983)).

If the nonlinear model is too complicated to be completely linearized, it may be possible to partially linearize it after obtaining rough estimates of some parameter values. To demonstrate this procedure consider Example 3.5:

First we estimate θ_1 by $\theta_1^{(0)} = \max(Y_i) + 0.2(\max(Y_i) - \min(Y_i)) = 41.08$. Then we estimate θ_2 and θ_3 by linear least squares applied to the model

$$\log(\theta_1^{(0)} - Y_i) = \log(\theta_2) + X\log(\theta_3) + (\theta_1^{(0)} - Y_i)^{-1}\varepsilon_i.$$

This gives the initial estimates $\theta_2^{(0)} = 15.329$ and $\theta_3^{(0)} = 0.95392$. Since these initial estimates are closer to the least squares estimates $\hat{\theta}_1, \hat{\theta}_2$ and $\hat{\theta}_3$ than that in (3.4.12) are, we can expect to find better approximations of $\hat{\theta}_1$, $\hat{\theta}_2$ and $\hat{\theta}_3$. Besides, the latter has a simple procedure.

3.4.3 RSNTO on a large domain

If only a few properties of the current model are known and we have no effective method to shrink D considerably, then we propose the following more conservative algorithm: Take a large rectangle $D^* = [\boldsymbol{a}, \boldsymbol{b}]$ such that $\hat{\boldsymbol{\theta}} \in D^* \subset D$, and then find the approximation of $\hat{\boldsymbol{\theta}}$ on D^* by RSNTO (section 3.2). We illustrate this method by the following example.

Example 3.6

This example is taken from Hartley (1961) and Wei (1989) for fertility experiments. We want to fit the data using the asymptotic regression model in (3.4.4), i.e.

$$EY = \theta_1 + \theta_2 \exp(-\theta_3 X).$$

In this example X stands for the normalizing quantity of fertilizer and Y represents wheat production. A set of observations is as follows:

X	−5	−3	−1	1	3	5
Y	127	151	379	421	460	426

We want to find the least squares estimates of θ_1, θ_2, and θ_3.

Suppose that we do not know how to construct the parameter space. So we take a large domain $D^* = [0, 1000] \times [0, 300] \times [0, 2]$. Then apply SNTO on D^* with $\delta = 1, n_1 = 1220$, and $n_2 = n_3 = \cdots = 418$ for $L(\boldsymbol{\theta})$ of the model. We obtain approximations of L and $\hat{\boldsymbol{\theta}}$ as follows

$$\hat{\boldsymbol{\theta}} \approx \hat{\boldsymbol{\theta}}^{(9)} = (570.73, 210.84, 0.15928), \quad L \approx L^{(9)} = 13776.2.$$

Since the initial region D^* is large, the errors of the above solutions may also be large. Thus we should search again by using SNTO. Now we set a rectangle

$$\bar{D} = [471, 771] \times [151, 271] \times [0.11, 0.21],$$

with the centre $\hat{\boldsymbol{\theta}}^{(9)}$. Applying SNTO on $\bar{D}$ with $\delta = 0.0001$, $n_1 = 1459$, and $n_2 = n_3 = \cdots = n_{13} = 266$, we have the respective minimum of L and the estimate of $\boldsymbol{\theta}$ as follows:

$$L \approx 13390.1 \qquad \text{and} \qquad \hat{\boldsymbol{\theta}} \approx (523.40, 157.04, 0.19957) = \hat{\boldsymbol{\theta}}^{(13)},$$

the latter is very close to the least squares estimate $\hat{\boldsymbol{\theta}}$:

$$\hat{\boldsymbol{\theta}} = (523.3, 157.1, 0.1994).$$

If we use SNTO again in a smaller rectangle with centre $\hat{\boldsymbol{\theta}}^{(13)}$, then we may obtain a solution which is closer to the least squares estimate $\hat{\boldsymbol{\theta}}$.

Fang and Zhang (1990) had used RSNTO for many sets of various nonlinear models in Ratkowsky (1983) to compare with the results obtained by classical nonlinear techniques. They found that the RSNTO is efficient for most cases and has the advantage that the algorithm is easy to program and can be universally used for different models with only minor modifications.

3.5 Robust regression model

Consider a general regression model (3.4.1). The robust estimate $\hat{\boldsymbol{\theta}}^*$ for $\boldsymbol{\theta}$ satisfies

$$L^*(\hat{\boldsymbol{\theta}}^*) = \min_{\boldsymbol{\theta}} L^*(\boldsymbol{\theta}), \tag{3.5.1}$$

where

$$L^*(\boldsymbol{\theta}) = \sum_{i=1}^{N} h(Y_i, g(\boldsymbol{x}_i, \boldsymbol{\theta})), \tag{3.5.2}$$

$\{Y_i, \boldsymbol{x}_i = (X_{i1}, \cdots, X_{ip})', i = 1, \cdots, N\}$ are observations, and $h(\cdot, \cdot)$ is a nonnegative function satisfying some regularity conditions, for example $h(u, v) = |u - v|$.

The least squares method has long dominated the literature of the applications of regression techniques. The research on robust estimation naturally leads statisticians to consider the corresponding improvements on regression. Many authors proposed various kinds of robust estimation of $\boldsymbol{\theta}$ in model (3.4.1) by choosing different h's in (3.5.2). The reader is refered to Li (1985) and Rousseeuw and Leroy (1987) for a good review of robust regression. We only discuss l_1-regression in this section. Now

$$L^*(\boldsymbol{\theta}) = \sum_{i=1}^{N} |Y_i - g(\boldsymbol{x}_i, \boldsymbol{\theta})|. \tag{3.5.3}$$

The regression model may be linear or nonlinear. Consider first the linear model

$$EY = \boldsymbol{X}\boldsymbol{\theta}, \tag{3.5.4}$$

where $\boldsymbol{X}$ and $\boldsymbol{\theta}$ are $N \times p$ and $p \times 1$ matrices respectively. We denote by $\hat{\boldsymbol{\theta}}_1$ the corresponding robust estimate of $\boldsymbol{\theta}$, which is called the ℓ_1-estimate of (3.5.4). The usual least squares estimate of $\boldsymbol{\theta}$ in (3.5.4) is denoted by $\hat{\boldsymbol{\theta}}_2$. We use the notation

$$\|\boldsymbol{a}\|_1 = \sum_i |a_i| \qquad \text{and} \qquad \|\boldsymbol{a}\|_2 = \Big(\sum_i a_i^2\Big)^{1/2}$$

which denote the l_1-norm and l_2-norm of $\boldsymbol{a}$ respectively. If $\boldsymbol{A} = (a_{ij})$ is a $n \times m$ matrix, then the l_1-norm of $\boldsymbol{A}$ is defined by

$$\|\boldsymbol{A}\|_1 = \max_j \Big(\sum_{i=1}^{n} |a_{ij}| \Big). \tag{3.5.5}$$

For any vector $\boldsymbol{a}$, we have

$$\|\boldsymbol{a}\|_1 \geq \|\boldsymbol{a}\|_2. \tag{3.5.6}$$

There are a lot of effective methods to find the l_1-estimate $\hat{\boldsymbol{\theta}}_1$ of $\boldsymbol{\theta}$, for example, the simplex method. Now we shall illustrate how to obtain $\hat{\boldsymbol{\theta}}_1$ by SNTO. Since the parameter space of $\boldsymbol{\theta}$ is usually too large, the result will be not good if SNTO is applied on D directly. Hence we need a method to shrink D to a smaller region D^* which is used instead of D. This problem is discussed as follows.

When the regression model is (3.5.4), then $L^*(\boldsymbol{\theta})$ is reduced to

$$L^*(\boldsymbol{\theta}) = \|\boldsymbol{y} - \boldsymbol{X\theta}\|_1, \tag{3.5.7}$$

where $\boldsymbol{y} = (Y_1, \cdots, Y_N)'$. Set $d = \min_{\boldsymbol{\theta}} L^*(\boldsymbol{\theta})$ and

$$\Theta = \{\boldsymbol{\theta} : L^*(\boldsymbol{\theta}) = d\}.$$

Then Θ is the set of all feasible solutions of $\hat{\boldsymbol{\theta}}_1$.

Theorem 3.3

$$\|\hat{\boldsymbol{\theta}}_1 - \hat{\boldsymbol{\theta}}_2\|_2 \leq \frac{2\|\boldsymbol{y} - \boldsymbol{X}\hat{\boldsymbol{\theta}}_2\|_1}{\sqrt{\lambda_{\min}}}, \tag{3.5.9}$$

where $\hat{\boldsymbol{\theta}}_1$ is any element of Θ and $\lambda_{\min}$ is the smallest eigenvalue of $\boldsymbol{X}'\boldsymbol{X}$.

PROOF Since $\hat{\boldsymbol{\theta}}_1 \in \Theta$, we have $\|\boldsymbol{y} - \boldsymbol{X}\hat{\boldsymbol{\theta}}_2\|_1 \geq \|\boldsymbol{y} - \boldsymbol{X}\hat{\boldsymbol{\theta}}_1\|_1$. Hence

$$\|\boldsymbol{X}(\hat{\boldsymbol{\theta}}_2 - \hat{\boldsymbol{\theta}}_1)\|_1 \leq \|\boldsymbol{y} - \boldsymbol{X}\hat{\boldsymbol{\theta}}_2\|_1 + \|\boldsymbol{y} - \boldsymbol{X}\hat{\boldsymbol{\theta}}_1\|_1 \leq 2\|\boldsymbol{y} - \boldsymbol{X}\hat{\boldsymbol{\theta}}_2\|_1. \tag{3.5.10}$$

On other hand we have

$$\begin{aligned}\|\boldsymbol{X}(\hat{\boldsymbol{\theta}}_2-\hat{\boldsymbol{\theta}}_1)\|_1 \geq \|\boldsymbol{X}(\hat{\boldsymbol{\theta}}_2-\hat{\boldsymbol{\theta}}_1)\|_2 &= ((\boldsymbol{\theta}_1-\boldsymbol{\theta}_2)'\boldsymbol{X}'\boldsymbol{X}(\boldsymbol{\theta}_1-\boldsymbol{\theta}_2))^{1/2} \\ &\geq \|\hat{\boldsymbol{\theta}}_2-\hat{\boldsymbol{\theta}}_1\|_2\sqrt{\lambda_{\min}}.\end{aligned}$$

Therefore, the assertion follows. □

From Theorem 3.3 we have the following algorithm for shrinking D into a smaller domain D^*:

Step 1 Find the least squares estimate $\hat{\boldsymbol{\theta}}_2 = \boldsymbol{X}(\boldsymbol{X}'\boldsymbol{X})^{-1}\boldsymbol{y}$, the least eigenvalue $\lambda_{\min}$ of $\boldsymbol{X}'\boldsymbol{X}$ and $\|\boldsymbol{y}-\boldsymbol{X}\hat{\boldsymbol{\theta}}_2\|_1$.

Step 2 Set $c = 2\|\boldsymbol{y}-\boldsymbol{X}\hat{\boldsymbol{\theta}}_2\|_1/\sqrt{\lambda_{\min}}$ and

$$\boldsymbol{D}^* = \{\boldsymbol{\theta} : \hat{\boldsymbol{\theta}}_2 - c\mathbf{1}_p \leq \boldsymbol{\theta} \leq \hat{\boldsymbol{\theta}}_2 + c\mathbf{1}_p\}, \tag{3.5.11}$$

where $\mathbf{1}_p = (1, \cdots, 1)'$.

Step 3 Find the minimum of $L^*(\boldsymbol{\theta})$ defined by (3.5.7) on D^* by SNTO or RSNTO which are approximations of d and $\hat{\boldsymbol{\theta}}_1$.

Remark 3.5

The inequality (3.5.9) was obtained by Cao (1992) and determines a ball

$$B = \left\{\boldsymbol{\theta} : \|\hat{\boldsymbol{\theta}}_2 - \boldsymbol{\theta}\|_2 \leq \frac{2\|\boldsymbol{y}-\boldsymbol{X}\hat{\boldsymbol{\theta}}_2\|_1}{\sqrt{\lambda_{\min}}}\right\} \tag{3.5.12}$$

and the D^* defined by (3.5.11) is the cube circumscribed about B. It is convenient to apply SNTO or RSNTO on a rectangle. Of course we may take B to replace D^*, and this case will be discussed in next section.

Remark 3.6

The condition rank$(\boldsymbol{X}) = p < N$ is necessary. If $\boldsymbol{X}$ is singular, then for any given large number M, there exist a $\boldsymbol{\theta}_1 \in \Theta$ such that $\|\boldsymbol{\theta}_1\|_1 > M$. This means that the set Θ of feasible solutions is unbounded. So the robustness property is lost. If the model is nonlinear, then $L^*(\boldsymbol{\theta})$ is given by (3.5.3) or (3.5.2). We usually cannot get an inequality similar to (3.5.9), and thus we have no analytic expression for the domain D^* as in the case of the linear model. Then we may determine D^* by the method stated in section

3.4. For safety, D^* should be enlarged so that $\hat{\boldsymbol{\theta}}_1$ is included in D^*. Then we find the minimum and minimum point of $L^*(\boldsymbol{\theta})$ ((3.5.2)) on D^* by SNTO, and the latter is a good approximation of $\hat{\boldsymbol{\theta}}_1$.

3.6 A version of SNTO in special regions, SNTO–D

We have discussed the optimization problem (3.1.1) in D when D is a rectangle. If D is not a rectangle, for example D is A_s, B_s, U_s, T_s or V_s as stated in Section 1.5, can we use also SNTO to find the approximations of M and $\boldsymbol{x}^*$? In this section we shall give a version of SNTO for more general regions D, which we denote by SNTO–D.

Let random vector $\boldsymbol{x}$ be uniformly distributed on D and $\boldsymbol{x}$ have a stochastic representation

$$\boldsymbol{x} = \boldsymbol{x}(U_1, \cdots, U_t) = \boldsymbol{x}(\boldsymbol{u}), \tag{3.6.1}$$

where $\boldsymbol{u} = (U_1, \cdots, U_t)' \in C^t$, $t \leq s$ and $U_1, \cdots, U_t$ are statistically independent. Given an NT-net $\{\boldsymbol{c}_k, k = 1, \cdots, n\}$ on C^t, we can obtain an NT-net $\mathcal{P} = \{\boldsymbol{y}_k, k = 1, \cdots, n\}$ on D where $\boldsymbol{y}_k = \boldsymbol{h}(\boldsymbol{c}_k)$ for some function $\boldsymbol{h}$ (section 1.5). Let $\boldsymbol{x}_0$ be a point of $\mathcal{P}$ where the function f achieves its maximum among the points of $\mathcal{P}$, i.e. $f(\boldsymbol{x}_0) = \max_k f(\boldsymbol{y}_k)$. Let $\boldsymbol{u}_0$ be the point of $\{\boldsymbol{c}_k\}$ corresponding to $\boldsymbol{x}_0$ by $\boldsymbol{x}_0 = \boldsymbol{h}(\boldsymbol{u}_0)$. Then a rectangle $[\boldsymbol{a}, \boldsymbol{b}]$ with center $\boldsymbol{u}_0$ corresponds a domain of D which includes $\boldsymbol{x}_0$. Consequently, we can define the process for contractions of D by means of the contractions on a rectangle $[\boldsymbol{a}, \boldsymbol{b}]$ in SNTO and get a version of SNTO on this kind of domain. The algorithm SNTO–D includes the following steps:

Step 0 **Initialization.** Set $q = 0$, $D^{(0)} = D$, $E^{(0)} = C^t = [\boldsymbol{0}, \boldsymbol{1}]$, $\boldsymbol{a}^{(0)} = \boldsymbol{a} = \boldsymbol{0}$ and $\boldsymbol{b}^{(0)} = \boldsymbol{b} = \boldsymbol{1}$.

Step 1 **Generate an NT-net on $D^{(q)}$.** Generate an NT-net $\{\boldsymbol{c}_k^{(q)}, k = 1, \cdots, n_q\}$ on $E^{(q)}$. Then we have an NT-net $\mathcal{P}^{(q)} = \{\boldsymbol{y}_k^{(q)} = \boldsymbol{h}(\boldsymbol{c}_k^{(q)}), k = 1, \cdots, n_q\}$ on $D^{(q)}$.

Step 2 **Compute new approximation.** Find $\boldsymbol{x}^{(q)} \in \mathcal{P}^{(q)} \cup \{\boldsymbol{x}^{(q-1)}\}$ and $M^{(q)}$ such that

$$M^{(q)} = f(\boldsymbol{x}^{(q)}) \geq f(\boldsymbol{y}), \qquad \forall \boldsymbol{y} \in \mathcal{P}^{(q)} \cup \{\boldsymbol{x}^{(q-1)}\}.$$

$\boldsymbol{x}^{(q)}$ and $M^{(q)}$ are the best approximations to $\boldsymbol{x}^*$ and M so far.

Step 3 **Termination criterion.** Let $\boldsymbol{c}^{(q)} = (\boldsymbol{b}^{(q)} - \boldsymbol{a}^{(q)})/2$. If $\max \boldsymbol{c}^{(q)} < \delta$, a pre-assigned small number, then $\boldsymbol{x}^{(q)}$ and $M^{(q)}$ are acceptable; terminate algorithm. Otherwise, proceed to next step.

Step 4 **Contract domain.** Let $\boldsymbol{u}^{(q)}$ be the point of C^t corresponding to $\boldsymbol{x}^{(q)}$ by $\boldsymbol{x}^{(q)} = \boldsymbol{h}(\boldsymbol{u}^{(q)})$. Form new domain $E^{(q+1)} = [\boldsymbol{a}^{(q+1)}, \boldsymbol{b}^{(q+1)}]$, where

$$\begin{aligned} a_i^{(q+1)} &= \max(u_i^{(q)} - \gamma c_i^{(q)}, 0), \\ & \qquad\qquad\qquad\qquad i = 1, \cdots, t, \\ b_i^{(q+1)} &= \min(u_i^{(q)} - \gamma c_i^{(q)}, 1), \end{aligned}$$

and γ is a predefined contraction ratio. Let $D^{(q+1)}$ be the subdomain of D which corresponds to $E^{(q+1)}$ by the relation $D^{(q+1)} = \{\boldsymbol{x} = \boldsymbol{h}(\boldsymbol{u}), \boldsymbol{u} \in E^{(q+1)}\}$. Set $q = q + 1$. Go to Step 1.

Remark 3.7

If D has no stochastic representation (3.6.1) and the dimension of D is s, then we may use SNTO on $[\boldsymbol{a}, \boldsymbol{b}]$ which is the rectangle circumscribed about D. Note that the points of the NT-net falling outside of D should be deleted. In the next section, we shall give some applications by SNTO–D.

3.7 A system of nonlinear equations

Many problems in statistics are often reduced to solving a system of nonlinear equations, for example, moment estimates of parameters, maximum likelihood estimation, and the quantizer of a univariate distribution (section 4.2).

Suppose that D is a bounded domain of R^s. We want to solve a system of equations

$$\begin{cases} f_1(\boldsymbol{x}) = f_1(x_1, \cdots, x_s) = 0 \\ \qquad \cdots \\ f_t(\boldsymbol{x}) = f_t(x_1, \cdots, x_s) = 0. \end{cases} \quad \boldsymbol{x} \in D, \qquad (3.7.1)$$

If there is no analytic expression for the solution of (3.7.1), there are a number of methods for the solution of (3.7.1), such as Newton's method, Brent's method and the quasi-Newton method (Brent (1973), Brown (1967), Feng (1989)). However, these methods require that the f_i have continuous derivatives of first or even higher orders, or satisfy certain properties of convexity in order that the convergence of these methods is ensured.

In fact the problem for solving system of equations such as (3.7.1) can be reduced to an optimization problem. Let $\boldsymbol{F}(\boldsymbol{x}) = (f_1(\boldsymbol{x}), \cdots, f_t(\boldsymbol{x}))$ and

$$f(\boldsymbol{x}) = \|\boldsymbol{F}(\boldsymbol{x})\|_p, \quad \boldsymbol{x} \in D \tag{3.7.2}$$

where the norm is the l_p-norm, $p > 0$ and the usual values of p are 1, 2 and $+\infty$. Then the problem of finding a solution $\boldsymbol{x}^*$ of (3.7.1) is equivatent to that of finding a point $\boldsymbol{x}^*$ such that $f(\boldsymbol{x})$ attains its minimum $M = 0$ for certain p. When the solution of (3.7.1) is not unique, our aim is to find one at least among them.

Remark 3.8

The use of $f(\boldsymbol{x})$ for solving system of equations (3.7.1) may not be the best way (Kahaner, Moler and Nash (1989)), but we still can find satisfactory results in most practical problems.

When D is a rectangle, we can use the SNTO to obtain a solution of (3.7.1). Otherwise SNTO–D can be applied if D has a stochastic representation (3.6.1). As an example for solving the system of nonlinear equations by a version of SNTO, we consider here a problem for finding a fixed point of a mapping from U_s to U_s.

Let $\boldsymbol{g}$ be a continuous mapping which maps D into D. A point $\boldsymbol{x}^* \in D$ is called a **fixed point $\boldsymbol{x}^*$** of $\boldsymbol{g}$ on D if

$$\boldsymbol{x}^* = \boldsymbol{g}(\boldsymbol{x}^*). \tag{3.7.3}$$

Let

$$f(\boldsymbol{x}) = \|\boldsymbol{x} - \boldsymbol{g}(\boldsymbol{x})\| = \sum_{i=1}^{s} |x_i - g_i(\boldsymbol{x})|, \tag{3.7.4}$$

where g_i and x_i are respective components of $\boldsymbol{g}$ and $\boldsymbol{x}$. Then finding a fixed point of (3.7.3) is a special case of (3.7.1).

Example 3.7
Let $D = U_3$ be the unit sphere in R^3. Let

$$\begin{cases} y_1 = (x_1 + 2x_2 + 3x_3)/S \\ y_2 = (4x_1 + 5x_2 + 6x_3)/S \\ y_3 = (7x_1 + 8x_2 + 9x_3)/S, \end{cases}$$

where

$$S = [(x_1+2x_2+3x_3)^2+(4x_1+5x_2+6x_3)^2+(7x_1+8x_2+9x_3)^2]^{1/2}.$$

Obviously, this is a continuous mapping which maps U_3 into U_3. There exists at least one fixed point by Brower's theorem.

Let $\{\boldsymbol{c}_k = (c_{k1}, c_{k2}), k = 1, \cdots, n\}$ be an NT-net on C^2. Then $\{\boldsymbol{x}_k = (x_{k1}, x_{k2}, x_{k3})'\}$, where

$$\begin{cases} x_{k1} = 1 - 2c_{k1} \\ x_{k2} = 2\sqrt{c_{k1}(1 - c_{k1})}\,\cos(2\pi c_{k2}) \\ x_{k3} = 2\sqrt{c_{k1}(1 - c_{k1})}\,\sin(2\pi c_{k2}), \end{cases}$$

is an NT-net on U_3 ((1.5.28)).

Generate an NT-net by the *glp* method with a generating vector $(233; 1, 144)$ and take $\delta = 10^{-7}$. With SNTO–D, in the first stage we obtain $M^{(1)} = 0.1368986$ and

$$\boldsymbol{x}^{(1)} = (0.2660945, 0.5760145, 0.7729173)'$$

which corresponds to $\boldsymbol{u}^{(1)} = (0.3669528, 0.1480687)$ ((3.6.1)). Then take $(144; 1, 89)$ as the generating vector and work on $E^{(1)} = [0.1169529, 0.6169528] \times [0, 0.3980687]$ which is generated by $\boldsymbol{u}^{(1)}$

and the rule of SNTO–D. We find that $M^{(2)} = 0.1092555$,

$$\begin{aligned}\boldsymbol{x}^{(2)} &= (0.3042889, 0.5017185, 0.8097449)\\ \boldsymbol{u}^{(2)} &= (0.3478556, 0.1617154)\\ E^{(2)} &= [0.2228556, 0.4728556] \times [0.00367154, 0.2867154],\end{aligned}$$

etc. and we finally obtain the approximations of M and $\boldsymbol{x}^*$ as follows:

$$\begin{aligned}M^{(21)} &= 1.043081 \times 10^{-7},\\ \boldsymbol{x}^{(21)} &= (0.2319707, 0.5253221, 0.8186734).\end{aligned}$$

The total number of points used in SNTO–D is 3113, and the precision of the result is high.

The following is another example finding a fixed point on T_s.

Example 3.8

Conside a mapping

$$\begin{cases} y_1 = (x_1 + 2x_2 + 3x_3)/S,\\ y_2 = (4x_1 + 5x_2 + 6x_3)/S,\\ y_3 = (7x_1 + 8x_2 + 9x_3)/S,\end{cases}$$

where $S = 12x_1 + 15x_2 + 18x_3$ and $\boldsymbol{x} = (x_1, x_2, x_3) \in D = T_3$. This is a continuous mapping which maps D into D with $y_2 \equiv 1/3$. We want to find a fixed point of this mapping. By SNTO–D with $n_1 = 233$, $n_2 = 144$ and 18 domain contractions, we have the approximations of M and $\boldsymbol{x}^*$ as follows:

$$M^{(18)} = 2.98\times^{-7}, \text{ and } \boldsymbol{x}^{(18)} = (0.1471927, 0.3333332, 0.5194746)'$$

which are very close to M=0 and $\boldsymbol{x}^*$=(0.1471927, 1/3, 0.5194742)'.

There is a rich literature on fixed point problems such as Dixon (1972), Todd (1976), Allgower and Georg (1980) and Wang (1986, 1987) in which the reader can find more detailed discussion.

Remark 3.9

The design of domain contraction in SNTO–D is to shorten each edge length of the original rectangle by the contraction ratio γ, and thus the volume of the resulting rectangle is γ^s time the original one. Now we shall use contraction ratios $1/2, 1/4, 1/8, 1/16$ in our example for comparison, and we see that the amount of calculation can be reduced by at least one-half for obtaining a result with still higher precision if $\gamma \leq 1/4$ is used instead of $\gamma = 1/2$ in the contraction. The results are given in Table 3.16, where T denotes the number of contractions. In general n_2 cannot be chosen too small if the contraction ratio γ is small.

Table 3.16 *Comparison of the different contraction ratios*

contraction ratio	n_1	n_2	T	$L(x)$
1/2	233	144	18	2.98E−7
1/4	233	144	9	9.54E−7
1/8	233	144	9	9.54E−7
1/16	233	144	9	5.96E−8

Methods for solving nonlinear equation of a scalar variable $F(x) = 0$ often begin by identifying an interval $[a, b]$ such that $F(a)$ and $F(b)$ have opposite signs. This insures a solution $x^* \in [a, b]$. This interval is then iteratively shrunk while continuing to bracket a solution until a pre-defined error tolerance is satisfied.

For a system of nonlinear equations $\boldsymbol{F}(\boldsymbol{x}) = \boldsymbol{0}$ it is impossible to bracket a solution in the same way as for a single equation. None the less, the signs of the functions f_i are still useful in locating the solution. If SNTO is used to solve a system of nonlinear equations by maximizing the function $f(\boldsymbol{x}) = -\|\boldsymbol{F}(\boldsymbol{x})\|_p$, then one might wish to guarantee that all the f_i change sign in each domain $D^{(t)}$. Doing so increases the probability that these domains all contain a solution to the nonlinear system.

In particular, at each step of the iteration there should be a set $Q^{(t+1)} \subseteq D^{(t+1)}$ with the property:

$$\begin{gathered}\forall i, \exists \boldsymbol{y}_+, \boldsymbol{y}_- \in Q^{(t+1)} \quad \text{such that} \\ \operatorname{sign}(f_i(\boldsymbol{y}_+)) = +1 \text{ and } \operatorname{sign}(f_i(\boldsymbol{y}_-)) = -1. \end{gathered} \tag{3.7.5}$$

To satisfy this condition Step 4 of SNTO is modified as follows:

Step 4* **Contract domain.** Set $Q^{(t+1)} = \{\boldsymbol{x}^{(t)}\}$. Continue to add points $\boldsymbol{y} \in \mathcal{P}^{(t)} \cup Q^{(t)}$ to $Q^{(t+1)}$ in decreasing order of $f(\boldsymbol{y})$ until (3.7.5) is satisfied. Let $[\boldsymbol{\alpha}^{(t+1)}, \boldsymbol{\beta}^{(t+1)}]$ be the smallest box enclosing $Q^{(t+1)}$. Form new domain $D^{(t+1)} = [\boldsymbol{a}^{(t+1)}, \boldsymbol{b}^{(t+1)}]$ as follows:

$$\begin{aligned} a_i^{(t+1)} &= \max(\min(x_i^{(t)} - \gamma c_i^{(t)}, \alpha_i^{(t+1)}), a_i) \\ b_i^{(t+1)} &= \min(\max(x_i^{(t)} + \gamma c_i^{(t)}, \beta_i^{(t+1)}), b_i) \end{aligned} \tag{3.7.6}$$

where γ is a pre-determined contraction ratio.

An alternative and more conservative condition to (3.7.5) is to require that

$$\begin{aligned} &\forall \boldsymbol{r} = (r_1, \cdots, r_s) \text{ where } r_i = \pm 1, \\ &\exists \boldsymbol{x}_r \in Q^{(t)} \text{ such that } \operatorname{sign}(f_i(\boldsymbol{x}_r)) = r_i \ \forall i. \end{aligned} \tag{3.7.7}$$

We have successfully applied this algorithm to many problems. The results show that this algorithm is more efficient than the use of the original SNTO. The key idea of this algorithm is due to Dr. F.J. Hickernell by personal communication.

3.8 Regression with constraints

In many practical problems we meet regression model with constraints. Consider a general regression model (3.4.1)

$$EY = g(\boldsymbol{x}; \boldsymbol{\theta}), \tag{3.8.1}$$

where $\boldsymbol{\theta} \in \Theta$, the parameter space, with constraints $\boldsymbol{v}(\boldsymbol{\theta}) \in D^* \subset R^t$, $t \leq s$. In this case, how can we get an estimate of $\boldsymbol{\theta}$? Let

$$D = \{\boldsymbol{\theta} : \boldsymbol{\theta} \in \Theta,\ \boldsymbol{v}(\boldsymbol{\theta}) \in D^*\}. \tag{3.8.2}$$

Then the least squares estimate $\hat{\boldsymbol{\theta}}$ satisfies

$$L(\hat{\boldsymbol{\theta}}) = \min_{\boldsymbol{\theta} \in D} L(\boldsymbol{\theta}), \tag{3.8.3}$$

where $L(\boldsymbol{\theta})$ is defined by (3.4.2). So we may use the version of SNTO stated in section 3.6 for finding $\hat{\boldsymbol{\theta}}$.

For illustration we discuss only a set of constraints which appear often in regression modelling with compositional data. Consider the linear regression model

$$\hat{Y} = a + b_1 x_1 + \cdots + b_s x_s \tag{3.8.4}$$

with constraints

$$b_1 + \cdots + b_s = 1, \quad b_i \geq 0, \; i = 1, \cdots, s, \tag{3.8.5}$$

i.e. $\boldsymbol{b} = (b_1, \cdots, b_s) \in T_s$.

Boot (1964) suggested that the above model can be considered as a problem of quadratic programming. Waterman (1974) proposed to find the least squares estimate by the use of all possible regression models among the s independent variables, and his method needs considerable computing time if s is large. Fang, Wang and Wu (1982) and Fang and He (1985) use the method of eliminating transformations to treat this problem. Although the latter is efficient the method is complicated. (The program can be found in SASD, a statistical package written by the Computing Centre, Academia Sinica.)

For simplicity, we assume $a = 0$ in (3.8.4), because the case $a \neq 0$ can be treated similarly. Given a set of observations $\{Y_i, X_{i1}, \cdots, X_{is},\ i = 1, \cdots, N\}$, let

$$L(\boldsymbol{b}) = L(b_1, \cdots, b_s) = \sum_{i=1}^{N} (Y_i - b_i X_{i1} - \cdots - b_s X_{is})^2. \tag{3.8.6}$$

Then the least squares estimate $\hat{\boldsymbol{b}} = (\hat{b}_1, \cdots, \hat{b}_s)$ satisfies

$$L(\hat{\boldsymbol{b}}) = \min_{\boldsymbol{b} \in T_s} L(\boldsymbol{b}). \tag{3.8.7}$$

Given an NT-net $\{\boldsymbol{c}_k = (c_{k1}, \cdots, c_{k,s-1})', k = 1, \cdots, n\}$ on C^{s-1}, then $\{\boldsymbol{x}_k = (x_{k1}, \cdots, x_{ks})', k = 1, \cdots, n\}$ defined in (1.5.33) is an NT-net on T_s. With this net and SNTO–D we can solve the following problem.

Example 3.9

This example was discussed by Fang, Wang and Wu (1982) which considered a specification for the model of making a bituminous concrete. The data is given in Table 3.17.

Table 3.17 *Data*

x_1	x_2	x_3	x_4	x_5	x_6	Y
100.0	100.0	100.0	100.0	100.0	100.0	100.0
10.15	72.25	100.0	100.0	100.0	100.0	85.0
0	0	87.5	98.4	94.7	100.0	60.0
0	0	32.8	65.55	88.4	100.0	45.0
0	0	27.3	52.45	86.8	100.0	37.0
0	0	12.0	15.05	78.0	95.3	29.0
0	0	4.7	0.95	5.2	77.9	22.0
0	0	4.2	0.55	5.2	71.9	19.0
0	0	1.5	0.20	0.4	56.4	15.0

The least squares estimate of $\boldsymbol{b}$ is

$$\hat{\boldsymbol{b}} = (0.0761, 0.2941, 0.2191, 0.1561, 0.0000, 0.2545)'$$

which is obtained by the use of 7 eliminating transformations (Fang, Wang and Wu (1982)). The coresponding sum of square residuals is 12.62245.

Now we use SNTO–D to treat this problem. Take $\delta = 10^{-6}$, $n_i = 1069$, $i = 1, 2, \cdots$. Then we have $M^{(14)} = 12.62459$ and

$$\boldsymbol{b}^{(14)} = (0.0764, 0.2933, 0.2210, 0.1546, 0.0000, 0.2546)'$$

which are quite close to $M = 12.62245$ and $\hat{\boldsymbol{b}}$. In order that the reader can easily check the above result, we list all $M^{(t)}$ for $1 \le t \le 14$ in Table 3.18.

Remark 3.10

This example shows that the magnitude of calculation in SNTO–D is larger than that of the method of eliminating transformation, but the SNTO–D program is simple. Hence it is still worth while for those people who perform regression analysis with constraints.

Table 3.18 *Results by SNTO–D*

t	1	2	3	4	5
$M^{(t)}$	46.08358	37.20098	18.56352	16.96464	13.60261
t	6	7	8	9	10
$M^{(t)}$	13.05130	12.74801	12.66653	12.63025	12.62659
t	11	12	13	14	
$M^{(t)}$	12.62525	12.62513	12.62464	12.62459	

3.9 Mode of a multivariate distribution

In some practical problems, the mode is a useful characteristic number of a density function, but there is no analytic expression for many multivariate distributions. Hence a numerical calculation is required to obtain its approximate value. Suppose that $p(\boldsymbol{x})$ is a p.d.f. of an s-dimensional distribution $F(\boldsymbol{x})$. Then $\boldsymbol{x}_0$ is called the **mode** of $F(\boldsymbol{x})$ (or $p(\boldsymbol{x})$) if $\boldsymbol{x}_0 \in R^s$ and

$$p(\boldsymbol{x}_0) = \max_{\boldsymbol{x} \in D} p(\boldsymbol{x}). \tag{3.9.1}$$

where D is the support of $p(\boldsymbol{x})$, i.e. $D = \{\boldsymbol{x} : p(\boldsymbol{x}) > 0\}$. This is a typical optimization problem for finding the mode $\boldsymbol{x}_0$, and we may obtain its approximate value by using SNTO–D. Now we discuss the following two cases.

(a) D is a large s-dimensional domain, even $D = R^s$ or $D = R^s_+$ the positive part of R^s. This will lead to difficulties in numerical calculation. It is well-known that there are some connections between the mean and mode. For univariate distribution, if the distribution is symmetric about some point and unimodal, the mean and the mode are the same. Otherwise they are distinct, and the absolute value of their difference depends roughly on the skewness of the distribution. Let $\boldsymbol{\mu}$ and $\boldsymbol{\sigma}$ be the respective mean and standard deviation vectors of $F(\boldsymbol{x})$. For most distributions, the mode $\boldsymbol{x}_0$ falls in the rectangle $[\boldsymbol{\mu} - 3\boldsymbol{\sigma},\ \boldsymbol{\mu} + 3\boldsymbol{\sigma}] \cap D$, which may be regarded as $D^{(1)}$ in the SNTO–D algorithm for finding the approximation of $\boldsymbol{x}_0$. If the distribution is almost symmetric, we may use $[\boldsymbol{\mu}-2\boldsymbol{\sigma},\ \boldsymbol{\mu}+2\boldsymbol{\sigma}]\cap D$ or even the rectangle $[\boldsymbol{\mu}-\boldsymbol{\sigma},\ \boldsymbol{\mu}+\boldsymbol{\sigma}]\cap D$ as $D^{(1)}$.

(b) Suppose that the dimension of D is less than s. To find the mode $\boldsymbol{x}_0$ of $F(\boldsymbol{x})$ by SNTO–D, we need an NT-net on D or on a subdomain of D containing $\boldsymbol{x}_0$. We give here an example to illustrate the algorithm.

Example 3.10

The additive logistic elliptical distributions were defined in section 2.6, and its density function is of the form

$$f(\boldsymbol{x}) = (\det\boldsymbol{\Sigma})^{-1/2}\prod_{i=1}^{s} x_i^{-1} g((\boldsymbol{y_x}-\boldsymbol{\mu})'\boldsymbol{\Sigma}^{-1}(\boldsymbol{y_x}-\boldsymbol{\mu})), \tag{3.9.2}$$

where $\boldsymbol{y_x}$ is the associated vector of $\boldsymbol{x}$ defined by (2.6.9). When $g(u) = (2\pi)^{-(s-1)/2}\exp(-\frac{1}{2}u)$, (3.9.2) reduces to the p.d.f. of the additive logistic normal distributions. When the function g in (3.9.2) has the form

$$g(u) = c(1+u/m)^{-p},\ p > \frac{s}{2},\ m > 0, \tag{3.9.3}$$

where c is the normalizing constant, the corresponding distribution is called **additive logistic elliptical Pearson Type VII distribution** (section 2.6). In this case to find the mode of $f(\boldsymbol{x})$ is equivalent to obtaining the maximum of

$$h(\boldsymbol{x}) = \prod_{i=1}^{s} x_i^{-1}\left[1+\frac{1}{m}(\boldsymbol{y_x}-\boldsymbol{\mu})'\boldsymbol{\Sigma}^{-1}(\boldsymbol{y_x}-\boldsymbol{\mu})\right]^{-p} \tag{3.9.4}$$

over T_s.

Now we give the algorithm for finding the approximation of $\boldsymbol{x}_0$ on T_s by SNTO–D, where the contraction is done on T_s directly.

Step 1 Take an NT-net $\mathcal{P}^{(1)} = \{\boldsymbol{y}_k^{(1)}, k = 1,\cdots,n_1\}$ on $D^{(1)} = T_s$ with suitable n_1. Find the maximum $M^{(1)}$ of $h(\boldsymbol{x})$ among these points, and assume that it is attained at $\boldsymbol{x}^{(1)} = (x_1^{(1)}, \cdots, x_s^{(1)})'$.

Step 2 Find a small subdomain $D^{(2)}$. For instance, $D^{(2)}$ is a domain with $\boldsymbol{x}^{(1)}$ located near the centre of gravity of $D^{(2)}$. More precisely, we choose a_i's such that

$$0 \le a_i < x_i^{(1)}, \ i = 1, \cdots, s.$$

Set $a = a_1 + \cdots + a_s$ and $b_i = a_i + 1 - a$, $i = 1, \cdots, s$. Then

$$\begin{aligned} 1 \ge b_i \ge a_i + \sum_{j=1}^{s} x_j^{(1)} - \sum_{k=1}^{s} a_k \\ = x_i^{(1)} + \sum_{\substack{j=1 \\ j \ne i}}^{s} (x_j^{(1)} - a_j) > x_i^{(1)}, \ i = 1, \cdots, s. \end{aligned}$$

Let

$$\begin{aligned} D^{(2)} = \{\boldsymbol{x} = (x_1, \cdots, x_s)' : a_1 \le x_i \le b_i, \\ i = 1, \cdots, s, \ \boldsymbol{x} \in D^{(1)}\}. \end{aligned}$$

Let $\{\boldsymbol{z}_k = (z_{k1}, \cdots, z_{ks})', k = 1, \cdots, n_1\}$ be an NT-net on T_s. Then we have a set $\mathcal{P}^{(2)} = \{\boldsymbol{y}_k^{(2)}, k = 1, \cdots, n_2\}$, where

$$y_{ki}^{(2)} = a_i + (1 - a) z_{ki}, \ i = 1, \cdots, s, \ k = 1, \cdots, n_2,$$

is an NT-net on $D^{(2)}$. Denote by $M^{(2)}$ the maximum of the function on $\mathcal{P}^{(1)} \cup \mathcal{P}^{(2)}$ and $M^{(2)}$ is attained at the point $\boldsymbol{x}^{(2)}$.

Step 3 Suppose that in the tth-step we have found the maximum $M^{(t)}$ of the function and the corresponding point $\boldsymbol{x}^{(t)}$. By a similar method we can reduce the domain $D^{(t)}$ to $D^{(t+1)}$, and make a set of points on $D^{(t+1)}$, by which we can find another maximum $M^{(t+1)}$ of the function and the corresponding point $\boldsymbol{x}^{(t+1)}$.

Repeat Step 3 until the search domain is small enough. The last maximum $M^{(t+1)}$ is expected to be close to the global maximum of the function.

Applying the above program to our problem with parameters $s = 3$, $p = 9$, $m = 5.5$,

$$\boldsymbol{\mu} = \begin{pmatrix} 0 \\ 0 \end{pmatrix} \quad \text{and} \quad \Sigma = \begin{pmatrix} 1 & -0.7 \\ 0.7 & 1 \end{pmatrix},$$

and taking the *glp set* with $n_1 = n_2 = \cdots = 233$, $h_1 = 1$ and $h_2 = 144$ for each step, we have the results listed in Table 3.19. The result is quite close to the real solution by 4 contractions only. Since we should check the result in each step, it is possible to accelerate the process of contraction according to results feedback.

Table 3.19

t	a_t	b_t	$M^{(t)}$	$x_1^{(t)}$	$x_2^{(t)}$	$x_3^{(t)}$
1	0.0000	1.0000	26.55203	0.3271035	0.3306723	0.3422242
2	0.3000	0.4000	26.97836	0.3304331	0.3343395	0.3352274
3	0.3300	0.3400	26.99543	0.3327104	0.3330673	0.3342224
4	0.3330	0.3340	26.99994	0.3332711	0.3333067	0.3334223
The global maximum			27.00000	0.333333	0.3333333	0.3333333

3.10 Mixture of SNTO and other methods

Many examples and the discussions in preceeding sections show that there are many advantages of SNTO: there is a higher chance of finding the global maximum of a multimodal function using SNTO than by using other methods; the program is easy to implement and can be universally used for different models with only minor modification; we need not calculate the derivatives of the function f on D. The SNTO has also some shortcomings: first, the domain for optimization problem should be bounded. Otherwise the problem should be reduced to an optimization problem on a bounded domain such that the maximum of the function f is outside the bounded domain. We have done such pretreatment in many problems in previous sections. Secondly, SNTO requires more computing time when compared with the Newton-like methods when the domain is large. However the enhancement of the convergence rate is still an open problem, discussed later in this

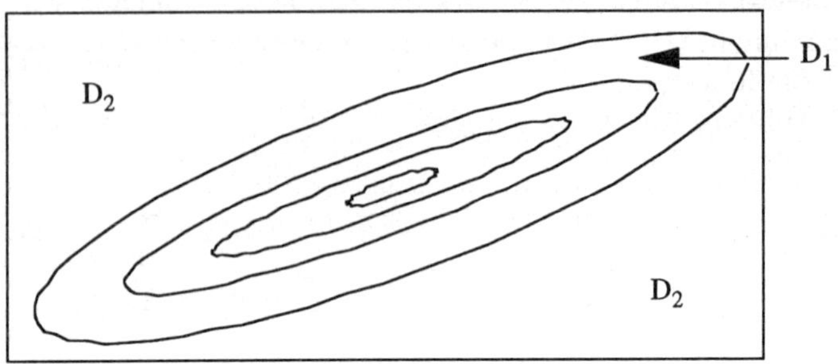

Figure 3.4 *Contour of f*

section. Thirdly, if the values of f are approximately zero in a large part of D, say D_2, as shown by Figure 3.4, we would still put many points on D_2 by the use of SNTO which is obviously not economic. Therefore SNTO and SNTO–D should be improved in the above points. Now we propose some suggestions for improving SNTO and SNTO–D as follows:

3.10.1 Mixtures of SNTO and Newton-like method

There are a lot of optimization methods to treat the unimodal function f with high convergence rate. But if the function is not unimodal, then we can get probably a local maximum. If we can use SNTO or SNTO–D first to shrink the domain D into a smaller domain D^* such that $f(\boldsymbol{x})$ is unimodal on D^* and D^* contains the maximum point $\boldsymbol{x}^*$ of D, then we may use a Newton-like method to get the maximum point $\boldsymbol{x}^*$ of $f(\boldsymbol{x})$ on D^*. This suggests that combining the two methods may lead to a hybrid which is both more efficient than SNTO and more robust than Newton-like methods. The following is an outline of such an algorithm.

Step 0 **Initialization.** Set $t = 0$, $D^{(0)} = D$.

Step 1 **Generate an NT-net.** Use a number-theoretic method to generate n_t points $\mathcal{P}^{(t)}$ uniformly scattered on $D^{(t)} = [\boldsymbol{a}^{(t)}, \boldsymbol{b}^{(t)}]$.

Step 2 **Compute new approximation from NT-net.** Find $\boldsymbol{x}^{(t)} \in \mathcal{P}^{(t)} \cup \{\boldsymbol{x}^{(t-1)}\}$ and $M^{(t)}$ such that $M^{(t)} = f(\boldsymbol{x}^{(t)}) \geq f(\boldsymbol{y}), \forall \boldsymbol{y} \in \mathcal{P}^{(t)} \cup \{\boldsymbol{x}^{(t-1)}\}$. $\boldsymbol{x}^{(t)}$ and $M^{(t)}$ are the best approximations to $\boldsymbol{x}^*$ and M so far.

Step 3 **NT Termination criterion.** Let $\boldsymbol{c}^{(t)} = (\boldsymbol{b}^{(t)} - \boldsymbol{a}^{(t)})/2$. If $\max(\boldsymbol{c}) < \delta$, then $D^{(t)}$ is small enough; $\boldsymbol{x}^{(t)}$ and $M^{(t)}$

are acceptable; terminate algorithm. Otherwise, proceed to next step.

Step 4 **Contract domain.** Form new domain $D^{(t+1)} = [\boldsymbol{a}^{(t+1)}, \boldsymbol{b}^{(t+1)}]$ as follows:

$$a_i^{(t+1)} = \max(x_i^{(t)} - \gamma c_i^{(t)}, a_i)$$

and

$$b_i^{(t+1)} = \max(x_i^{(t)} + \gamma c_i^{(t)}, b_i)$$

where γ is a pre-determined contraction ratio.

Step 5 **Compute new approximation by Newton-like method.** Set $t = t+1$. Use one step of a Newton-like method to compute a new approximation $\boldsymbol{x}^{(t)}$ to $\boldsymbol{x}^*$. Set $M^{(t)} = f(\boldsymbol{x}^{(t)})$.

Step 6 **NT/Newton-like branching criterion.** If $M^{(t)} < M^{(t-1)}$, the Newton-like method is not useful. Reject this approximation and go to step 1. Otherwise, proceed to Step 7.

Step 7 **Newton-like termination criterion.** If $\|\boldsymbol{x}^{(t)} - \boldsymbol{x}^{(t-1)}\| < \delta$, then $\boldsymbol{x}^{(t)}$ and $M^{(t)}$ are acceptable; terminate algorithm. Otherwise, proceed to Step 5.

This hybrid algorithm is also valid if Newton-like method is replaced by any other optimization method and is similar to probit analysis where one uses the simplex method followed by the Newton-Raphson method. Many examples show that the convergence rate is enhanced by this algorithm (Lai (1992)). Lai tests the hybrid algorithm with many typical examples and shows that the hybrid one dominates both SNTO and the conjugate direction method.

3.10.2 Mixtures of SNTO and Monte Carlo optimization

One reason that the efficiency of SNTO is not high is that the properties of the function are not considered. For example, if there are more points of an NT-net falling on D_1 compared with D_2 according to the property of f (Fig. 3.3) and if the domains for optimization are shrinked steadily, then the efficiency of the algorithm can be highly enhanced. Therefore, another way to improve the efficiency of SNTO is to generate sets of points $\mathcal{P}^{(t)}$ that are representative points with respect to a distribution function $F^{(t)}(\boldsymbol{x})$

other than the uniform distribution. For example, $F^{(t)}$ might be a multivariate normal distribution centered at $\boldsymbol{x}^{(t-1)}$. The advantage of such a modification is that one can cluster points near to where the maximum is assumed to be. Some implementations of Monte Carlo methods for optimization use this approach. This modification to SNTO or SNTO–D can be implemented easily by modifying Step 1 as follows:

Step 1 **Generate an NT-net.** Generate an NT-net $\mathcal{P}^{(t)}$ of n_t points on $D^{(t)}$ and then produce a set of rep-points of $F^{(t)}(\boldsymbol{x})$ (section 4.2).

The above modification assumes that $F^{(t)}$ is known. In practice, we may wish to estimate $F^{(t)}$ from information gained from previous iterations. One way to do so is proposed by Tang, M.C. (1992). First, let $\bar{f}^{(t-1)}$ be the sample mean of $f(\boldsymbol{x})$ for $\boldsymbol{x} \in \mathcal{P}^{(t-1)}$, and let $\mathcal{P}_*^{(t-1)} = \{\boldsymbol{x} \in \mathcal{P}^{(t-1)} : f(\boldsymbol{x}) > \bar{f}^{(t-1)}\}$. The set $\mathcal{P}_*^{(t-1)}$ consists of points which are likely to be closer to $\boldsymbol{x}^*$. The sample mean and covariance matrix of the points in $\mathcal{P}_*^{(t-1)}$ is then used to define a multivariate normal distribution $F^{(t)}$.

There are a number of methods of Monte Carlo optimization, such as the nonadaptive random search, adaptive random search, random search double-trial, acceptance-rejection random search, statistical gradient random search algorithms (Rubinstein (1986)). In picking up the ideas in these methods and applying them to the NTM for optimization we might find more efficient optimization algorithms.

Exercise

3.1 Draw the contours of the function f in Example 3.1 and write a computer program to reproduce the results in Table 3.2.

3.2 Let $f_1(\boldsymbol{x}) = \exp(\sum_{j=1}^s x_j)\sin(\sum_{j=1}^s x_j)$, $\boldsymbol{x} \in C^s$. Draw the contours of function f_1 for $s = 2$ and write a computer program to find the maximum and minimum point of f_1.

3.3 Let

$$f_2(\boldsymbol{x}) = \exp\Big(\prod_{j=1}^{s} x_j^j\Big)\sin\Big(\sum_{j=1}^{s} x_j\Big), \quad \boldsymbol{x} \in C^s$$

and

$$f_3(\boldsymbol{x}) = -\exp\Big(\prod_{j=1}^{s} x_j\Big)\sin\Big(\sum_{j=1}^{s} jx_j\Big), \quad \boldsymbol{x} \in C^s.$$

Write a program to find the maximum and minimum point for $s = 3, 4$ by the following methods:

(a) Generate an infinite NT-net on C^s and use the procedure (3.1.8). For example, generate the NT-nets by the *gp*–1 and *gp*–2 methods, H-W, Haber, and Hammersley methods defined in section 1.3.

(b) Apply SNTO.

3.4 Let $f_1(x, y)$ and $f_2(x, y)$ by the so-called **Rosenbrock functions:**

$$\begin{aligned} f_1(x, y) &= 100(y - x)^2 + (x - 1)^2 \\ & \qquad\qquad\qquad\qquad -3 \le x, y \le 2. \\ f_2(x, y) &= 10000(y - x)^2 + (x - 1)^2, \end{aligned}$$

Draw the contours of the above two functions and apply Newton's method to find their maximum and minimum points.

3.5 Let $f_4(\boldsymbol{x}) = -(\boldsymbol{x}-\boldsymbol{a})'\boldsymbol{A}(\boldsymbol{x}-\boldsymbol{a})$, where $\boldsymbol{x}, \boldsymbol{a} \in R^s$, $\boldsymbol{A} : s \times s$, $\boldsymbol{A} > 0$ are given. Obviously, the function f_4 has zero maximum value at the point $\boldsymbol{a}$. Suppose that $\boldsymbol{a} \sim U(C^s)$ and is generated by the Monte Carlo method. If we require the precision to be $\|\boldsymbol{a} - \tilde{\boldsymbol{a}}\| \le 10^{-7}$, where $\tilde{\boldsymbol{a}}$ is the approximation of $\boldsymbol{a}$ by some numerical method, give the comparisons among the Newton's method, the truncated Newton method, the simplex method and SNTO in the following way. Let $\boldsymbol{A}$ have all the diagonal elements one and all the other elements ρ ($|\rho| < 1$). Consider $\rho = -0.9, -0.3, 0.3, 0.9$. Then generate $\boldsymbol{a}_1, \cdots, \boldsymbol{a}_{100}$ by the Monte Carlo method and record the computer time needed to meet the required precision.

3.6 Let

$$f_5(\boldsymbol{x}) = -[\lambda_1(\boldsymbol{x} - \boldsymbol{a})'\boldsymbol{A}(\boldsymbol{x} - \boldsymbol{a}) + \lambda_2(\boldsymbol{x} - \boldsymbol{b})'\boldsymbol{B}(\boldsymbol{x} - \boldsymbol{a})],$$

where $\lambda_1 > 0, \lambda_2 > 0, \lambda_1 + \lambda_2 = 1, \boldsymbol{B} > 0, \boldsymbol{B} : s \times s, \boldsymbol{b} \sim U(D)$ with $D = [\boldsymbol{a}, \boldsymbol{a} + \boldsymbol{1}_s]$, and $\boldsymbol{a}$ and $\boldsymbol{A}$ are the same with Exercise

3.5. Apply the same comparison method using in Exercise 3.5 to the function $f_5(\boldsymbol{x})$.

3.7 The following data set is from a Weibull distribution with three parameters γ, α and β. Give the MLEs of γ, α, β by the NTM (section 3.3).

5.37	5.92	4.81	4.97	5.97	4.24	4.96	4.45	4.93	4.42
5.57	6.06	5.59	5.31	5.58	6.64	5.36	4.64	4.64	5.41
4.79	6.25	5.14	5.56	4.50	5.62	5.77	4.53	4.85	6.08
4.50	5.84	4.90	5.09	6.44	3.98	3.88	5.66	6.58	6.10
7.14	6.15	1.44	3.82	5.71	3.93	4.78	5.05	3.94	5.28
5.66	7.83	4.52	2.91	6.67	4.14	4.70	4.53	5.04	6.37
6.92	6.55	6.06	5.89	7.71	4.82	6.03	4.98	7.31	5.60
5.84	6.92	6.65	6.52	6.42	5.24	7.49	3.42	5.11	5.35
4.34	5.64	4.07	6.64	5.14	5.07	7.75	6.02	7.01	8.01
4.62	6.56	6.74	4.22	6.52	6.34	6.30	5.15	6.86	5.38

3.8 The following data set is from a beta distribution with parameters α and β. Give the MLEs of α and β by the NTM (section 3.3).

0.21	0.18	0.41	0.52	0.49	0.66	0.24	0.38	0.20	0.30
0.49	0.32	0.43	0.31	0.15	0.18	0.28	0.45	0.62	0.36
0.37	0.61	0.38	0.42	0.50	0.35	0.31	0.37	0.21	0.46
0.43	0.47	0.26	0.53	0.49	0.28	0.74	0.43	0.51	0.32
0.63	0.34	0.39	0.61	0.25	0.39	0.37	0.39	0.28	0.27
0.60	0.42	0.66	0.11	0.41	0.31	0.33	0.28	0.34	0.20
0.52	0.17	0.27	0.29	0.53	0.23	0.33	0.45	0.44	0.41
0.49	0.30	0.09	0.37	0.28	0.55	0.57	0.42	0.45	0.30
0.16	0.40	0.23	0.32	0.29	0.52	0.33	0.48	0.36	0.23
0.45	0.52	0.35	0.57	0.37	0.44	0.60	0.49	0.49	0.53

3.9 The following dat set is taken from Example 5.2 from the steel industry (Fang, Quan and Chen (1989)):

X	2	3	4	5	7	8
Y	106.42	108.20	109.58	110.00	109.93	110.49

X	11	14	15	16	18	19
Y	110.59	110.60	110.90	110.76	111.00	111.20

Use the following models to fit the data by the LNL algorithm:

$$EY = \frac{1}{a + b/X}$$

$$EY = ae^{b/X}$$

$$EY = \frac{X}{aX + b}$$

and

$$EY = \frac{1}{a + be^{-X}}.$$

The estimates of a and b are to be obtained by the LSE and the robust estimation with $h(u, v) = |u - v|$ (section 3.5).

3.10 Use the following data sets

Set I		Set II		Set III		Set IV	
x	y	x	y	x	y	x	y
9	8.93	1	16.08	0	1.23	0.5	1.3
14	10.80	2	33.83	1	1.52	1.5	1.3
21	18.59	3	65.80	2	2.95	2.5	1.9
28	22.33	4	97.20	3	4.34	3.5	3.4
42	39.35	5	191.55	4	5.26	4.5	5.3
57	56.11	6	326.20	5	5.84	5.5	7.1
63	61.73	7	386.87	6	6.21	6.7	10.6
70	64.62	8	520.55	8	6.50	7.5	16.0
79	67.08	9	590.03	10	6.83	8.5	16.4
		10	651.92			9.5	18.3
		11	724.93			10.5	20.9
		12	699.56			11.5	20.5
		13	689.96			12.5	21.3
		14	637.56			13.5	21.2
		15	717.41			14.5	20.9

to fit the five models:

$$EY = \alpha \exp[-\exp(\beta - \gamma x)]$$

$$EY = \frac{\alpha}{1 + \exp(\beta - \gamma x)}$$

$$EY = \frac{\alpha}{[1 + \exp(\beta - \gamma x)]^{1/\delta}}$$

$$EY = \frac{\beta\gamma + \alpha x^\delta}{\gamma + x^\delta}$$

and

$$EY = \alpha + \beta \exp(-\gamma x^\delta)$$

by the partial linearization method and RSNTO algorithm.

3.11 Apply the RSNTO algorithm to Example 3.3.

3.12 Apply the RSNTO algorithm to Example 3.6 and find the approximate value of $\hat{\boldsymbol{\theta}}$ with higher precision than that shown in Example 3.6.

CHAPTER 4

Representative points of a multivariate distribution

In the previous chapters we have discussed many applications of an NT-net to a closed and bounded domain D in statistics. The points in an NT-net on D can be considered representative points of the distribution $U(D)$. In this chapter we shall consider representative points, or rep-points for short, of $F(\boldsymbol{x})$, where $F(\boldsymbol{x})$ is a given multivariate continuous distribution function and pay attention to the following two criteria: F-discrepency (or Kolmogorov-Smirnov distance) and mean square error (MSE) for measuring the closeness of representation. In this chapter these two kinds of rep-points will be given for many useful distributions. Some applications of rep-points in evaluating probabilities and in geometric probability will also be given.

4.1 F-discrepancy criterion

Let $F(\boldsymbol{x}) = F(x_1, \cdots, x_s)$ be a given s-dimensional continuous distribution function. Let $\mathcal{P} = \{\boldsymbol{x}_k, k = 1, \cdots, n\}$ be a set of n points of R^s. The F-discrepancy $D_F(n, \mathcal{P})$ defined by (1.2.2) is a measure of the representation of $\mathcal{P}$ to $F(\boldsymbol{x})$. If we can find a set $\mathcal{P}^* = \{\boldsymbol{x}_k^*, k = 1, \cdots, n\}$ such that

$$D_F(n, \mathcal{P}^*) = \min_{\mathcal{P}} D_F(n, \mathcal{P}), \tag{4.1.1}$$

where $\mathcal{P}$ runs over all sets of n points in R^s, then $\mathcal{P}^*$ is called a set of **cdf-rep-points** of $F(\boldsymbol{x})$.

When $s = 1$, the problem for finding a set of cdf-rep-points for continuous distribution functions is solved (Example 1.4). For completeness, we record the conclusion below.

Theorem 4.1
Let $F(x)$ be a univariate continuous c.d.f. and $F^{-1}(y)$ be its inverse function. Then the set

$$\mathcal{Q}_F = \{F^{-1}((2i-1)/2n),\ i = 1, \cdots, n\} \qquad (4.1.2)$$

contains the cdf-rep-points of size n of $F(x)$ with F–discrepancy $n/2$.

When $s > 1$, it is difficult to find a set $\mathcal{P}^*$ of n points in R^s such that (4.1.1) holds even in the simplest case that $F(\boldsymbol{x})$ is the uniform distribution over C^s. Hence we shall be satisfied with a set $\mathcal{P}$ with low F-discrepancy, and regard $\mathcal{P}$ to be the rep-points of $F(\boldsymbol{x})$. How low is "low discrepancy"? We made some suggestions in Definition 1.3, and we may similarly give the following

Definition 4.1
Let $\boldsymbol{N}$ be an infinite subset of natural numbers and $\{\mathcal{P}_n, n \in \boldsymbol{N}\}$ be a sequence of sets of points on C^s, where $\mathcal{P}_n$ has n points with a certain structure. If

$$D_F(n, \mathcal{P}_n) = o(n^{-\frac{1}{2}}),$$

then $\{\mathcal{P}_n\}$ is said to be a **sequence of rep-points** of $F(\boldsymbol{x})$. If for any given $\varepsilon > 0$ we have

$$D_F(n, \mathcal{P}_F) = O(n^{-1+\varepsilon}),$$

then $\{\mathcal{P}_n\}$ is called a **sequence of good rep-points** of $F(\boldsymbol{x})$.

When $F(\boldsymbol{x})$ is the uniform distribution $U(C^s)$, we have introduced several effective methods for obtaining sequences of good rep-points of $U(C^s)$, and we may obtain the rep-points of a multivariate distribution with independent components by using the transformation method.

Definition 4.2
Let T be an one-to-one continuous transformation from R^s to R^s and $\boldsymbol{y} = T(\boldsymbol{x})$. Let $F(\boldsymbol{x})$ and $G(\boldsymbol{y})$ be respective c.d.f's of $\boldsymbol{x}$ and

$\boldsymbol{y}$. Given a set of points $\mathcal{P} = \{\boldsymbol{x}_k, k = 1, \cdots, n\}$ we have a set $\mathcal{P}^* = \{\boldsymbol{y}_k = T(\boldsymbol{x}_k), k = 1, \cdots, n\}$. We shall call T a **discrepancy-preserving transformation (DPT)** if

$$D_F(n, \mathcal{P}) = D_G(n, \mathcal{P}^*) \tag{4.1.3}$$

for any $\mathcal{P}$ and its image $\mathcal{P}^*$.

Example 4.1
Let $F(\boldsymbol{x})$ be a continuous multivariate distribution and

$$F(\boldsymbol{x}) = F(x_1, \cdots, x_s) = \prod_{i=1}^{s} F_i(x_i), \tag{4.1.4}$$

where $F_i(x_i)$ $(i = 1, \cdots, s)$ are the marginal distribution functions of $\boldsymbol{x}$. If $\{\boldsymbol{c}_k = (c_{k1}, \cdots, c_{ks}),\ k = 1, \cdots, n\}$ is a set of n points in C^s with discrepancy d, then the set

$$\{\boldsymbol{x}_k = (F_1^{-1}(c_{k1}), \cdots, F_s^{-1}(c_{ks})),\ k = 1, \cdots, n\} \tag{4.1.5}$$

has F–discrepancy d with respect to $F(\boldsymbol{x})$ (Example 1.5). Therefore, this independent inverse transformation of a distribution function is a DPT.

Example 4.2
Let $\boldsymbol{x}$ be a random vector in R^s and

$$\boldsymbol{y} = \boldsymbol{B}\boldsymbol{x} + \boldsymbol{a}$$

be its linear transformation, where $\boldsymbol{a}$ is a constant vector and $\boldsymbol{B}$ is a diagonal matrix with positive diagonal elements. Then the transformation is a DPT. The proof is left to the reader.

But, there are no independent structures in most multivariate distributions. How do we find their rep-points? Note that we have obtained the low discrepancy rep-points of the uniform distributions $U(A_s)$, $U(B_s)$, $U(U_s)$, $U(T_s)$ and $U(V_s)$ by the stochastic representation method. For the convenience we prefer to use the terminology "NT-net" or "uniformly scattered" for the various uniform distributions instead of rep-points which is used for other distributions.

Suppose that $\boldsymbol{x} = (X_1, \cdots, X_s)$ is a random vector with continuous c.d.f. $F(\boldsymbol{x})$ and has a stochastic representation

$$\boldsymbol{x} \stackrel{d}{=} \boldsymbol{x}(\boldsymbol{\phi}) = \boldsymbol{x}(\phi_1, \cdots, \phi_t), \tag{4.1.6}$$

where $\phi_1, \cdots, \phi_t$ are independent random variables with continuous c.d.f.'s. From (4.1.5) we can obtain a set $\{\boldsymbol{y}_k, k = 1, \cdots, n\}$ of R^t with low F–discrepancy d. Then the set

$$\{\boldsymbol{x}_k = \boldsymbol{x}(\boldsymbol{y}_k), k = 1, \cdots, n\} \tag{4.1.7}$$

can be considered as a set of rep-points of $F(\boldsymbol{x})$, where $\{\boldsymbol{x}_k\}$ has quasi F-discrepancy d (Definition 1.9).

We have already discussed in section 1.5 that the rep-points obtained by (4.1.7) are reasonably good. Since the stochastic representation (4.1.6) could not be unique, we may obtain different sets of rep-points for the same $F(\boldsymbol{x})$. For example, we have discussed this for the uniform distributions $U(U_s)$ and $U(T_s)$ in section 1.6. It is similar to Monte Carlo simulation, in that there are many methods for generating a random variate from the same distribution. Different methods may yield different efficiencies, but there is no common criterion to judge the superiority of the methods.

It is interesting to note that many useful multivariate distributions have the following stochastic representation:

$$\boldsymbol{x} \stackrel{d}{=} R\boldsymbol{y} \tag{4.1.8}$$

where $\boldsymbol{x} \sim F(\boldsymbol{x})$, an s-dimensional multivariate distribution, $R > 0$ a positive random variable, $\boldsymbol{y} \sim U(D)$ where D is an $(s-1)$-dimensional bounded domain in R^s, and R and $\boldsymbol{y}$ are independent. If an NT-net on D can be generated we propose the following algorithm to generate a set of rep-points of $F(\boldsymbol{x})$ by the independence of R and $\boldsymbol{y}$:

Step 1 Generate an NT-net $\{\boldsymbol{c}_k = (c_{k1}, \cdots, c_{ks})', k = 1, \cdots, n\}$ on C^s.

Step 2 Denote the c.d.f. of R by $F_R(r)$ and let F_R^{-1} be its inverse function. Compute $r_k = F_R^{-1}(c_{ks}), k = 1, \cdots, n$.

Step 3 Generate an NT-net $\{\boldsymbol{y}_k, k = 1, \cdots, n\}$ on D with the first $(s-1)$-components of $\boldsymbol{c}_k$, $k = 1, \cdots, n$.

Step 4 Then $\{\boldsymbol{x}_k = r_k \boldsymbol{y}_k, k = 1, \cdots, n\}$ is a set of rep-points of $F(\boldsymbol{x})$.

This algorithm is called the **NTSR algorithm**, and we shall use it to produce the rep-points of many multivariate distributions in the next section with $D = U_s$ or T_s.

4.2 Rep-points of some classes of multivariate distributions

In this section we shall use the NTSR algorithm to generate the rep-points of some classes of multivariate distributions such as spherical distributions, multivariate l_1-norm symmetric distributions, and multivariate Liouville distributions.

4.2.1 Classes of spherically and elliptically symmetric distributions

The spherically and elliptically symmetric distributions have been defined in section 2.2. It is known that $\boldsymbol{x}$ has a spherical distribution if and only if

$$\boldsymbol{x} \stackrel{d}{=} R\boldsymbol{u}^{(s)}, \tag{4.2.1}$$

where $R \geq 0$ is independent of $\boldsymbol{u}^{(s)}$ and $\boldsymbol{u}^{(s)} \sim U(U_s)$ (Theorem 2.2).

We have discussed the spherical coordinate transformation method in section 1.5 for generating an NT-net on U_s. In the next section we shall introduce a more efficient method for generating an NT-net on U_s based on an NT-net on C^{s-1}. Applying the NTSR algorithm to (4.2.1) we have the following algorithm for generating rep-points of a given spherical distribution:

Step 1 Generate an NT-net $\{\boldsymbol{c}_k, k = 1, \cdots, n\}$ on C^s.

Step 2 Generate an NT-net on U_s, $\{\boldsymbol{y}_k, k = 1, \cdots, n\}$, based on the first $(s-1)$ components of $\{\boldsymbol{c}_k\}$.

Step 3 Denote the c.d.f. of R defined in (4.2.1) by $F_R(r)$. Compute

$$r_k = F_R^{-1}(c_{ks}), k = 1, \cdots, n. \tag{4.2.2}$$

Step 4 Then $\{x_k = r_k \boldsymbol{y}_k, k = 1, \cdots, n\}$ is a set of rep-points of $\boldsymbol{x}$ given in (4.2.1).

For convenience we list below some of the $F_R(r)$ which generate different families of spherically symmetric distributions.

(a) The uniform distribution on B_s

If $\boldsymbol{x}$ has a uniform distribution on B_s, the c.d.f. of its generating variate R is given by

$$F_R(r) = \begin{cases} 0, & r \leq 0, \\ r^s, & 0 < r < 1, \\ 1, & r \geq 1, \end{cases} \tag{4.2.3}$$

((2.2.24)).

(b) The multivariate standard normal distribution

Let $\boldsymbol{x} \sim N_s(\boldsymbol{0}, \boldsymbol{I}_s)$. Then the c.d.f. of the generating variate R is given by the chi-distribution with s degrees of freedom:

$$F_R(r) = \int_0^r \frac{t^{s-1} e^{-t^2/2}}{\Gamma(s/2) 2^{s/2-1}} dt.$$

In this case, there is a simpler method for generating rep-points is based on the independence structure of $\boldsymbol{x}$, i.e.

$$F(\boldsymbol{x}) = \prod_{i=1}^{s} \Phi(x_i).$$

where $\Phi(x)$ is the c.d.f. of $N(0,1)$. By Example 4.1, we can obtain a set of rep-points $\{\boldsymbol{x}_k, k = 1, \cdots, n\}$ of $\boldsymbol{x}$, where

$$\boldsymbol{x}_k = (\Phi^{-1}(c_{k1}), \cdots, \Phi^{-1}(c_{ks})), k = 1, \cdots, n,$$

and $\{\boldsymbol{c}_k = (c_{k1}, \cdots, c_{ks}), k = 1, \cdots, n\}$ is an NT-net on C^s.

(c) The symmetric Kotz type distributions

This subclass of spherical distributions is defined in Example 2.2. The corresponding distribution of R can be obtained by (2.2.4) and (2.2.14) as follows:

$$F_R(r) = \frac{2tb^{(2N+s-2)/(2t)}}{\Gamma((2N+s-2)/(2t))} \int_0^r u^{2N+s-3} e^{-bu^{2t}} du. \qquad (4.2.4)$$

(d) The symmetric multivariate Pearson Type VII distributions

As defined in Example 2.3, the corresponding p.d.f. of $U = R^2$ can be obtained by (2.2.4) and (2.2.15) as follows:

$$\frac{1}{B(s/2, N-s/2)} m^{-s/2} u^{s/2-1} (1+u/m)^{-N}, \quad u > 0, \qquad (4.2.5)$$

i.e. R^2/m has the beta type II distribution with parameters $s/2$ and $N - s/2$. In this case, it is better to obtain the rep-points of R indirectly via the c.d.f. of a beta distribution. This shall be explained after the following remark.

Remark 4.1

A random variable B_2 is said to have the beta type II distribution with parameters α and β if it has density

$$\frac{1}{B(\alpha,\beta)} b_2^{\alpha-1} (1+b_2)^{-(\alpha+\beta)}, \quad b_2 > 0 \qquad (4.2.6)$$

and we denote it by $B_2 \sim Be\text{II}(\alpha,\beta)$. It is easy to see that if $B_2 \sim Be\text{II}(\alpha,\beta)$, then $B_1 = \frac{B_2}{1+B_2} \sim Be(\alpha,\beta)$, the beta distribution with parameters α and β.

From Remark 4.1, we have

$$B \equiv \frac{R^2/m}{1+R^2/m} = \frac{R^2}{m+R^2} \sim Be\Big(\frac{s}{2}, \frac{N-s/2}{2}\Big), \qquad (4.2.7)$$

where R^2 has the density (4.2.5). Using the inverse of beta distribution, we may obtain the r_k in (4.2.2). Let $G(b)$ be the c.d.f.

of $Be(\frac{s}{2}, \frac{N-s/2}{2})$ and $b_k = G^{-1}(c_{ks})$. From (4.2.7), we have $R = [mB/(1-B)]^{1/2}$, and so

$$r_k = \left(\frac{mb_k}{1-b_k}\right)^{1/2}. \tag{4.2.8}$$

As we have mentioned in Example 2.3, the multivariate t-distribution and the multivariate Cauchy distribution are special cases of the symmetric multivariate Pearson Type VII distributions. Therefore, the above method can applied to these two kinds of distributions.

(e) The symmetric multivariate Pearson Type II distributions
These distributions are defined in Example 2.4. From (2.2.4) and (2.2.20) the density of the corresponding R is

$$\frac{2\Gamma(s/2+m+1)}{\Gamma(s/2)\Gamma(m+1)} r^{s-1}(1-r^2)^m, \qquad 0 \le r \le 1 \tag{4.2.9}$$

and the density of $U = R^2$ is thus

$$\frac{2\Gamma(s/2+m+1)}{\Gamma(s/2)\Gamma(m+1)} u^{s/2-1}(1-u)^m, \qquad 0 \le u \le 1 \tag{4.2.10}$$

which implies that $R^2 \sim Be(s/2, m+1)$. So that the rep-points of R can be generated by using the inverse of a beta c.d.f. with a square root adjustment.

Finally we shall illustrate how to produce the rep-points of elliptically symmetric distributions. If a random vector $\boldsymbol{y} \sim S_s(g)$, then from Definition 2.1 we obtain

$$\boldsymbol{x} \stackrel{d}{=} \boldsymbol{\mu} + \boldsymbol{\Sigma}^{1/2}\boldsymbol{y} \sim EC_s(\boldsymbol{\mu}, \boldsymbol{\Sigma}, g),$$

where $\boldsymbol{\mu} : s \times 1$, and $\boldsymbol{\Sigma} > 0 : s \times s$. If $\{\boldsymbol{y}_k\}$ is a set of rep-points of $S_s(g)$, then $\{\boldsymbol{x}_k\}$ are rep-points of $EC_s(\boldsymbol{\mu}, \boldsymbol{\Sigma}, g)$, where

$$\boldsymbol{x}_k = \boldsymbol{\mu} + \boldsymbol{\Sigma}^{1/2}\boldsymbol{y}_k, \quad k = 1, \cdots, n. \tag{4.2.11}$$

Note that $\Sigma^{1/2}$ can be replaced by any $s \times s$ matrix $\boldsymbol{A}$ such that $\Sigma = \boldsymbol{A}\boldsymbol{A}'$. It is most convenient to take $\boldsymbol{A}$ as an upper triangular matrix with positive diagonal elements. It is known as the **Cholesky root** of Σ. With this choice of $\boldsymbol{A}$, we may write (4.2.11) as

$$\boldsymbol{x}_k = \boldsymbol{\mu} + \boldsymbol{A}\boldsymbol{y}_k, \quad k = 1, \cdots, n.$$

4.2.2 The class of multivariate l_1–norm symmetric distributions

Definition 4.3

An s–dimensional random vector $\boldsymbol{x}$ is said to have a multivariate l_1–norm symmetric distribution if and only if

$$\boldsymbol{x} \stackrel{d}{=} R\boldsymbol{u}, \tag{4.2.12}$$

where $R \geq 0$ is independent of $\boldsymbol{u}$, and $\boldsymbol{u} \sim U(T_s)$ (Fang and Fang (1988) or Fang, Kotz and Ng (1990)).

If $R \sim Ga(s, 1)$, the gamma distribution (section 2.6), then the components of $\boldsymbol{x}$ defined by (4.2.12) are i.i.d. and each has the standard exponential distribution. Therefore the class of multivariate l_1–norm symmetric distributions has many properties similar to those of the exponential distribution, and it may be regarded as a multivariate version of the exponential distribution.

Note that (4.2.12) has the structure of (4.1.8). Thus we can use the NTSR algorithm to generate the rep-points of multivariate l_1–norm symmetric distributions.

Let $F_R(x)$ be the c.d.f. of R in (4.2.12) and $F_R^{-1}(x)$ be its inverse function. Given an NT-net $\{\boldsymbol{c}_k = (c_{k1}, \cdots, c_{ks}),\ k = 1, \cdots, n\}$ on C^s, then the set $\{\boldsymbol{x}_k = (x_{k1}, \cdots, x_{ks}),\ k = 1, \cdots, n\}$, where

$$\begin{aligned} x_{ki} &= r_k \prod_{j=1}^{i-1} c_{kj}^{\frac{1}{s-j}} (1 - c_{ki}^{\frac{1}{s-i}}), \quad i = 1, \cdots, s-1, \\ x_{ks} &= r_k \prod_{j=1}^{s-1} c_{kj}^{\frac{1}{s-j}}, \quad r_k = F_R^{-1}(c_{ks}), \quad k = 1, \cdots, n, \end{aligned} \tag{4.2.13}$$

is the required set of rep-points ((1.5.33)). It is possible to find some recurrence formulae to calculate $\{\boldsymbol{x}_k\}$ for saving computing time ((4.2.19)).

4.2.3 The class of multivariate Liouville distributions

Definition 4.4
An s-dimensional random vector $\boldsymbol{x}$ is said to have a **multivariate Liouville distribution** if and only if

$$\boldsymbol{x} \stackrel{d}{=} R\boldsymbol{y}, \tag{4.2.14}$$

where $R \geq 0$ is independent of $\boldsymbol{y}$, and $\boldsymbol{y}$ has a Dirichlet distribution (Definition 2.3). If $\boldsymbol{y} \sim D_s(\alpha_1, \cdots, \alpha_s)$, then we write $\boldsymbol{x} \sim L(\alpha_1, \cdots, \alpha_s, F_R)$, where F_R is the c.d.f. of R in (4.2.14).

When $\alpha_1 = \cdots = \alpha_s = 1$, $\boldsymbol{y}$ reduces to the uniform distribution $U(T_s)$. Hence the class of multivariate l_1-norm symmetric distributions is a special class of multivariate Liouville distribution (Fang, Kotz and Ng (1990), Chap. 6).

Since (4.2.14) has the structure of (4.1.8), we can obtain the rep-points of the Liouville distribution by the NTSR algorithm if we can obtain the rep-points of a Dirichlet distribution, and of course we may use (2.6.3) to generate the rep-points of a Dirichlet distribution. Let F_j be the c.d.f. of $Ga(\alpha_j)$, $j = 1, \cdots, s$. If $\{\boldsymbol{c}_k = (c_{k1}, \cdots, c_{ks})\}$ is an NT-net on C^s and $b_{ki} = F_i^{-1}(c_{ki})$, then $\{\boldsymbol{y}_k = (y_{k1}, \cdots, y_{ks})\}$ is a set of rep-points of $\boldsymbol{y} \sim D_s(\alpha_1, \cdots, \alpha_s)$, where

$$y_{ki} = \frac{b_{ki}}{b_{k1} + \cdots + b_{ks}}, \quad i = 1, \cdots, s, \quad k = 1, \cdots, n. \tag{4.2.15}$$

Since $\boldsymbol{y} \in T_s$, this is a random vector with $(s-1)$ independent variables. Now we generate the rep-points by s independent variables. The representation may be not very good when n is small (Examples 1.11 and 1.12). Therefore we need a more effective method.

Let $B_1, \cdots, B_{s-1}$, be independent and $B_i \sim Be(\sum_{j=1}^{i} \alpha_j, \alpha_{i+1})$, $i = 1, \cdots, s-1$, and let $\boldsymbol{y} = (Y_1, \cdots, Y_s) \sim D_s(\alpha_1, \cdots, \alpha_s)$. Then by (2.6.6) we have

$$(Y_1, \cdots, Y_s) \stackrel{d}{=} \Big(\prod_{i=1}^{s-1} B_i, (1 - B_1) \prod_{i=2}^{s-1} B_i, \cdots, 1 - B_{s-1} \Big). \quad (4.2.16)$$

The rep-points of the multivariate Liouville distribution can be easily obtained from the stochastic representation (4.2.16).

Let F_i be the c.d.f. of $Be(\sum_{j=1}^{i} \alpha_i, \alpha_{i+1})$, $i = 1, \cdots, s$, and F_R be the c.d.f. of R defined in (4.2.14). Given an NT-net $\{\boldsymbol{c}_k\}$ on C^s, then $\{\boldsymbol{x}_k\}$ is a set of rep-points of the multivariate Liouville distribution $L(\alpha_1, \cdots, \alpha_s, F_R)$, where

$$\begin{aligned} x_{ki} &= r_k(1 - b_{k,i-1}) \prod_{j=i}^{s-1} b_{kj}, \\ r_k &= F_R^{-1}(c_{ks}), \qquad b_{k0} = 0, \\ b_{ki} &= F_i^{-1}(c_{ki}), \qquad i = 1, \cdots, s-1, \; k = 1, \cdots, n. \end{aligned} \quad (4.2.17)$$

Remark 4.2

When $\alpha_1 = \cdots = \alpha_s = 1$, $D_s(1, \cdots, 1)$ reduces to $U(T_s)$ and then $B_i \sim Be(i, 1)$, $i = 1, \cdots, s - 1$. Since the p.d.f. of $Be(i, 1)$ is $i^{-1}x^{i-1}$ $(0 < x < 1)$, and its c.d.f. is

$$F_i(x) = x^i, \quad 0 < x < 1.$$

Since $P(R = 1) = 1$ for the uniform distribution $U(T_s)$, we obtain $r_k \equiv 1$ and so (4.2.17) reduces to

$$\begin{cases} x_{k1} = \prod_{j=1}^{s-1} c_{kj}^{1/j}, \\ x_{ki} = (1 - c_{k,i-1}^{1/(i-1)}) \prod_{j=i}^{s-1} c_{kj}^{1/j}. \end{cases} \quad (4.2.18)$$

$$i = 2, \cdots, s, \quad k = 1, \cdots, n.$$

This gives another formula for obtaining the NT-net set on T_s. Comparing (4.2.18) and (1.5.33), we can see that they are essentially the same although they are obtained in different ways.

It is better to use the following recurrence formula for writing a computer program for (4.2.18):

(a) Let $g_{ks} = 1$ and $g_{k0} = 0, k = 1, \cdots, n$.

(b) Recursively compute for $k = 1, \cdots, n$

$$g_{kj} = g_{k,j+1} c_{kj}^{1/j}, j = s-1, s-2, \cdots, 1. \qquad (4.2.19)$$

(c)

$$x_{kj} = \sqrt{g_{kj} - g_{k,j-1}}, j = 1, \cdots, s, k = 1, \cdots, n.$$

The NTSR algorithm can be applied similarly to the classes of **α–symmetric distributions, generalized symmetric Dirichlet distributions**, rotationally invariant distributions (Fang and Anderson (1990)), and additive logistic elliptical distributions (section 2.6).

4.3 An efficient method for generating an NT-net on U_s

The uniform distribution $U(U_s)$ is the basis of spherical and elliptical distributions as well as of directional statistics. We have introduced in section 1.5.3 the method for generating an NT-net on U_s by the use of a spherical coordinate transformation (SCT). In this method, the values on a set of points of the inverse functions of the distribution functions

$$F_j(\phi) = \frac{\pi}{B(\frac{1}{2}, \frac{s-j}{2})} \int_0^\phi (\sin(\pi t))^{s-j-1} dt, \qquad j = 1, \cdots, s \quad (4.3.1)$$

should be evaluated. It is pointed out in Appendix B.2 that $F_j(\phi)$, and therefore its inverse function, can be expressed by the incomplete beta function. When n is large it can take a comparatively long time to obtain an NT-net of size n on U_s because computing the incomplete beta function is time consuming. We often use the Box-Muller transformation to generate the s-dimensional standard normal variate $\boldsymbol{x}$, and then obtain the uniform distribution $\boldsymbol{x}/\|\boldsymbol{x}\|$

in order to save computer time (Example 1.11). But the uniformity of the set obtained by this method is usually inferior to that of the former method when n is small. In this section we shall introduce an algorithm for generating an NT-net on U_s of which the uniformity of net is not inferior than that given by spherical coordinate tranformation, and the computer time is only slightly longer than that of Box-Muller tranformation.

Lemma 4.1
Suppose that $\boldsymbol{u} \sim U(U_s)$ and is partitioned into m parts

$$\boldsymbol{u} = \begin{pmatrix} \boldsymbol{u}_1 \\ \vdots \\ \boldsymbol{u}_m \end{pmatrix}, \quad \boldsymbol{u}_i : t_i \times 1, \quad t_1 + \cdots + t_m = s.$$

Then

$$\boldsymbol{u} \stackrel{d}{=} \begin{pmatrix} d_1 \boldsymbol{u}^{(1)} \\ \vdots \\ d_m \boldsymbol{u}^{(m)} \end{pmatrix}, \tag{4.3.2}$$

where

(1) $\boldsymbol{u}^{(j)} \sim U(U_{t_j})$, $j = 1, \cdots, m$.

(2) $(d_1, \cdots, d_m)$, $\boldsymbol{u}^{(1)}, \cdots, \boldsymbol{u}^{(m)}$ are independent.

(3) $d_j > 0$, $j = 1, \cdots, m$, and $(d_1^2, \cdots, d_m^2) \sim D_m(t_1/2, \cdots, t_m/2)$.

(Fang, Kotz and Ng (1990), Theorem 2.6).

In particular, when $s = 2m$ is even and $t_1 = \cdots = t_m = 2$, we have $\boldsymbol{u}^{(j)} \sim U(U_2)$ and $(d_1^2, \cdots, d_m^2) \sim D_m(1, \cdots, 1) = U(T_m)$. Consequently, the stochastic representation (4.3.2) can be expressed in the more explicit form

$$\boldsymbol{u} \stackrel{d}{=} \begin{pmatrix} d_1 \cos(2\pi\phi_1) \\ d_1 \sin(2\pi\phi_1) \\ d_2 \cos(2\pi\phi_2) \\ d_2 \sin(2\pi\phi_2) \\ \vdots \\ d_m \cos(2\pi\phi_m) \\ d_m \sin(2\pi\phi_m) \end{pmatrix}, \tag{4.3.3}$$

where

(1) $(d_1^2, \cdots, d_m^2) \sim U(T_m)$ and $d_j > 0$, $j = 1, \cdots, m$.
(2) $(\phi_1, \cdots, \phi_m)$ and $(d_1, \cdots, d_m)$ are independent.
(3) $(\phi_1, \cdots, \phi_m) \sim U(C^m)$.

With this fact and (4.2.19), we propose the following algorithm:

Step 1 Generate an NT-net $\{\boldsymbol{c}_k = (c_{k1}, \cdots, c_{k,s-1}),\ k = 1, \cdots, n\}$ on $C^{(s-1)}$.

Step 2 Set $g_{km} = 1$ and $g_{k0} = 0, k = 1, \cdots, n$.

Step 3 Recursively compute for $k = 1, \cdots, n$

$$g_{kj} = g_{k,j+1} c_{kj}^{1/j}, j = m-1, m-2, \cdots, 1. \tag{4.3.4}$$

Step 4 Compute $d_{kl} = \sqrt{g_{kl} - g_{k,l-1}}$ and

$$\begin{cases} x_{k,2l-1} = d_{kl} \cos(2\pi, c_{k,m+l-1}), & l = 1, \cdots, m \\ x_{k,2l} = d_{kl} \sin(2\pi, c_{k,m+l-1}), & k = 1, \cdots, m \end{cases} \tag{4.3.5}$$

Then $\{\boldsymbol{x}_k = (x_{k1}, \cdots, x_{ks}),\ k = 1, \cdots, n\}$ is an NT-net on U_s.

When $s = 2t + 3 (t \geq 0)$ is odd, we choose $t_1 = 3, t_2 = \cdots = t_m = 2$, and $m = t + 1$ in Lemma 4.1, and we have $(d_1^2, \cdots, d_m^2) \sim D_m(3/2, 1, \cdots, 1)$. Therefore, with (4.2.16) we obtain

$$(d_1^2, \cdots, d_m^2) \stackrel{d}{=} \Big(\prod_{i=1}^{t} B_i, (1 - B_1) \prod_{i=2}^{t} B_i, \cdots, 1 - B_t \Big), \tag{4.3.6}$$

where $B_1, \cdots, B_t$ are independent, $B_i \sim B_e((2i+1)/2, 1), i = 1, \cdots, t$. It is easy to see that the c.d.f. of B_i is $x^{(2i+1)/2}, 0 < x < 1$. The stochastic representation (4.3.2) now can be expressed as (cf.(1.5.28))

$$\boldsymbol{u} \stackrel{d}{=} \begin{pmatrix} d_1(1-2\phi_1) \\ d_1\sqrt{\phi_1(1-\phi_1)}\cos(2\pi\phi_2) \\ d_1\sqrt{\phi_1(1-\phi_1)}\sin(2\pi\phi_2) \\ d_2\cos(2\pi\phi_3) \\ d_2\sin(2\pi\phi_3) \\ \vdots \\ d_m\cos(2\pi\phi_{m+1}) \\ d_m\sin(2\pi\phi_{m+1}) \end{pmatrix} \tag{4.3.7}$$

where

(1) $(d_1^2, \cdots, d_m^2) \sim D_m(3/2, 1, \cdots, 1)$ and has the stochastic representation (4.3.6).
(2) $(\phi_1, \cdots, \phi_{m+1})$, and $(d_1^2, \cdots, d_m^2)$ are independent.
(3) $(\phi_1, \cdots, \phi_{m+1}) \sim U(C^{m+1})$.

Therefore we have the following algorithm

Step 1 Generate an NT-net $\{\boldsymbol{c}_k = (c_{k1}, \cdots, c_{k,s-1}), k = 1, \cdots, n\}$ on $C^{(s-1)}$.

Step 2 Set $m = (s-1)/2, g_{km} = 1$ and $g_{ko} = 0, k = 1, \cdots, n$.

Step 3 Recursively compute for $k = 1, \cdots, n$

$$g_{kj} = g_{k,j+1}c_{kj}^{2/(2j+1)}, \quad j = m-1, m-2, \cdots, 1.$$

Step 4 Compute

$$d_{kj} = \sqrt{g_{kj} - g_{k,j-1}}, \quad j = 1, \cdots, m, k = 1, \cdots, m.$$

Step 5 For $k = 1, \cdots, n$ compute

$$\begin{aligned} x_{k1} &= d_{k1}(1-2c_{km}) \\ x_{k2} &= d_{k1}\sqrt{c_{km}(1-c_{km}}\cos(2\pi c_{k,m+1}), \\ x_{k3} &= d_{k1}\sqrt{c_{km}(1-c_{km}}\sin(2\pi c_{k,m+1}), \\ x_{k,2l} &= d_{kl}\cos(2\pi c_{k,2l}), \\ x_{k,2l+1} &= d_{kl}\sin(2\pi c_{k,2l}), \\ & \qquad l = 2, 3, \cdots, m. \end{aligned}$$

Then $\{\boldsymbol{x}_k = (x_{k1}, \cdots, x_{ks}), k = 1, \cdots, n\}$ is an NT-net on U_s.

The above two algorithms are due to Tashiro (1977) and Fang, Wang and Wong (1992), and therefore we will call them TFWW algorithms. The numerical results given by Fang, Wang and Wong (1992) show that the computer time used by the TFWW algorithm is much shorter than that used by the SCT. The NT-nets on U_s generated by TFWW and SCT have the same level of uniformity. Therefore the TFWW algorithm is to be recommended.

4.4 MSE criterion (univariate case)

Let X be a random variable with p.d.f. $p(x)$ and $\sigma^2 = \mathrm{Var}(X)$. For any given n numbers $-\infty < x_1 < x_2 < \cdots < x_n < \infty$, the n–level quantizer $Q_n(\cdot)$ is defined as the step function:

$$Q_n(x) = x_k, \quad \text{if} \quad a_k < x \leq a_{k+1}, \quad k = 1, \cdots, n, \tag{4.4.1}$$

where

$$a_1 = -\infty, \; a_{n+1} = \infty, \; a_k = (x_k + x_{k-1})/2, \; k = 2, \cdots, n.$$

We use the following mean square error (MSE)

$$\begin{aligned} \mathrm{MSE}(\boldsymbol{x}) = \mathrm{MSE}(x_1, \cdots, x_n) &= \frac{1}{\sigma^2} E(X - Q_n(X))^2 \\ &= \frac{1}{\sigma^2} \int_{-\infty}^{\infty} \min_i (x - x_i)^2 p(x) dx \\ &= \frac{1}{\sigma^2} \sum_{i=1}^{n} \int_{a_i}^{a_{i+1}} (x - x_i)^2 p(x) dx \end{aligned} \tag{4.4.2}$$

to measure the distortion between X and $Q_n(X)$ for a given $\boldsymbol{x} = (x_1, \cdots, x_n)'$.

The **optimum quantizer** $Q_n^*(X)$ minimizes $\mathrm{MSE}(\boldsymbol{x})$ over $\boldsymbol{x}$ in the region

$$TR_n = \{(x_1, \cdots, x_n) : x_1 < x_2 < \cdots < x_n\}.$$

That is, let $\boldsymbol{x}^* \in TR_n$ and

$$\text{MSE}(\boldsymbol{x}^*) = \min_{\boldsymbol{x} \in TR_n} \text{MSE}(\boldsymbol{x}),$$

then $Q_n^*(X)$ is the associate n–level quantizer of $\boldsymbol{x}^*$. We shall call $\boldsymbol{x}^*$ a set of **mse-rep-points** of X (or $p(x)$). The problem for obtaining a set of mse-rep-points has appeared in many fields, such as information theory, clustering analysis, theory of quantization, standardization and theory of stochastic simulation (Cox (1957), Bofinger (1970), Anderberg (1973) and Fang and He (1982)). For example, let $F(x)$ be the c.d.f. of the height of male adults in a country. Suppose that we want to design the trousers with three different sizes (S,M,L) for all male adults. How do we choose three models who are good representation of the population? A suggestion is to choose the three mse-rep-points of $F(x)$ as the model's heights. Max (1960), Lloyd (1982), Fang and He (1982) and Cambanis and Gerr (1983) have proposed the numerical methods for finding the mse-rep-points. Their methods are essentially the same. Without loss of generality, we may assume that $\sigma = 1$.

Let $\{x_{n1}, x_{n2}, \cdots, x_{nn}\}$ $(n = 2, 3, \cdots)$ be a sequence of mse-rep-points for $p(x)$, and let

$$L_n = L(x_{n1}, \cdots, x_{nn}) = \text{MSE}(x_{n1}, \cdots, x_{nn})$$

be the minimum of MSE over TR_n. The following facts can easily be verified.

(a) The probability distribution of the optimum quantizer $Q_n(X)$ is

$$P(Q_n(X) = x_{ni}) = p_{ni},$$

where

$$p_{n1} = \int_{-\infty}^{(x_{n1}+x_{n2})/2} p(x)dx,$$

$$p_{ni} = \int_{(x_{n,i-1}+x_{ni})/2}^{(x_{ni}+x_{n,i+1})/2} p(x)dx, \quad i = 2, \cdots, n-1,$$

and

$$p_{nn} = \int_{(x_{n,n-1}+x_{nn})/2}^{+\infty} p(x)dx.$$

(b) $0 < L_{n+1} \leq L_n \leq 1$.
(c) $L_n \to 0$ as $n \to \infty$.
(d)

$$\begin{aligned} &E(Q_n(X)) = E(X), \\ &\mathrm{Var}(Q_n(X)) = (1 - L_n)\mathrm{Var}(X), \\ &\mathrm{Var}(Q_n(X)) \leq \mathrm{Var}(Q_{n+1}(X)) \end{aligned}$$

and

$$\mathrm{Var}(Q_n(X)) \to \mathrm{Var}(X) \quad \text{as } n \to \infty.$$

(e) $E(X - Q_n(X))^2 = L_n \mathrm{Var}(X) \to 0$ as $n \to \infty$.

PROOF (a) follows from (4.4.1) immediately. Now we proceed to prove (b). Since

$$\begin{aligned} L_{n+1} &= L(x_{n+1,1}, \cdots, x_{n+1,n+1}) \\ &\leq L(x_{n1}, \cdots, x_{nn}, x_{nn} + 1) \leq L_n, \end{aligned}$$

it is sufficient to show that $L_1 \leq 1$. We have

$$L_1 = \frac{1}{\sigma^2} \int_{-\infty}^{\infty} (x - x_1)^2 p(x)dx,$$

which attains its minimum at $x_1 = E(x)$. Hence $x_{11} = E(x)$ and $L_1 = 1$. We omit the proofs of (c), (d) and (e). □

Taking $\partial \text{MSE}/\partial x_i = 0$ in (4.4.2), $i = 1, \cdots, n$, we have the following system of non-linear equations

$$\begin{cases} f_1(\boldsymbol{x}) = \displaystyle\int_{-\infty}^{(x_1+x_2)/2} (x - x_1)p(x)dx = 0, \\ f_2(\boldsymbol{x}) = \displaystyle\int_{(x_1+x_2)/2}^{(x_2+x_3)/2} (x - x_2)p(x)dx = 0, \\ \quad \cdots\cdots \\ f_n(\boldsymbol{x}) = \displaystyle\int_{(x_{n-1}+x_n)/2}^{\infty} (x - x_n)p(x)dx = 0, \end{cases} \tag{4.4.4}$$

where $\boldsymbol{x} \in TR_n$. Let

$$L(\boldsymbol{x}) = \sum_{i=1}^{n} |f_i(\boldsymbol{x})|, \quad \boldsymbol{x} \in TR_n. \tag{4.4.5}$$

Then the problem for finding the mse-rep-points is equivalent to the problem for finding an $\boldsymbol{x}^*$ such that $L(\boldsymbol{x})$ attains its minimum zero. This is a typical optimization problem which can be solved by the SNTO (section 3.7).

Example 4.3 The standard normal distribution

Suppose that $X \sim N(0,1)$. Since the p.d.f. is symmetric with respect to origin, we have $x_i^* = -x_{n+1-i}^*$, $i = 1, 2, \cdots, (n+1)/2$, and therefore $x_{(n+1)/2}^* = 0$ if n is odd. So we need only to find $0 \le x_1^* < \cdots < x_s^*$, where $s = [n/2]$. Hence (4.4.4) becomes

$$\begin{cases} f_1(\boldsymbol{x}) = \phi(v) - \phi(\frac{x_1 + x_2}{2}) - x_1[\Phi(\frac{x_1 + x_2}{2} - \Phi(v)] = 0, \\ f_2(\boldsymbol{x}) = \phi(\frac{x_1 + x_2}{2}) - \phi(\frac{x_2 + x_3}{2}) \\ \qquad -x_2[\Phi(\frac{x_2 + x_3}{2}) - \Phi(\frac{x_1 + x_2}{2})] = 0, \\ f_s(\boldsymbol{x}) = \phi(\frac{x_{s-1} + x_s}{2}) - x_3[1 - \Phi(\frac{x_{s-1} + x_s}{2})] = 0, \end{cases} \tag{4.4.6}$$

where

$$\phi(x) = \frac{1}{\sqrt{2\pi}} e^{-x^2/2}, \qquad \Phi(x) = \int_{-\infty}^{x} \phi(t)dt$$

and

$$v = \begin{cases} 0, & \text{if } n \text{ even}, \\ x_1/2, & \text{otherwise}. \end{cases}$$

The following results show that the advantages of SNTO are not only for its simple algorithm but also for its precision compared with some other known methods. Let

$$L(\boldsymbol{x}) = \sum_{i=1}^{s} |f_i(\boldsymbol{x})|, \quad \boldsymbol{x} \in TR_s^+, \tag{4.4.7}$$

where

$$TR_s^+ = \{(x_1, \cdots, x_s) : 0 \le x_1 < \cdots < x_s\}.$$

The problem for finding a mse-rep-point $\boldsymbol{x}^*$ is equivalent to the problem for finding a minimum point of $L(\boldsymbol{x})$ on TR_s^+. It is known by the properties of the standard normal distribution that $\boldsymbol{x}^*$ must fall in the region

$$TR_s^+(B) = \{(x_1, \cdots, x_s) : 0 \le x_1 < \cdots < x_s < B\}, \tag{4.4.8}$$

where B is a constant, for examples, we may take $B = 3.5$ if $10 < n \le 30$ and $B = 2 \sim 3$ if $n \le 10$. Let $D = TR_s^+(B)$. We can obtain the approximation of a mse-rep-point $\boldsymbol{x}^*$ by SNTO with $n_2 = n_3 = \cdots$. The results are given in Tables 4.1 and 4.2. In Table 4.1, we use N to denote the total number of points for the calculation, T the number of contractions for the domain $TR_s^+(B)$, $\boldsymbol{x}^{**}$ a mse-rep-point obtained by SNTO, $\boldsymbol{x}^*$ a mse-rep-point given in Fang and He (1982) which is slightly better than those obtained by Max (1960), and $L(\boldsymbol{x}^{**})$ and $L(\boldsymbol{x}^*)$ the respective values of the function $L(\boldsymbol{x})$ at $\boldsymbol{x}^{**}$ and $\boldsymbol{x}^*$. Table 4.1 shows that the $\boldsymbol{x}^{**}$ has higher precision than those obtained by the methods given in Fang and He (1982) and Max (1960). Table 4.2 gives the values of $\boldsymbol{x}^{**}$ for $n \le 10$.

The above method can be applied to any continuous univariate distributions. For example, Chong (1991) and Lai (1992) obtained the mse-rep-points of the lognormal and beta distributions.

Table 4.1 *Comparison between SNTO and Fang-He's methods*

n	n_1	n_2	T	N	$L(\boldsymbol{x}^{**})$	$L(\boldsymbol{x}^*)$
4	114	89	20	1835	1.490E−8	3.308E−6
5	114	89	20	1835	1.490E−8	6.124E−6
6	701	199	22	4880	1.490E−8	1.305E−5
7	701	199	23	5079	1.490E−8	1.882E−5
8	1069	523	22	12052	7.078E−8	2.424E−5
9	1069	523	22	12052	2.235E−8	7.964E−3
10	4001	1069	22	26450	1.576E−6	3.400E−5

Table 4.2 *Mse-rep-points of* $N(0,1)$

n	x_1	x_2	x_3	x_4	x_5
4	0.452780	1.51042			
5	0.764567	1.72415			
6	0.317717	1.00011	1.89360		
7	0.560579	1.18814	2.03336		
8	0.245096	0.756014	1.34392	2.15195	
9	0.443636	0.918791	1.47639	2.25465	
10	0.199614	0.609828	1.05777	1.59125	2.34488

4.5 MSE criterion (multivariate case)

We now consider the mse-rep-points and quantizers of multivariate distributions. A special issue of IEEE Transactions on Information Theory on this subject appeared in March 1982.

Let $\boldsymbol{x} = (X_1, \cdots, X_s)$ be an s–dimensional random vector with p.d.f. $p(\boldsymbol{x})$ and $\text{Var}(X_i) = 1$, $i = 1, \cdots, s$. The following is the usual definition of the multivariate quantizer in the literature.

Definition 4.5

An n**-level quantizer** $Q = (\mathcal{Y}, \mathcal{S})$ of an s-dimensional vector $\boldsymbol{x}$ consists of a set of output vectors $\mathcal{Y} = \{\boldsymbol{y}_1, \cdots, \boldsymbol{y}_n\}$ of R^s, a partition $\mathcal{S} = (S_1, \cdots, S_n)$ of the space R^s with n disjoint and exhaustive regions and a mapping $Q : R^s \to \mathcal{S}$ defined by $Q(\boldsymbol{x}) = \boldsymbol{y}_i$, if $\boldsymbol{x} \in S_i$.

When $s > 1$ we often call it a **block quantizer**, **vector quantizer**, or an **s-dimensional quantizer**. The **optimal n-level quantizer** minimizes the mean square error

$$\mathrm{MSE}(Q) = \frac{1}{s} E\|\boldsymbol{x} - Q(\boldsymbol{x})\|^2, \tag{4.5.1}$$

over all n-level quantizers. The corresponding $\{\boldsymbol{y}_i\}$ of the optimal n-level quantizer are called the **mse-rep-points** of $\boldsymbol{x}$.

It is easy to obtain the following necessary conditions for an optimal quantizer: Associated with each quantizer output $\boldsymbol{y}_i$ a partition cell S_i, $i = 1, \cdots, n$, should satisfy

$$S_i = \{\boldsymbol{x} : \|\boldsymbol{x} - \boldsymbol{y}_i\| < \|\boldsymbol{x} - \boldsymbol{y}_j\|, \ j \neq i\}, \tag{4.5.2}$$

if we neglect the boundary of S_i; and for each partition cell S_i, $\boldsymbol{y}_i$ should be the conditional mean

$$\boldsymbol{y}_i = E[\boldsymbol{x} \,|\, \boldsymbol{x} \in S_i]. \tag{4.5.3}$$

Since the mse-rep-points are sorted in order for univariate distributions, its MSE can be expressed as a sum (4.4.2) in which each term is easily calculated. This property does not hold for multivariate distributions, and so it is difficult to find the mse-rep-points for multivariate distributions. Although there are several numerical methods that claim to generate mse-rep-points, no one has shown that these methods can produce the real mse-rep-points. A theoretical basis for the asymptotically optimum vector quantization has been established by Zador (1982) and Gersho (1979, 1982). Zador found that for the asymptotic case of high quality quantization

$$\mathrm{MSE}(Q) = A(s, 2) n^{-2/s} \|p(\boldsymbol{x})\|_{s/(s+r)}, \tag{4.5.4}$$

where

$$\|p(\boldsymbol{x})\|_t = \left[\int |p(\boldsymbol{x})|^t d\boldsymbol{x} \right]^{1/t}$$

and $A(s, 2)$ is independent of the density $p(\boldsymbol{x})$. Toth (1959) pointed out that $A(2, 2) = 5\sqrt{3}/108 \cong 0.080187537$. When n is not large Linde, Buzo and Gray (1982) suggested an iterative vector quantizer algorithm LBG based on a training sequence. This approach has been used to calculate low bit-rate quantizers for multinormal, multivariate Laplace, multivariate gamma and multivariate uniform distributions (Abut, Gray and Rebolledo (1982), Gray and Linde (1982) and Gray (1984)). The necessary conditions (4.5.2) and (4.5.3) provide the basis of the LBG algorithm, and the details of which are given in Linde, Buzo and Gray (1982).

The LBG algorithm has the overwhelming advantage and is the most useful one for generating vector quantizers. But there are some drawbacks with the approach: The resulting output vectors are only locally optimal and depend on the initial set of output vectors. The training sequence and numerical evaluation of the MSE are based on the Monte Carlo method that has poor efficiency.

A number of authors propose various kinds of initial output vectors. Wilson (1980), Fischer and Dicharry (1984) and Fischer (1989) gave some geometric designs. An alternative approach is the lattice quantizer which only deals with the uniform distribution. Fischer (1986) provided a source coding theorem using Shannon's entropy theory (1948). This approach is only effective for very high dimensional quantizers.

In order to overcome the shortcomings in the LBG algorithm stated above, Fang, Yuan and Bentler (1990) modified the LBG algorithm by using the NTM to generate the initial vector and training sequence.

Let $F(\boldsymbol{x})$ be a given continuous c.d.f. in R^s and random vector $\boldsymbol{x} \sim F(\boldsymbol{x})$. Suppose that a set of cdf-rep-points of $F(\boldsymbol{x})$ can be generated by an NT-net on C^s. The so-called **NTLBG algorithm** involves the following steps:

Step 1 Set $t = 0$. Generate a set of cdf-rep-points of $F(\boldsymbol{x})$ based on a *glp* set on C^s as an initial set of output vectors $\mathcal{Y}_t = \{\boldsymbol{y}_{t1}, \cdots, \boldsymbol{y}_{tn}\}$.

Step 2 Generate a set $\{\boldsymbol{x}_1, \cdots, \boldsymbol{x}_N\}$ of cdf-rep-points of $F(\boldsymbol{x})$ as a training sequence.

Step 3 Form a partition $\{S_j\}$ of $\{\boldsymbol{x}_k, k = 1, \cdots, N\}$ such that each $\boldsymbol{x}_i$ assigned to cells S_j of a partition, i.e. $\boldsymbol{x}_i \in S_j$ if

$$\|\boldsymbol{x}_i - \boldsymbol{y}_{tj}\| < \|\boldsymbol{x}_i - \boldsymbol{y}_{tk}\|, \quad k \neq j.$$

Step 4 Calculate the sample conditional means (4.5.3) and form a new set of output vectors $\mathcal{Y}_{t+1} = \{\boldsymbol{y}_{t+1,1}, \cdots, \boldsymbol{y}_{t+1,n}\}$, where

$$\boldsymbol{y}_{t+1,j} = \frac{1}{n_{t+1,j}} \sum_{k \in S_j} \boldsymbol{x}_k,$$

and $n_{t+1,j}$ is the number of $\boldsymbol{x}_k$ falling in S_j. These output vectors are in general different from the initial output vectors.

Step 5 If $\mathcal{Y}_{t+1} = \mathcal{Y}_t$ terminate the algorithm and find the final output $\mathcal{Y}_t$. Otherwise let $t = t + 1$ and return to Step 3.

This process provides a nonincreasing MSE, and the algorithm eventually converges to a locally optimal solution $\mathcal{Y}=\{\boldsymbol{y}_1,\cdots,\boldsymbol{y}_n\}$. The NTLBG algorithm is essentially based on the k–means method of clustering analysis. Different proofs for the convergence of the k-means algorithm are given by MacQueen (1967), Pollard (1981, 1982), Gray, Kieffer and Linde (1980), Gray and Karnin (1982) and Zhang and Fang (1982).

In the next section applying the NTLBG algorithm to spherical distributions can obtain better results than that by the literatures.

Remark 4.3

Why can the NTLBG algorithm improve the LBG algorithm? The key point is the use of cdf-rep-points for the initial set of output vectors.

Consider the simplest case C^s first. We take a sequence of NT-nets $\mathcal{P}_n = \{\boldsymbol{x}_k^{(n)}, k = 1, \cdots, n\}$ on C^s with discrepancy $D(\mathcal{P}_n) = O(n^{-1} \log^s n)$, where n belongs to a subset of the natural numbers. Then we have

$$\max_{\boldsymbol{x} \in C^s} \min_{1 \le k \le n} \|\boldsymbol{x} - \boldsymbol{x}_k^{(n)}\| = O(n^{-1/s} \log n),$$

where $\|\boldsymbol{y}\|$ is the l_2–norm of $\boldsymbol{y}$ and

$$\text{MSE} = \frac{1}{s} \int_{C^s} \min \|\boldsymbol{x} - \boldsymbol{x}_k^{(n)}\|^2 d\boldsymbol{x} = O(n^{-2/s} \log^2 n)$$

which is close to the order $O(n^{-2/s})$ in (4.5.4). The result may be better if we use the mse-rep-points of $U(C^s)$ to produce the initial points in step 1 (Fang, Yuan and Bentler (1990)). This is available only for small s.

Remark 4.4
We would naturally use the NTM to generate a training sequence. The simplest way is to use the methods described in sections 4.1 and 4.2. But the symmetrization method, introduced in section 2.3 to enhance the convergence rate of quadrature formula, can also be applied to the problem of generating a training sequence. For example, if $\{\boldsymbol{c}_k = (c_{k1}, c_{k2})\}$ is an NT-net on C^2, then the symmetrization NT-net is given by

$$\{(c_{k1}, c_{k2}), (c_{k1}, 1-c_{k2}), (1-c_{k1}, c_{k2}), (1-c_{k1}, 1-c_{k2}), k=1, \cdots, n\}$$

and the corresponding rep-points $\{\boldsymbol{x}_k\}$ of $N_2(\boldsymbol{0}, \boldsymbol{I})$ can be obtained by the methods introduced in section 4.2.

4.6 Mse-rep-points of spherical distributions

In this section we shall apply the NTLBG algorithm to obtain mse-rep-points for spherically symmetric distributions.

Example 4.4 The multivariate normal distribution
Without loss of generality, we consider only the bivariate normal distribution $N_2(\boldsymbol{0}, \boldsymbol{I})$. The same method can be applied to any standard and hence general multivariate normal distribution. The algorithm for generating an initial point is given in the previous section, and will not be repeated here. There are two ways to produce the training sequence. One is to use the property that $N_2(\boldsymbol{0}, \boldsymbol{I})$ has independent components, and the other is to regard the multivariate normal distribution as a spherically symmetric distribution and use the method stated in section 4.2. Table 4.3 gives the MSE for $3 \leq n \leq 32$ obtained by the former method, and the result is represented by Figure 4.1 with '•', where the size of a training sample is $N = 70,844$ including the points obtained by symmetrization. We see that the MSE is slightly large at $n = 6$, and the point curve decreases smothly.

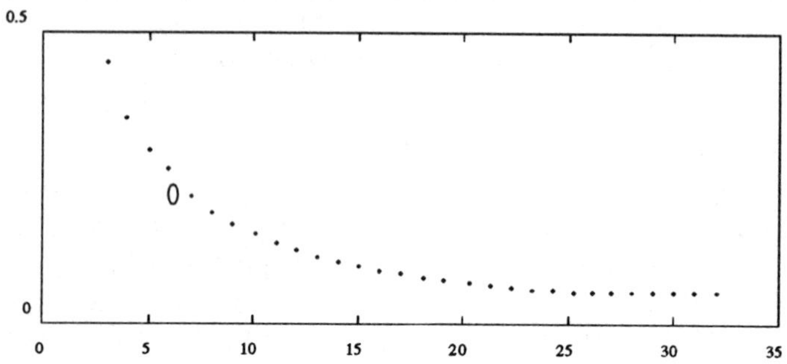

Figure 4.1 *MSE's of mse-rep-points of* $N_2(\mathbf{0}, \boldsymbol{I})$

Table 4.3 *MSE's of mse-rep-points of normal distribution*

n	MSE	n	MSE	n	MSE
3	0.4618390	13	0.1301080	23	0.0771025
4	0.3623600	14	0.1227065	24	0.0744985
5	0.3063070	15	0.1172445	25	0.0714630
6	0.2711950	16	0.1067880	26	0.0689305
7	0.2216000	17	0.1020085	27	0.0659590
8	0.2002523	18	0.0967775	28	0.0639525
9	0.1806665	19	0.0922800	29	0.0620685
10	0.1635840	20	0.0875885	30	0.0600280
11	0.1513470	21	0.0835925	31	0.0582365
12	0.1387165	22	0.0800835	32	0.0569415

If we use the latter method to generate a training sequence, the result has no substantial difference when compared with the former one. For example, the MSE is equal to 0.200268, 0.106794 and 0.056556 for $n = 8, 16$ and 32, and the MSE is not improved for $n = 6$. Figure 4.2 provides a plot of a part of mse-rep-points, and we see that the results are reasonable. Since $N_2(\mathbf{0}, \boldsymbol{I})$ is invariant under the orthogonal transformation, the points in the figure have the same MSE under rotation.

Remark 4.5

For symmetrical distributions, as pointed out by Sharma (1978), the optimal quantizer need not be symmetric. From Figure 4.2 we see that most of the output vectors are not symmetric. But there

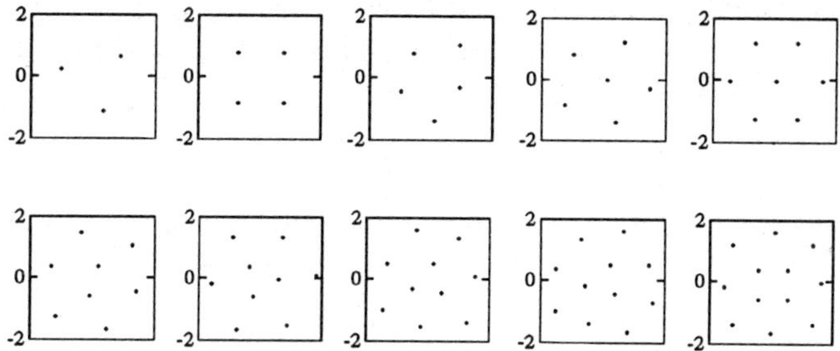

Figure 4.2 *Representative points of* $N_2(\mathbf{0}, \boldsymbol{I})$

are some symmetric relationships in some cases ($n = 3, 4, 7, 8$, etc). For $n = 7$ we guess that one point should be at the origin and others equally located on a circle with centre 0 and a radius r to be determined as follows.

When $n = 7$, the r values of polar coordinates of the output vectors are 0, 1.43501, 1.43713, 1.43529, 1.43503, 1.43422, 1.43320 in Figure 4.2. The average of these six numbers is 1.435647 which can be regarded as an estimate of r. Then we choose 0 and six points on the circle with $r = 1.435647$ as the initial set of output vectors in the NTLBG algorithm, the final vectors have r values 0, 1.43502, 1.43555, 1.43444, 1.43503, 1.43556, 1.43556 and the associated MSE is 0.221597. This technique can be applied to other cases and improves the results obtained by NTLBG algorithm. For example, when $n = 8$, the improved result has r values 0.48077, 0.48110, 1.37299, 1.37299, and four 1.63416 and the associated MSE is 0.200251.

Remark 4.6

Wang and Fang (1981) proposed a method to obtain a set $\mathcal{P}_n(a)$ with low discrepancy by deleting the last point of $\mathcal{P}_{n+1}(a)$ which has a low discrepancy. The idea provides another way to design the initial set of output vectors. For example, when $n = 6$, the output vectors with MSE$= 0.2522955$ are plotted in Figure 4.3(a) and the MSE value is plotted by '0' in Figure 4.1 which is better than that with MSE$= 0.2711950$ in Figure 4.3(b). Similarly we can obtain a set of output vectors in the case of $n = 32$ with a MSE$= 0.056895$

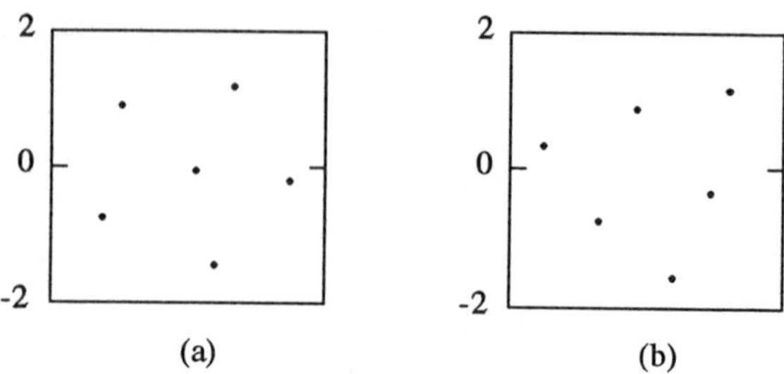

Figure 4.3 *Two sets of mse-rep-points of* $N_2(\mathbf{0}, \boldsymbol{I})$

which gives a slightly better result than MSE$= 0.0569415$ in Table 4.3.

We have seen here the importance of the choice of initial set of output vectors. The NTM gives a general algorithm to generate an initial set of output vectors, and it seems that it is better than other methods in literature, but the results are not always good if only a single NTM is applied. Therefore we suggest the use of several methods, for example, the methods stated in Remarks 4.5 and 4.6, and compare their results so that we can find the best one.

Example 4.5 The uniform distribution on B_s

We have introduced in detail the method for obtaining the rep-points of $U(B_s)$ with lower quasi F–discrepancy stated in section 1.5. Now we proceed to give the mse-rep-points of $U(B_s)$ by the MSE criterion.

The uniform distribution $U(B_s)$ is spherically symmetric with stochastic decomposition (4.2.1), where the c.d.f. of R is

$$F_s(r) = r^s$$

and $r_k = c_{ks}^{1/s}$ in (4.2.2).

The mse-rep-points of $U(B_s)$ for $n = 8$, 16 and 32 are presented in Figure 4.4 by the NTLBG algorithm and their corresponding MSE values are 0.032936, 0.016591 and 0.008116. Compared with Figure 1.11, we see that the mse-rep-points obtained by these two criteria are different. The mse-rep-points obtained by MSE criterion are located approximately symmetrically on the circle, for

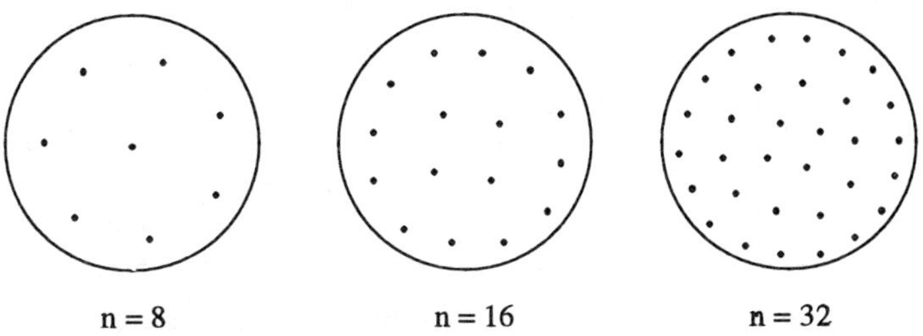

Figure 4.4 *The mse-rep-points of* $U(B_2)$

example, one point is at the origin and the other 7 points are located uniformly on a concentric circle for $n = 8$. When $n = 16$, four points are uniformly scattered on a concentric circle and the other twelve points are on another concentric circle. If the rep-points are obtained by the F-discrepancy criterion and are represented by polar coordinates (r_i, θ_i), then it is possible that there are no two points with the same r_i or same θ_i.

Example 4.6 Symmetric multivariate Pearson type VII distributions

These distributions have been defined in Example 2.4. The p.d.f. of a symmetric multivariate Pearson type VII distribution has the form

$$\frac{\Gamma(N)}{(\pi m)^{-s/2}\Gamma(N-\frac{s}{2})}\left(1+\frac{1}{m}\boldsymbol{x}'\boldsymbol{x}\right)^{-N}, \quad N > \frac{s}{2}, \ m > 0. \qquad (4.6.1)$$

We have suggested an efficient method (4.2.8) for calulating r_k in (4.2.2) for (4.5.1). Thus the mse-rep-points of a symmetric multivariate Pearson type VII distribution can be obtained by using the NTLBG algorithm. For example, when $m = 4$, $N = 15$ and $s = 2$, we use the NTLBG algorithm to generate the sets of output vectors for $n = 8$, 16 and 32 which are shown in Figure 4.5 and have the respective MSE of 0.032294, 0.017586 and 0.009286.

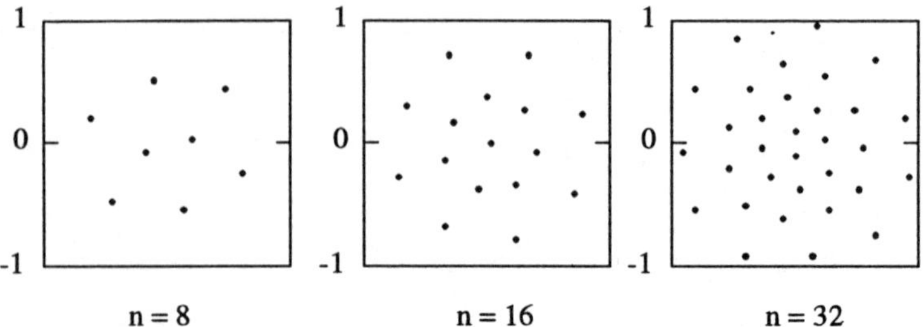

Figure 4.5 *The mse-rep-points*

4.7 Remarks and applications of rep-points to integration

We have introduced the methods for generating two kinds of rep-points. One is based on the F–discrepancy or quasi F–discrepancy criterion, and the other on the MSE criterion. It is difficult to conclude which method has the advantage in general because each has its special applications. We would address the following remarks according to our experiences as a reference for the reader. For simplicity, we use rep-points (I) to denote the rep-points of the first kind, and rep-points (II) similarly.

The rep-points (I) are easily generated and the rep-points (II) are not. Therefore the latter is suitable for the cases which need only a small number of rep-points, for instances, experimental designs (Chapter 5), industrial standardization and theory of information. For numerical evaluation of multiple integrals, optimization and statistical simulation, we recommend the use of rep-points (I).

Since the principle for choosing the rep-points (I) is that the difference between empirical distribution and population distribution is small, the stochastic property of rep-points (I) is much better than those of rep-points (II) if a set of rep-points is to be regarded as a sample. Hence the discrepancy between the sample moments obtained by rep-points (I) and population moments is small. (Figure 1.9 and Tables 1.1 and 1.2).

These two kinds of rep-points can all be used in experimental designs (Chapter 5). However one expects that the rep-points are uniformly scattered in experimental domain. For example, put 5 components on a disc and then heat them in a oven. Of course we desire that these 5 components are located in a good place of the disc. This is equivalent to finding a set of rep-points uniformly

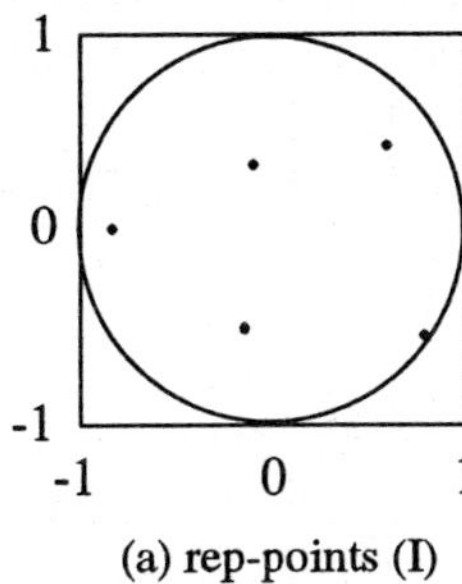

(a) rep-points (I)

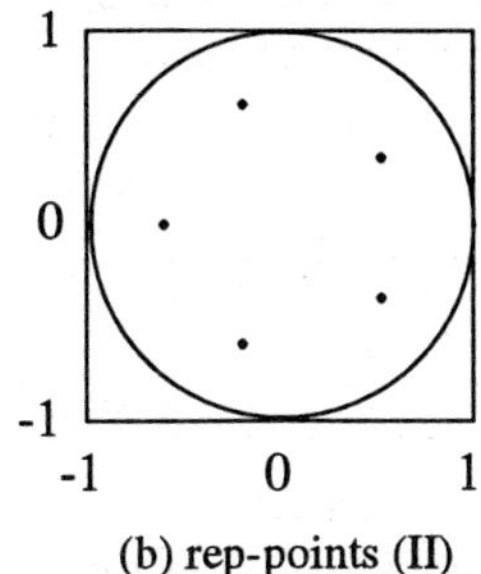

(b) rep-points (II)

Figure 4.6 *Two kinds of rep-points*

scattered on disc. Usually, the rep-points (I) are not readily acceptable to engineers since they are not visually uniform. They prefer to use rep-points (II) (Figure 4.6).

We use only the rep-points (I) of the uniform distribution for numerical evaluation of multiple integrals in Chapter 2. Of course the rep-points (I) of other distributions can also be used for numerical integration. Let $\boldsymbol{x}$ be an s–dimensional random vector with p.d.f. $p(\boldsymbol{x})$ and let $f(\boldsymbol{x})$ be an integrable function. Then the integral

$$I \equiv E(f(\boldsymbol{x})) = \int f(\boldsymbol{x})p(\boldsymbol{x})d\boldsymbol{x} \tag{4.7.1}$$

can be approximated by

$$I \cong \frac{1}{n}\sum_{i=1}^{n} f(\boldsymbol{x}_i), \tag{4.7.2}$$

where $\{\boldsymbol{x}_1, \cdots, \boldsymbol{x}_n\}$ is a set of rep–points of $p(\boldsymbol{x})$. For example, the integral

$$I = \int_D p(\boldsymbol{x})d\boldsymbol{x} \cong \frac{1}{n}\sum_{i=1}^{n} I(\boldsymbol{x}_i \in D), \tag{4.7.3}$$

where $I(\,\cdot\,)$ is the indicator function of D, is the approximate value of the probability $P(\boldsymbol{x} \in D)$, $D \in R^s$. This method is similar to the **Hit or Miss method** in Monte Carlo methodology (Rubinstein (1981) P.115) and is called the **NT-hit-miss method** (**NTHM**). The **NT-mean method** given in Chapter 2 is similar to the sample mean method in Monte Carlo methodology. The latter method is better than the former in Monte Carlo theory, and the same conclusion is also true for the NTM; that is, (4.7.2) is inferior to the NT-mean method. However (4.7.2) is still useful in some cases. For instance, suppose that $\{D_i, i = 1, \cdots, m\}$ is a partition of R^s. Then the probabilities

$$p_i = P(\boldsymbol{x} \in D_i), \quad i = 1, \cdots, m.$$

can be approximated by (4.7.3), i.e.

$$p_i \cong \frac{1}{n} \sum_{j=1}^{n} I(\boldsymbol{x}_j \in D_i), \qquad i = 1, \cdots, m. \tag{4.7.4}$$

We generate here only a set of rep-points $\{\boldsymbol{x}_k\}$ and then calculate the number of $\{\boldsymbol{x}_k\}$ that fall into D_i which is denoted by n_i for $1 \leq i \leq m$. Then by (4.7.4)

$$p_i \cong n_i/n, \qquad i = 1, \cdots, m.$$

This means that the approximate values of these m probabilities are obtained simultaneously.

Example 4.7 Orthant probability

We have introduced the method for calculating the orthant probabilities in Example 2.6. Now we proceed to calculate the probabilities of all eight octants. Set

$$\begin{aligned}
D_1 &= \{(x_1, x_2, x_3) : x_1 > 0,\ x_2 > 0,\ x_3 > 0\},\\
D_2 &= \{(x_1, x_2, x_3) : x_1 > 0,\ x_2 > 0,\ x_3 < 0\},\\
D_3 &= \{(x_1, x_2, x_3) : x_1 > 0,\ x_2 < 0,\ x_3 > 0\},\\
D_4 &= \{(x_1, x_2, x_3) : x_1 > 0,\ x_2 < 0,\ x_3 < 0\},\\
D_5 &= \{(x_1, x_2, x_3) : x_1 < 0,\ x_2 > 0,\ x_3 > 0\},\\
D_6 &= \{(x_1, x_2, x_3) : x_1 < 0,\ x_2 > 0,\ x_3 < 0\},\\
D_7 &= \{(x_1, x_2, x_3) : x_1 < 0,\ x_2 < 0,\ x_3 > 0\},\\
D_8 &= \{(x_1, x_2, x_3) : x_1 < 0,\ x_2 < 0,\ x_3 < 0\}
\end{aligned}$$

and

$$p_i = P(\boldsymbol{x} \in D_i), \qquad 1 \le i \le 8,$$

where $\boldsymbol{x} \sim Mt_3(5, \boldsymbol{0}, \boldsymbol{R})$ with

$$\boldsymbol{R} = \begin{pmatrix} 1 & 0.5 & 0.5 \\ 0.5 & 1 & 0.5 \\ 0.5 & 0.5 & 1 \end{pmatrix}$$

which is the same as in Example 2.7. We know already $p_1 = 0.25$ and the other p_i can be obtained by the use of (2.3.2) and a transformation of D_i into D_1. For instance,

$$p_2 = \int_0^\infty \int_0^\infty \int_{-\infty}^0 p(\boldsymbol{y}, \boldsymbol{0}, \boldsymbol{R}) d\boldsymbol{y}.$$

Set $z_1 = y_1$, $z_2 = y_2$ and $z_3 = -y_3$. Then

$$p_2 = \int_0^\infty \int_0^\infty \int_0^\infty p(\boldsymbol{z}; \boldsymbol{0}, \boldsymbol{R}_2) d\boldsymbol{z},$$

where

$$\boldsymbol{R}_2 = \begin{pmatrix} 1 & 0.5 & -0.5 \\ 0.5 & 1 & -0.5 \\ -0.5 & -0.5 & 1 \end{pmatrix}.$$

Since $\det(\boldsymbol{R}_2) = \det(\boldsymbol{R})$, we obtain $p_2 = 1/12$ by (2.3.1). Similarly we have $p_i = 1/12$, $i = 3, 4, 5, 6, 7$ and $p_8 = 0.25$.

Let p_{ni} be the approximate value of p_i obtained by (4.7.4) and let

$$T_n = \sum_{i=1}^{8} |p_{ni} - p_i|$$

be the total sum of absolute errors of these eight probabilities p_{ni} $(1 \le i \le n)$, where n is the number of rep-points used for calculating p_{ni}. The values of T_n are given in Table 4.4, where the third column is obtained by the symmetrization of integrand (section 2.1), and so the size of sample is $8n$. We conclude that the precision of NTHM is inferior to the method given by Example

2.7, and that the process of the symmetrization of integrand has no significant advantages here.

Table 4.4 *Errores of two methods*

n	NTHM	Symmetrization of integrand ($8n$)
135	0.04444	0.007407
597	0.01256	0.010050
1010	0.00693	0.003630
2440	0.001639	0.001776
5037	0.003077	0.003971
10007	0.0013324	0.000333
39029	0.000675	0.000521

Since we do not need to reduce an infinite domain into a finite domain in the NTHM, it is really an advantage that the determination of the value A as shown in Example 2.7 is neglected. If we do not require a high precision in the approximate evaluation of probability, this method is good enough. Besides, this method can also be applied to the distributions over a sphere or a simplex. We demonstrate this by the following example:

Example 4.8

The additive logistic normal distribution, where the p.d.f. of $\boldsymbol{x}$ is given by (2.6.12), and denoted by $\boldsymbol{x} \sim ALN_s(\boldsymbol{\mu}, \boldsymbol{\sigma})$. Let

$$D_i = \{\boldsymbol{x} \in T_s : x_j > x_i, \quad j \neq i\}.$$

when $s = 3$, D_1, D_2 and D_3 are shown by Figure 4.7. Now we proceed to calculate the probabilities $p_i = P\{\boldsymbol{x} \in D_i)$, $i = 1, 2, 3$. Tables 4.5 and 4.6 give the values of $\{p_i\}$ for $\mu = 0, 1, 2, 3, 4$ when $\boldsymbol{x} \sim ALN_3\left(\begin{pmatrix}\mu\\ \mu\end{pmatrix}, \boldsymbol{I}_2\right)$ and $\boldsymbol{x} \sim ALN_3\left(\begin{pmatrix}\mu\\ \mu\end{pmatrix}, \begin{pmatrix}1 & 0.5\\ 0.5 & 1\end{pmatrix}\right)$ respectively. It is interesting to see that p_1, p_2 and p_3 are not the same even when $\boldsymbol{\mu} = \mathbf{0}$ and $\boldsymbol{\Sigma} = \boldsymbol{I}_2$, because the components of the ALN-distribution are not symmetric.

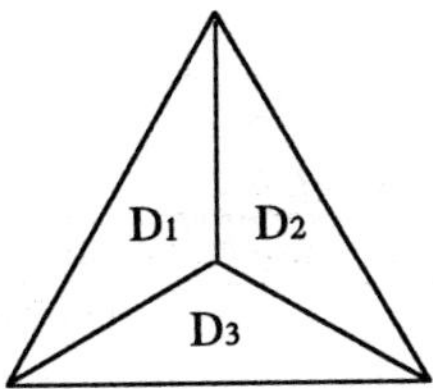

Figure 4.7

Table 4.5 *Probabilities of* $ALN_3\left(\begin{pmatrix}\mu\\\mu\end{pmatrix}, \boldsymbol{I}_2\right)$

μ	D_1	D_2	D_3
0	0.3750228	0.3750228	0.2500457
1	0.1461721	0.1461721	0.7076558
2	0.0226567	0.0222912	0.9550521
3	0.0013704	0.0014617	0.9971679
4	0.0000000	0.0000000	1.0000000

Table 4.6 *Probabilities of* $ALN_3\left(\begin{pmatrix}\mu\\\mu\end{pmatrix}, \begin{pmatrix}1 & 0.5\\0.5 & 1\end{pmatrix}\right)$

μ	D_1	D_2	D_3
0	0.3703636	0.3058652	0.3237712
1	0.1659967	0.0894391	0.7445642
2	0.0358122	0.0086790	0.9555089
3	0.0034716	0.0002741	0.9962543
4	0.0001827	0.0000000	0.9998173

In fact the idea of (4.7.1) and (4.7.2) can be applied to evaluate the integral of any continuous function. Consider an s-dimensional integral

$$I = \int f(\boldsymbol{x})d\boldsymbol{x}.$$

Let $p(\boldsymbol{x})$ be a p.d.f. with support R^s, i.e., $p(\boldsymbol{x}) > 0$ for any $\boldsymbol{x} \in R^s$. Rewrite

$$I = \int \frac{f(\boldsymbol{x})}{p(\boldsymbol{x})} p(\boldsymbol{x}) d\boldsymbol{x}. \tag{4.7.5}$$

Applying (4.7.1) and (4.7.2) to (4.7.5), we have

$$I \cong \frac{1}{n} \sum_{i=1}^{m} \frac{f(\boldsymbol{x}_i)}{p(\boldsymbol{x}_i)}, \tag{4.7.6}$$

where $\{\boldsymbol{x}_i\}$ is a set of rep-points of $p(\boldsymbol{x})$. It follows from the theory of Monte Carlo method that the efficiency of (4.7.6) will be high if $p(\boldsymbol{x})$ and $f(\boldsymbol{x})$ are similar, i.e.

$$f(\boldsymbol{x})/p(\boldsymbol{x}) \approx 1 \tag{4.7.7}$$

(Rubinstein (1981) section 4.3.1).

The approximation (4.7.6) may be applied to the class of elliptically symmetric distributions $EC_s(\boldsymbol{\mu}, \boldsymbol{\Sigma}, g)$, since the contour of any element of the class is ellipsoidal, therefore (4.7.7) is easily satisfied. More precisely, let $f(\boldsymbol{x}) = g((\boldsymbol{x} - \boldsymbol{\mu})'\boldsymbol{\Sigma}^{-1}(\boldsymbol{x} - \boldsymbol{\mu}))$ be the p.d.f. of an $EC_s(\boldsymbol{\mu}, \boldsymbol{\Sigma}, g)$ and $p(\boldsymbol{x})$ the p.d.f. of multivariate normal distribution $N_s(\boldsymbol{\mu}, \boldsymbol{\Sigma})$. We want to evaluate the probability

$$I = \int_D f(\boldsymbol{x}) d\boldsymbol{x}. \tag{4.7.8}$$

Write

$$I = \int_D \frac{f(\boldsymbol{x})}{p(\boldsymbol{x})} p(\boldsymbol{x}) d\boldsymbol{x}. \tag{4.7.9}$$

Although the contours of $f(\boldsymbol{x})$ and $p(\boldsymbol{x})$ are all ellipsoids, the condition (4.7.7) may still not be satisfied. In fact, let $\boldsymbol{y} \sim EC_s(\boldsymbol{\mu}, \boldsymbol{\Sigma}, g)$ and $\boldsymbol{z} \sim N_s(\boldsymbol{\mu}, \boldsymbol{\Sigma})$ with respective stochastic decompositions

$$\boldsymbol{y} = \boldsymbol{\mu} + R_1 \boldsymbol{\Sigma}^{1/2} \boldsymbol{u}$$

and

$$\boldsymbol{z} = \boldsymbol{\mu} + R_2 \boldsymbol{\Sigma}^{1/2} \boldsymbol{u},$$

where $R_1 > 0, R_2 > 0$, R_1, R_2 and $\boldsymbol{u}$ are independent, and $\boldsymbol{u} \sim U(U_s)$. Since R_1 and R_2 have distinct distributious, (4.7.7) is not usually satisfied. Suppose that the second moment of R_1 is finite. Let

$$c = \sqrt{\text{Var}(R_2)/\text{Var}(R_1)}. \tag{4.7.10}$$

Since

$$E(\boldsymbol{y}) = E(\boldsymbol{z}) = \boldsymbol{\mu},$$

the densities of $c(\boldsymbol{y} - \boldsymbol{\mu})$ and $(\boldsymbol{z} - \boldsymbol{\mu})$ are similar. Therefore, the probability (4.7.8) can be approximated by

$$I_n \cong \frac{1}{nc} \sum_{i=1}^{n} \frac{g(\frac{1}{c}(\boldsymbol{x}_i - \boldsymbol{\mu})'\Sigma^{-1}(\boldsymbol{x}_i - \boldsymbol{\mu}))}{p(\boldsymbol{x})} I_D(\boldsymbol{x}_i) \tag{4.7.11}$$

where $\{\boldsymbol{x}_i, i = 1, \cdots, n\}$ is a set of rep-points of $N_s(\boldsymbol{\mu}, \boldsymbol{\Sigma})$, and I_D is the index function of D.

The advantage of using (4.7.11) to evaluate the probability of an elliptically symmetric distribution is that only the rep-points of normal distribution are needed in the program. In general, the number of points n is comparatively large, and the computational time will be reasonable if we use the NTM and Box-Muller transformation to generate a set of rep-points for a normal distribution (section 4.3). Since the use of rep-points of $EC_s(\boldsymbol{\mu}, \boldsymbol{\Sigma}, g)$ is avoided, the program becomes simple and universal, and many examples show that this method is effective both in computational time and also in accuracy.

4.8 Applications of rep-points in simulation

Statistical simulation is an important tool, because many problems in statistics have no analytic solution. Let $\boldsymbol{x}, \cdots, \boldsymbol{x}_N$ be a sample from a population with c.d.f. $F(\boldsymbol{x})$ and T be a statistic of $\boldsymbol{x}_1, \cdots, \boldsymbol{x}_N$. If the distribution of T has no analytic expression, statistical simulation suggests generation of a sample $\boldsymbol{x}_1, \cdots, \boldsymbol{x}_N$ by a computer and then finding a sample T_1 of T. Independently repeating this process m times, we obtain a sample, $T_1, \cdots, T_m$, of T. When m is large the empirical distribution of $T_1, \cdots, T_m$ approximates the distribution of T. Therefore, we can get approximations of the critical points of T and some related statistics based on $T_1, \cdots, T_m$. Statistical simulation has been widely used for a long time, but it has some shortcomings: 1) the convergence rate of the empirical distribution to the underlying distribution is slow; 2) if we want to use random variates $\boldsymbol{x}_1, \cdots, \boldsymbol{x}_N$ which are from the uniform distribution on D, a closed and bounded domain, to stand for D, the uniformity of $\boldsymbol{x}_1, \cdots, \boldsymbol{x}_N$ is poor. In this section we study only the second problem and illustrate the NTM approach by a real problem.

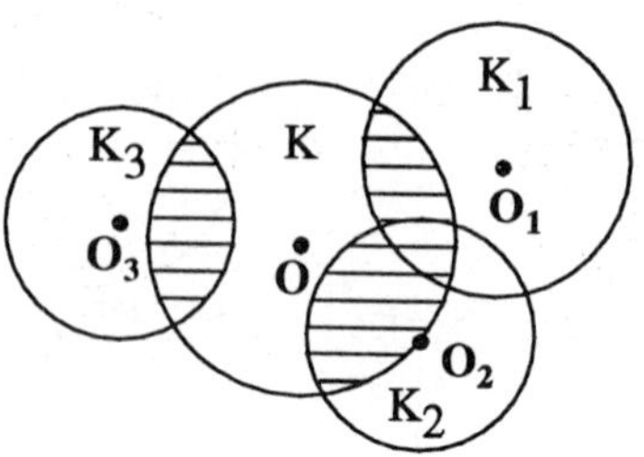

Figure 4.8 The overlapping area

Let D be a domain where a simulation is to be worked on. We need to find a set of points on D to stand for D. When D is a rectangle the simulation is usually done by using a set of equi-distributed points to stand for D. When D is not a rectangle, we propose using an NT-net on D. In this section we shall consider a practical problem and demonstrate that the NTM is really helpful in simulations. The problem is about the distribution of overlapping areas of a fixed circle and several random circles.

Given a unit circle K with centre at the origin $\mathbf{0}$ in R^2. Suppose that there are m random circles $K_1, \cdots, K_m$ with centres at location vectors $\boldsymbol{O}_1, \cdots, \boldsymbol{O}_m$ and radii $r_1, \cdots, r_m$, respectively, where the r_i are given constants. The m vectors $\mathbf{0}_1, \cdots, \mathbf{0}_m$ are independent and each has a bivariate normal distribution $\mathbf{0}_i \sim N_2(\mathbf{0}, \sigma_i^2 \boldsymbol{I}_2)$ in which $\sigma_i > 0$ is known and $\boldsymbol{I}_2$ denotes the 2×2 identity matrix. Let S be the overlapping area between K and the union of all random circles, i.e. the area of

$$K \cap (K_1 \cup \cdots \cup K_m). \tag{4.8.1}$$

Figure 4.8 gives an illustration for the case $m = 3$, where S denotes the shaded area. We want to find the distribution of S.

It is easily to find the distribution of S when $m = 1$, since the overlap area of two circles can be expressed explicitly in terms of the distance between their centres and the radii of two circles.

When $m = 2$, Figure 4.9 shows several possibilities of the location of K_1 and K_2, and it is difficult to find a simple formula for the distribution of S. Hence we need an effective method for finding the distributionof S. This is a problem of geometrical probability.

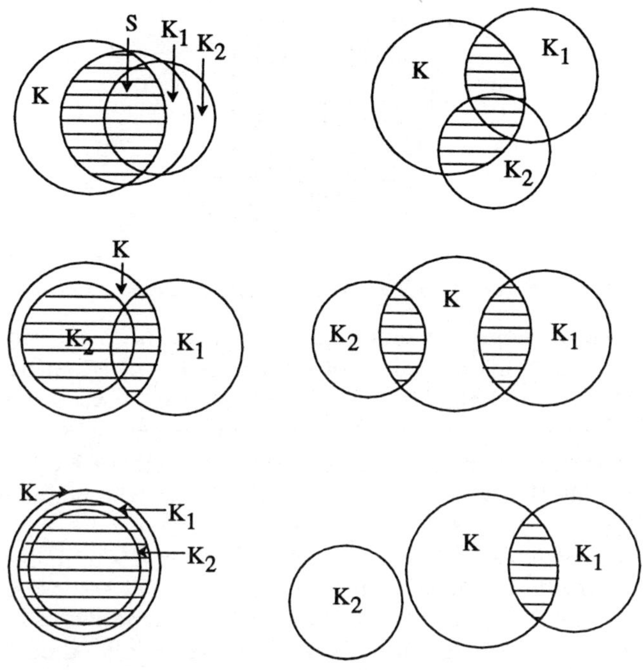

Figure 4.9

A natural way is to use simulation. The classical method is the so-called **lattice points method**. Let ABCD be the circumscribed square of the unit circle K as shown by Figure 4.10. Divide the square ABCD into n^2 equal subsquares of side length $2/n$, and we obtain the following n^2 lattice points in ABCD:

$$\left(-1+\frac{2i}{n},\ -1+\frac{2j}{n}\right),\quad 0\le i,j\le n-1$$

Suppose that there are N lattice points lying in K. We now generate m random circles with centres $\boldsymbol{O}_i \sim N_2(\boldsymbol{0}, \sigma_i^2 \boldsymbol{I}_2)$ and radii r_i $(i = 1, \cdots, m)$ by the Monte Carlo method. Suppose that M points among the N lattice points are covered by these m random circles. Then we get an observation $\pi M/N$ for the distribution of S. Repeating this process, we generate other m random circles and obtain another observation from the distribution of S. Continuing

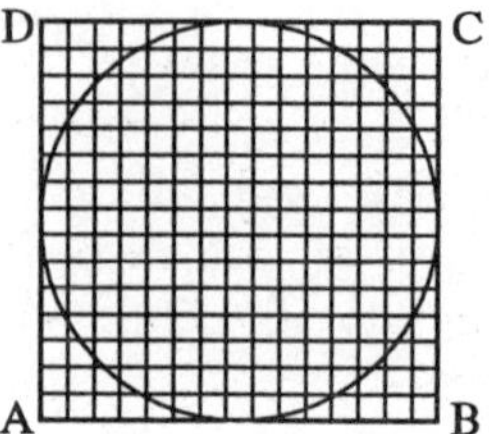

Figure 4.10 The lattice points method

this process we obtain an empirical distribution of S. For simplicity we shall call this procedure of simulating the distribution of S Method I. We find that its convergence rate is slow. The more serious problem with this method is that its accuracy is low even if N is large for example $N \approx 100,00$, because there are $O(\sqrt{n}\ )$ points among the N lattice points located nearly the boundary of K. Therefore, Method I is not acceptable.

An alternative procedure, called Method II here, is to replace the n^2 lattice points by an NT-net, for example the *glp* set, on ABCD instead of the n^2 lattice points and to do the simulation as before. It is surprising that Method II is much better than Method I, and we can get more accurate result than Method I by the following comparison: Since we know the exact distribution of S in the case of $m = 1$, the comparion of the two methods is performed in this case. For example, it takes more than 180 minutes using a PC/XT and method I to get a sample of size 1000 with an error 0.15, but it needs only 4 minutes using the same computer and Method II to obtain a sample of size 1500 with an error 0.02. The reason for the large computing time with Method I is that we must put more points on ABCD in the Method I than in Method II, otherwise the error of Method I would be too big.

These two methods are all based on uniformly scattered sets of points on ABCD but not on K. Of course we can do also the simulation by using an NT-net on K as shown in section 1.4. This leads to an even better result compared to that of method II. Jarrett and Morgan (1984) discussed the patch-gap model which is somehow related to the problem studied in this section and used a different approach.

4.9 Applications of rep-points in geometric probability

The problem discussed in the previous section can be considered as an application of NTM in geometric probability. Some basic knowledge of the geometric probability can be found in Kendall and Moran (1963). In this section we shall discuss another problem in geometric probability: the life distribution of a rotary ball roller. This problem comes from the production line of steel rolling and there has been no satisfactory solution for a long time (Cheng (1983)). People wish to increase the life of the roller by using a randomly rotary ball roller instead of the usual roller. Its mathematical model can be stated as follows: Let S be a unit sphere in R^3 and be covered by independent random belts with fixed thickness successively. Each belt is symmetric about a great circle of S and is uniformly distributed on S. Denote by $G_h(\boldsymbol{x})$ the random belt on S with thickness $2h$ and the normal direction $\boldsymbol{x}$. It is easy to see that

$$G_h(\boldsymbol{x}) = \{\boldsymbol{a} : |\boldsymbol{a}'\boldsymbol{x}| \leq h\}, \tag{4.9.1}$$

where $\boldsymbol{x} \sim U(S)$. In practice, we usually have $0 < h < 0.3$. Let $G_h(\boldsymbol{x}_1)$, $G_h(\boldsymbol{x}_2)$, $\cdots$ be a sequential samples from the population (4.9.1). For any $\boldsymbol{x} \in S$ we denote by $K_N(\boldsymbol{x})$ the number of belts which cover $\boldsymbol{x}$ in the first N random belts. The roller, i.e. S, is discarded if some point on S has been covered by m times, where m is a given positive integer. For given m and h, let $T_m(h)$ be the life of the roller, i.e. the minimum of N such that $K_N(\boldsymbol{x}) \geq m$ for some $\boldsymbol{x} \in S$. Obviously, we have

$$\begin{aligned} T_m(h) &= \min\{N : K_N(\boldsymbol{x}) \geq m, \quad \text{for some } \boldsymbol{x} \in S\} \\ &= \min\Big\{N : \sup_{\boldsymbol{x}\in S} \sum_{j=1}^{N} I_{G_h(\boldsymbol{x}_j)}(\boldsymbol{x}) \geq m, \Big\}, \end{aligned} \tag{4.9.2}$$

where $I_A(\cdot)$ is the indicator function of set A. We are required to obtain the distribution of $T_m(h)$ and $E(T_m(h))$, and to find some ways to increase the life of the roller.

It is easily seen that $I_{G_h(\boldsymbol{x}_j)}(\boldsymbol{x}) = I_{G_h(\boldsymbol{x})}(\boldsymbol{x}_j)$, and

$$E(I_{G_h(\boldsymbol{x})}(\boldsymbol{x}_j)) = h \quad \text{and} \quad \mathrm{Var}(I_{G_h(\boldsymbol{x})}(\boldsymbol{x}_j) = h(1-h),$$

Cheng *et al.* (1990) find some properties of the limiting distribution of $T_m(h)$:

(a) $\lim_{m\to\infty}(\frac{1}{m}T_m(h) - \frac{1}{n}) = 0$, a.s.

(b) $\lim_{m\to\infty} \frac{h}{\sqrt{1-h}} m^{1/2}(\frac{1}{m}T_m(h) - \frac{1}{h}) = \inf_{\boldsymbol{a}\in S} w(\boldsymbol{a})$, in distribution, where $\{w(\boldsymbol{a}), \boldsymbol{a} \in S\}$ is a Gaussian process with zero mean and covariance

$$E(w(\boldsymbol{a})w(\boldsymbol{b})) = \{EI_{G_h(\boldsymbol{a})}(\boldsymbol{x})I_{G_h(\boldsymbol{b})}(\boldsymbol{x}) - h^2\}/h(1-h),$$

where $\boldsymbol{x} \sim U(S)$.

(c) $\lim_{m\to\infty} \sup \frac{h}{\sqrt{1-h}} \{m(2\log\log m)^{-1}\}^{1/2}|\frac{1}{m}T_m(h) - \frac{1}{h}| = 1$, a.s.

The above theoretical results are interesting but we have difficulty when m is not large. For example, if $h = 0.1$ and $m = 20$, Fang and Wei (1989a) showed by simulation that the sample mean and the standard deviation are 99.7 and 9.8 respectively. But it follows from (a) and (c) that the mean of limiting distribution is $\frac{h}{m} = 200$ and the asymptotic standard deviation is $\sqrt{m(1-h)}\,/h \approx 42.4$, and so there is a big gap of the asymptotic results from the simulation values. Therefore, it is necessary to use simulation to find the distribution, mean and standard deviation of $T_m(h)$. Fang and Wei (1989) propose the following simulation algorithm:

Step 1 Give m and h.

Step 2 Generate an NT-net of n points $\{\boldsymbol{x}_k\}$, for example $n = 1067$. Set $N = 1$.

Step 3 Generate a point $\boldsymbol{a}_N$ from the population $U(s)$ by the Monte Carlo method and consequently we have the corresponding belt $G_h(\boldsymbol{a}_N)$.

Step 4 Count the number of belts which cover $\boldsymbol{x}_k$ in the first N belts and denote it by $K_N(\boldsymbol{x}_k)$. If $K_N(\boldsymbol{x}_k) = m$ for some k, go to Step 5, otherwise let $N+1$ replace N and return to Step 3.

Step 5 The number N is an observation of $T_m(h)$.

Repeat the above process n_0 times and obtain a sample of size n_0 of $T_m(h)$. For exapmle, when $m = 20$, $h = 0.1$ and $n_0 = 5000$, Fang and Wei's simulation shows the sample mean and sample standard deviation are

$$\bar{T}_m(h) = 99.7 \quad \text{and} \quad \sigma(T_m(h)) = 9.8,$$

respectively. By the same way, they obtain 20 samples of size 5000 (total 100,000 observations) and find that the results are very close each other. They also find that the distribution of $T_m(h)$ is close to a normal distribution, but it is not normal. If m is large, the distribution of $T_m(h)$ can be considered a normal distribution.

Furthermore, they note that the largest value of $T_m(h)$ among the 100,000 observations is 125. We denote by $G_h(\boldsymbol{a}_1^*), \cdots, G_h(\boldsymbol{a}_{125}^*)$ the correstponding belts such that the rotary ball roller has the longest life. This means that if $\boldsymbol{a}_i = \boldsymbol{a}_i^*, i = 1, 2, \cdots, 125$ are fixed, we always have $T_m(h) = 125$ in the case of $h = 0.1$ and $m = 20$, which is better than the random choices of $\{\boldsymbol{a}_i\}$. We may ask: is it possible to improve this result further? We find that $\boldsymbol{a}_1^*, \cdots, \boldsymbol{a}_{125}^*$ are not very uniformly scattered on S, and guess that the answer should be positive.

Since we can generate by the NTM a more uniformly scattered set of points on S than those by the Monte Carlo method, we use the *glp* set $\{\boldsymbol{c}_k = (c_{k1}, c_{k2}),\ k = 1, \cdots, n\}$ of generating vector $(n; 1, b)$ on C^2 and then have a set $\{\boldsymbol{a}_k^{**} = (a_{k1}, a_{k2}, a_{k3})',\ k = 1, \cdots, n\}$ on S, where

$$\begin{cases} a_{k1} = 1 - 2c_{k1}, \\ a_{k2} = 2\sqrt{c_{k1}(1 - c_{k1})}\cos(2\pi c_{k2}), \\ a_{k3} = 2\sqrt{c_{k1}(1 - c_{k1})}\sin(2\pi c_{k2}). \end{cases}$$

Finally we find that the largest $T_m^*(h) = 155$ if $(n; 1, b) = (155; 1, 20)$. This means that the life of the roller can be 155!

This example indicates that the NTM is significantly better than 100,000 experiments by the Monte Carlo method.

Exercises

4.1 Let $X \sim N(170, 9^2)$ be the distribution of height (cm) of men in a city. Give 7 rep-points for X by both criteria: F-discrepancy and MSE. If we want to choose 7 representative men's heights of the city for designing men's clothes, which kind of rep-points would you prefer to choose?

4.2 Write a computer program to generate an NT-net on U_s by the TFWW algorithm in section 4.3.

4.3 With Exercise 4.2 write a computer program to generate rep-points of the following distributions:
(a) the multivariate normal distribution $N_x(\mathbf{0}, \boldsymbol{I}_s)$;
(b) the symmetric Kotz type distribution;
(c) the symmetric multivariate Pearson Type VII distribution;
(d) the symmetric multivariate Pearson Type II distribution.

4.4 Write a computer program to generate rep-points of multivariate l_1-norm symmetric distributions with generating densities (4.2.4), (4.2.5) and (4.2.9).

4.5 Let $X \sim F(x)$ with finite second moment and let $Q_n(x)$ be its optimum quantizer and $(x_{n1}, \cdots, x_{nn})$ be the corresponding mse-rep-points. Prove that
(a) $L_n \to 0$ as $n \to 0$, where $L_n = \text{MSE}(x_{n1}, \cdots, x_{nn})$;
(b) $\text{Var}(Q_n(X)) = (1 - L_n)\text{Var}(X)$;
(c) $\text{Var}(Q_n(X)) \leq \text{Var}(Q_{n+1}(X))$;
(d) $\text{Var}(Q_n(X)) \to \text{Var}(X)$ as $n \to \infty$.

4.6 Let $\{\boldsymbol{y}_i, i = 1, \cdots, n\}$ be n points in R^s and $\{S_i, i = 1, \cdots, n\}$ be a partition of R^s. Let $p(\boldsymbol{x})$ be a p.d.f. in R^s and let $Q(\boldsymbol{x}) = \boldsymbol{y}_i$ if $\boldsymbol{x} \in S_i$ for any $\boldsymbol{x} \in R^s$. Assume the random vector $\boldsymbol{x} \sim p(\boldsymbol{x})$. Prove that
(a) $E\|\boldsymbol{x} - Q(\boldsymbol{x})\|^2 \geq E\|\boldsymbol{x} - Q^*(\boldsymbol{x})\|^2$,
where $Q^*(\boldsymbol{x}) = \boldsymbol{y}_i$ if $\boldsymbol{x} \in S_i^*$ for any $\boldsymbol{x} \in R^s$ and

$$S_i^* = \{\boldsymbol{x} : \|\boldsymbol{x} - \boldsymbol{y}_i\| < \|\boldsymbol{x} - \boldsymbol{y}_j\|, i \neq j\};$$

(b) $E\|\boldsymbol{x} - Q(\boldsymbol{x})\|^2 \geq E\|\boldsymbol{x} - Q_*(\boldsymbol{x})\|^2$,
where $Q_*(\boldsymbol{x}) = \boldsymbol{y}_i^*$ if $\boldsymbol{x} \in S_i$ for any $\boldsymbol{x} \in R^s$ and $y_i^* = E(\boldsymbol{x}|\boldsymbol{x} \in S_i)$.

4.7 Let $\boldsymbol{x}$ have a symmetric Kotz Type distribution with

$$\boldsymbol{\mu} = \mathbf{0} \text{ and } \Sigma = \begin{pmatrix} 1 & 0.5 & 0.5 \\ 0.5 & 1 & 0.5 \\ 0.5 & 0.5 & 1 \end{pmatrix}.$$

Apply (4.7.11) to calculate the probabilities of all eight octants (Example 4.7).

4.8 Let $\boldsymbol{x} \sim MT_3(5, \mathbf{0}, \boldsymbol{R})$, where $\boldsymbol{R}$ is given in Example 4.7. Apply (4.7.11) to calculate the probabilities of all eight octants and compare the results with Example 4.7.

4.9 With an NT-net on the unit circle K (section 4.8) and using simulation for $m = 3, r_1 = 0.1, r_2 = 0.2$ and $r_3 = 1.2$, find the distribution of S, the overlapping area between K and the union of three random circles.

CHAPTER 5

Experimental design and design of computer experiments

In this chapter we shall describe the applications of the number-theoretic method to experimental design. With the *glp* set we proposed a new kind of experimental designs in 1980, which we called **the uniform design** and denoted by UD. The number of experiments is significantly decreased by a large factor compared with the orthogonal design when the number of levels of factors is large. Besides the experimental design with independent factors, the UD can be also applied to experiments with dependent factors, for instance, **experiments with mixtures**. The **design of computer experiments** has becomes an interesting topic in recent years, and it has a certain ties with the UD. We shall give also a brief introduction on the design of computer experiments in this chapter.

5.1 Introduction

Experimental design is a branch of statistics which is extremely important in agriculture, industry and natural sciences. Let us look at an example first.

Example 5.1

Consider three variables which may affect the yield of a chemical product: temperature (A), time (B), and concentration of alkali (C), at the following values:

Temperature(A) :	80^oC,	85^oC,	90^oC
Time(B) :	$90m$,	$120m$,	$150m$
Concentration(C) :	5%,	6%,	7%

We shall denote the different values of the variables as $A_1, A_2, A_3, \cdots, C_2, C_3$, respectively. Here the temperature, time and concentration are called the **factors** of the experiment and $80^oC, 85^oC$, and 90^oC are called **levels** of factor A.

The purpose of the experiments is to study the influence of each factor on the response and to find the best combination of levels. A good experimental design should minimize the number of experiments to get the most amount of information. The following are some ways to design the experiment:

(1) Consider all combinations of levels. Totally there are $3 \times 3 \times 3 = 27$ combinations in Example 5.1. This method can be used only for small numbers of factors and levels. For example, suppose that there are six factors and each factor has five levels. The total number of combinations is $5^6 = 15625$ which is too large.

(2) Do a one-factor-experiment (one-way-experiment) several times. This way is effective only for the case of factors having no interactions.

(3) The orthogonal design. The so-called **orthogonal design** is the most popular one in practice and has a long history. It provides a series of the **orthogonal tables** for arrangement of experiments. For example, we may use the orthogonal table $L_9(3^4)$ (Table 5.1) to arrange the experiments of Example 5.1.

Table 5.1 $L_9(3^4)$

No	1	2	3	4
1	1	1	1	1
2	1	2	2	2
3	1	3	3	3
4	2	1	2	3
5	2	2	3	1
6	2	3	1	2
7	3	1	3	2
8	3	2	1	3
9	3	3	2	1

In notation $L_9(3^4)$, "L" denotes Latin square, "9" the number of experiments, "3" the number of levels, and "4" the maximum number of independent factors in the use of the table. We can choose any three columns of $L_9(3^4)$ for factors A, B and C, and then obtain the following experimental design in Table 5.2.

Table 5.2 *The design of experiments*

No	A	B	C
1	80^oC	90m	5%
2	80^oC	120m	6%
3	80^oC	150m	7%
4	85^oC	90m	6%
5	85^oC	120m	7%
6	85^oC	150m	5%
7	90^oC	90m	7%
8	90^oC	120m	5%
9	90^oC	150m	6%

This experimental design chooses 9 representative experiments from the 27 possible experiments and gives equal status to all factors and levels: (a) Each level of any factor is duplicated 3 times. (b) Each combination of all levels of any two factors has appeared the same number of times. We call (a) and (b) the properties of **equilibrium** and **regularity** respectively. We can see also that the experimental points designed by the orthogonal design are uniformly scattered on the experimental domain. The following example shows that the orthogonal design sometimes requests too many experiments.

Example 5.2

Quantitative risk assessment of a mixture of toxic chemicals present in the environment is a complex problem. The varying concentration of each chemical in the environment and the routes of exposure make any prediction difficult. An experiment considers the various concentrations of each of cadmium(Cd), copper(Cu), zinc(Zn), nickel(Ni), chromium(Cr), and lead(Pb) as follows (ppm): 0.01, 0.05, 0.1, 0.2, 0.4, 0.8, 1, 2, 4, 5, 8, 10, 12, 14, 16, 18, 20. The combined effect of six different trace metals on mortality of the Reuber H-35 rat hepatoma cell are studied.

The total number of combinations of this experiments is $17^6 = 24137569$ which is too many for any laboratory. It was known that there exist interactions among six metals. So we can not use the above second method, "do a one-factor-experiment several times". If we want to use the orthogonal design, the number of experiments is at least $17^2 = 289$ which is too large also. Therefore Fang (1980),

Wang and Fang (1981) proposed a new design called the uniform design (UD) and provided many tables of UD. For example, Table 5.3 shows a table of UD, $U_{17}(17^{16})$, where the notation $U_n(q^t)$ has a similar meaning to the orthogonal table: "U" denotes the UD, "n" the number of experiments, "q" the number of levels of each factor, and "t" the maximum number of columns of the table. We need choose 6 columns among 16 columns of $U_{17}(17^{16})$. It will be shown that different choices of 6 columns have different effects. Therefore, a recommendation of 6 columns is needed. According to Table 5.12 the recommended columns are 1, 4, 6, 10, 14 and 15 in $U_{17}(17^{16})$, and the corresponding design is given in Table 5.4.

Table 5.3 $L_{17}(17^{16})$

No	1	2	3	4	5	6	7	8
1	1	2	3	4	5	6	7	8
2	2	4	6	8	10	12	14	16
3	3	6	9	12	15	1	4	7
4	4	8	12	16	3	7	11	15
5	5	10	15	3	8	13	1	6
6	6	12	1	7	13	2	8	14
7	7	14	4	11	1	8	15	5
8	8	16	7	15	6	14	5	13
9	9	1	10	2	11	3	12	4
10	10	3	13	6	16	9	2	12
11	11	5	16	10	4	15	9	3
12	12	7	2	14	9	4	16	11
13	13	9	5	1	14	10	6	2
14	14	11	8	5	2	16	13	10
15	15	13	11	9	7	5	3	1
16	16	15	14	13	12	11	10	9
17	17	17	17	17	17	17	17	17

Table 5.3 $L_{17}(17^{16})$ *(continuation)*

No	9	10	11	12	13	14	15	16
1	9	10	11	12	13	14	15	16
2	1	3	5	7	9	11	13	15
3	10	13	16	2	5	8	11	14
4	2	6	10	14	1	5	9	13
5	11	16	4	9	14	2	7	12
6	3	9	15	4	10	16	5	11
7	12	2	9	16	6	13	3	10
8	4	12	3	11	2	10	1	9
9	13	5	14	6	15	7	16	8
10	5	15	8	1	11	4	14	7
11	14	8	2	13	7	1	12	6
12	6	1	13	8	3	15	10	5
13	15	11	7	3	16	12	8	4
14	7	4	1	15	12	9	6	3
15	16	14	12	10	8	6	4	2
16	8	7	6	5	4	3	2	1
17	17	17	17	17	17	17	17	17

In this chapter we shall introduce the idea of UD, how to find the tables of UD and discuss criteria for uniformity of the experimental points on the domain. We also introduce the uniform design for experiments with mixtures (**UDEM**).

The UD has been applied satisfactorily to problems in the textile industry, watch industry, metallurgy industry, software design, and military sciences inside China during the past ten years.

5.2 Uniform design

Suppose that there are s factors and each factor has q levels. Then the number of possible experiments is q^s. The orthogonal design is to choose q^2 experiments with best representation among these q^s experiments. We assume without loss of generality that the domain for experiments is C^s. The q^2 points corresponding to these q^2 experiments should be scattered uniformly and regularly on C^s as stated in section 5.1. Hence the q^2 points may be regarded as the rep-points of $U(C^s)$. However, if the uniformity of these q^2

Table 5.4 *Experimental design of Example 5.2*

No	Cd	Cu	Zn	Ni	Cr	Pb
1	0.01	0.2	0.8	5.0	14.0	16.0
2	0.05	2.0	10.0	0.1	8.0	12.0
3	0.1	10.0	0.01	12.0	2.0	8.0
4	0.2	18.0	1.0	0.8	0.4	4.0
5	0.4	0.1	12.0	18.0	0.05	1.0
6	0.8	1.0	0.05	4.0	18.0	0.4
7	1.0	8.0	2.0	0.05	12.0	0.1
8	2.0	16.0	14.0	10.0	5.0	0.01
9	4.0	0.05	0.1	0.4	1.0	18.0
10	5.0	0.8	4.0	16.0	0.2	14.0
11	8.0	5.0	16.0	2.0	0.01	10.0
12	10.0	14.0	0.2	0.01	16.0	5.0
13	12.0	0.01	5.0	8.0	10.0	2.0
14	14.0	0.4	18.0	0.2	4.0	0.8
15	16.0	4.0	0.4	14.0	0.8	0.2
16	18.0	12.0	8.0	1.0	0.1	0.05
17	20.0	20.0	20.0	20.0	20.0	20.0

points is measured by discrepancy, the result is poor (Example 1.2). On the other hand orthogonal design requires that the design for any two factors considers all the combinations of their levels (regularity). Hence, the number of experiments is at least q^2. The requirement of regular arrangement is for the sake of analysis of variance, so that the estimations of main effects and interaction effects among factors, and the related testing hypotheses can be obtained. When $q > 2$, if we want to obtain the estimates of all interaction effects, the number of experiments is at least q^s. Hence we cannot get the estimates of all interaction effects from the q^2 experiments using the orthogonal design. For example, we use the first 3 columns of $L_9(3^4)$ (Table 5.1) to arrange the experiments for the case $s = 3$ and $q = 3$, and use α_A, α_B, α_C to denote the effects of factors A, B, C and α_{AB}, α_{AC}, α_{BC} the interaction effects of second order. Then the four columns of $L_9(3^4)$ represent the following effects:

1	2	3	4
α_A	α_B	α_C	
α_{BC}	α_{AC}	α_{AB}	$\alpha_{AB}, \alpha_{AC}, \alpha_{BC}$

We see that each column is a mixture of at least two effects, and so the estimation for any effect is impossible. If the interaction effect of third order α_{ABC} is considered, the mixture among effects is more serious. Hence the analysis of variance is meaningless for orthogonal design when $n = q^2$ and $q > 2$ when considering interactions in the above sense. Since the analysis of variance is ineffective in the above case, it is not necessary to arrange the experiments according the requirement of analysis of variance, in particular, when q is large (Example 5.2). Hence it is reasonable that the experiments are arranged such that their corresponding representations on C^s are scattered uniformly.

The so-called UD is a design such that the experimental points are uniformly scattered on C^s. In virtue of the uniformity of NT-nets on C^s the UD is just to choose an NT-net on C^s, i.e. a set of rep-points for $U(C^s)$. In Chapter 4, we proposed two criteria, the F-discrepancy and MSE, to choose the rep-points for a given distribution. Both can be employed for experimental design. We discuss first the F-discrepancy criterion for UD.

Similar to the orthogonal design the UD provides many tables (Table 5.3, for example) which are generated by *glp* sets. We now introduce the method for constructing the tables of UD.

5.2.1 The tables of uniform design

A table of UD, $U_n(n^t)$, is obtained by a generating vector $(n; h_1, \cdots, h_t)$ (section 1.3) of a *glp* set, where $1 = h_1 < h_2 < \cdots < h_t < n$ and the g.c.d. $(n, h_i) = 1,\ i = 1, \cdots, t$. Let

$$q_{ki} \equiv kh_i (\operatorname{mod} n),\ k = 1, \cdots, n,\ i = 1, \cdots, t, \qquad (5.2.1)$$

where $0 < q_{ki} \leq n$ as the modified multiplication modulo n is applied. The table $U_n(n^t)$ is formed from the (q_{ki}).

Although the *glp* set has been discussed in section 1.3 and appeared in the previous chapters, a more detailed discussion is still necessary in this chapter. For example, the condition that g.c.d. $(n, h_i) = 1, i = 1, \cdots, t$ can be removed for applications of the *glp* set in integration, but it is necessary for UD.

What is the maximum number of h_i's for given n? Number theory (Hua (1956)) shows that the number of possible h_i is given by the Euler function $\varphi(n)$:

$$\varphi(n) = n \prod_{p|n} \left(1 - \frac{1}{p}\right),$$

where p runs over the prime divisors of n. For examples,

$$\varphi(12) = 12\left(1 - \frac{1}{2}\right)\left(1 - \frac{1}{3}\right) = 4$$

because $12 = 3 \times 2^2$, and the possibly associated h_i with $(h_i, n) = 1$ are 1, 5, 7 and 11;

$$\varphi(9) = 9\left(1 - \frac{1}{3}\right) = 6$$

and the possibly associated h_i are 1, 2, 4, 5, 7, 8; and

$$\varphi(7) = 7\left(1 - \frac{1}{7}\right) = 6$$

with 1, 2, 3, 4, 5, 6 as h_i's. Therefore for given n, we have $t = \varphi(n)$. The $U_7(7^6)$ and $U_9(9^6)$ are given by Tables 5.5 and 5.6.

Table 5.5 *Table* $U_7(7^6)$

No	1	2	3	4	5	6
1	1	2	3	4	5	6
2	2	4	6	1	3	5
3	3	6	2	5	1	4
4	4	1	5	2	6	3
5	5	3	1	6	4	2
6	6	5	4	3	2	1
7	7	7	7	7	7	7

Since $1 + 6 = 7$, $2 + 5 = 7$ and $3 + 4 = 7$ in $U_7(7^6)$, the rank of matrix (q_{ij}) is at most 4. In general the number of factors must be $\leq \varphi(n)/2+1$, and so the number of independent factors cannot exceed $\frac{\varphi(n)}{2}+1$ in $U_n(n^t)$ (Ding(1986)).

Table 5.6 *Table* $U_9(9^6)$

No	1	2	3	4	5	6
1	1	2	4	5	7	8
2	2	4	8	1	5	7
3	3	6	3	6	3	6
4	4	8	7	2	1	5
5	5	1	2	7	8	4
6	6	3	6	3	6	3
7	7	5	1	8	4	2
8	8	7	5	4	2	1
9	9	9	9	9	9	9

Suppose that there are s independent factors in a practical problem and each factor has n levels. If $t=(\varphi(n)+1)/2 \geq s$, we may use UD table $U_n(n^t)$ to arrange the experiments. It is well-known that any s columns in orthogonal table $L_n(q^t)$, $s<t$, are equivalent. But the efficiencies may be different for distinct choices of columns of $U_n(n^t)$ (Figures 1.6 (c) and (d)). Hence we need to consider the uniformity of designs and to choose s columns with the best uniformity among all possible choices of s columns.

5.2.2 *The Equivalence of design matrices*

Definition 5.1
An $n \times s$ matrix

$$\boldsymbol{A}=(a_{ij})=(\boldsymbol{a}_{(1)},\cdots,\boldsymbol{a}_{(n)})'=(\boldsymbol{a}_1,\cdots,\boldsymbol{a}_s) \tag{5.2.2}$$

is called a **UD-design matrix** if each column of $\boldsymbol{A}$ is a pemutation of $(1,\cdots,n)$ and $\text{rank}(\boldsymbol{A})=s$. Let

$$x_{ki}=(2a_{ki}-1)/2n, \quad i=1,\cdots,s; \quad k=1,\cdots,n.$$

The matrix $\boldsymbol{X}=(x_{ki})$ is called the **induced matrix** of $\boldsymbol{A}$. The set of all $n\times s$ UD-design matrices is denoted by $\mathcal{A}_{n\times s}$ or $\mathcal{A}$.

It is evident that for any $\boldsymbol{A} \in \mathcal{A}$, we have

$$\left.\begin{aligned} \sum_{k=1}^{n} a_{kj} &= n(n+1)/2, \\ \sum_{k=1}^{n} a_{jk}^2 &= n(n+1)(2n+1)/6, \end{aligned}\right. \qquad j = 1, \cdots, s. \tag{5.2.3}$$

Definition 5.2
Two design matrices $\boldsymbol{A}_1$ and $\boldsymbol{A}_2$ are called equivalent if there exist two permutation matrices $\boldsymbol{P}_1 : n \times n$ and $\boldsymbol{P}_2 : s \times s$ such that

$$\boldsymbol{A}_1 = \boldsymbol{P}_1 \boldsymbol{A}_2 \boldsymbol{P}_2. \tag{5.2.4}$$

We denote by $\boldsymbol{A}_1 \approx \boldsymbol{A}_2$.

Note that a matrix $\boldsymbol{P}$ is called a permutation matrix if every row (and also column) has only an element 1, and the other elements of $\boldsymbol{P}$ are zero. The matrix $\boldsymbol{Q} = (q_{ki})$ defined by (5.2.1) is obviously a UD-design matrix which is denoted by $\boldsymbol{Q}_n(h_1, \cdots, h_s)$. When $(h_1, \cdots, h_s) = (1, a, \cdots, a^{s-1})(\bmod n)$, then we denote it by $\boldsymbol{Q}_n(a)$. Recall that an integer b is called the inverse of a if $ab \equiv 1(\bmod n)$ and write $b = a^{-1}(\bmod n)$. We have the following:

Theorem 5.1
Suppose that $(l_1, \cdots, l_s)$ is a permutation of $(1, \cdots, s)$, $h_j^{(i)} = h_i^{-1} h_j(\bmod n)$, and $1, a, a^2, \cdots, a^{s-1}(\bmod n)$ are distinct integers. Then

(a) $\boldsymbol{Q}_n(h_1, \cdots, h_s) \approx \boldsymbol{Q}_n(h_{l_1}, \cdots, h_{l_s})$;
(b) $\boldsymbol{Q}_n(h_1, \cdots, h_s) \approx \boldsymbol{Q}_n(h_1^{(i)}, \cdots, h_s^{(i)})$, $1 \le i \le s$;
(c) $\boldsymbol{Q}_n(a) \approx \boldsymbol{Q}_n(a^{-1})$, $(\bmod n)$.

PROOF Let $\boldsymbol{P}_{m_i} = (p_{ij})$ be an $s \times s$ permutation matrix with elements zero except $p_{m_i i} = p_{i m_i} = 1$ and $p_{qq} = 1$ $(q \ne i, m_i)$. We multiply the right hand side of $\boldsymbol{Q}_n(h_1, \cdots, h_s)$ by $\boldsymbol{P}_{l_1} \cdots \boldsymbol{P}_{l}$, and then we have (a). Now we proceed to prove (b): Since $(h_i^{-1} h_1 k, \cdots, h_i^{-1} h_s k) = (h_1(h_i^{-1} k), \cdots, h_s(h_i^{-1} k))$ $(k = 1, \cdots, n)$, we know that $\boldsymbol{Q}_n(h_1^{(i)}, \cdots, h_s^{(i)})$ is obtained by a permutation of rows of

$\boldsymbol{Q}_n(h_1, \cdots, h_s)$. (c) follows by taking $h_j = a^{j-1}$ $(1 \leq j \leq s)$ and $i = s$ in (b). The theorem is proved. □

5.2.3 *Uniformity of design*

The so-called measure of uniformity for a design matrix $\boldsymbol{A}$ is a nonnegative function $U : \mathcal{A}(\text{ or } \mathcal{X}) \to R_+ = \{x : x \geq 0\}$. For convenience the measure of uniformity in this chapter is called the **U-criterion**. It is clear that the U-criterion must satisfy

$$U(\boldsymbol{A}) = U(\boldsymbol{B}) \tag{5.2.5}$$

provided $\boldsymbol{A} \approx \boldsymbol{B}$.

We shall introduce several U-criteria in this chapter. They are defined by the NTM, the geometrical method and the statistical method. In this subsection we illustrate only the NTM, and the other methods will be stated in section 5.4.

Given an induced matrix $\boldsymbol{X}$ and regarding n rows of $\boldsymbol{X}$ as a sample, let $F_n(\boldsymbol{x})$ denote its empirical distribution and $F(\boldsymbol{x})$ be the c.d.f. of $U(C^s)$. The so-called **l_p-discrepancy** of $\boldsymbol{X}$ is defined by

$$D_p(\boldsymbol{X}) = \left[\int_{C^s} |F_n(\boldsymbol{x}) - F(\boldsymbol{x})|^p d\boldsymbol{x}\right]^{1/p}. \tag{5.2.6}$$

When $p = 1, 2, \infty$, we have respective l_1-discrepancy, l_2-discrepancy and l_∞-discrepancy, where the latter is the well-known discrepancy (section 1.4). It is obvious that all the D_p's are U-criteria.

For l_2-discrepancy, Warnock (Niederreiter (1973b)) gives the following formula:

$$\begin{aligned} D_2(\boldsymbol{X}) =& 3^{-s} - \frac{2^{1-s}}{n} \sum_{k=1}^{n} \prod_{l=1}^{s} (1 - x_{kl}^2) \\ &+ \frac{1}{n^2} \sum_{k=1}^{n} \sum_{j=1}^{n} \prod_{i=1}^{s} (1 - \max(x_{ki}, x_{ji})). \end{aligned} \tag{5.2.7}$$

Since the calculation is heavy if we compute the discrepancy (i.e. D_∞) for the *glp* set directly, we reduce the problem by using some number-theoretic techniques: find a best vector $\boldsymbol{h}^* = (h_1^*, \cdots, h_s^*)$

such that $D(\boldsymbol{h}^*)$ minimizes

$$D(\boldsymbol{h}) = \frac{1}{n}\sum_{k=1}^{n}\prod_{r=1}^{s}\left(1 - \frac{1}{n}\ln\left(2\sin\left(\frac{\pi q_{kr}}{n+1}\right)\right)\right) \tag{5.2.8}$$

among all possible $\boldsymbol{h}$. Tables of best vector $\boldsymbol{h}^*$ for $4 \leq n \leq 31$ and $2 \leq s \leq 16$ are given by Wang and Fang (1981).

When n is large, the cost of a finding the best $\boldsymbol{h}^*$ is large. Therefore we suggest using sets of points of the form

$$\boldsymbol{x}_k = (k, kb, \cdots, kb^{s-1})(\bmod n), \; 1 \leq k \leq n, \tag{5.2.9}$$

to be the experimental points, where b is an integer satisfying $1 < b < n$ and $b^i \not\equiv b^j (\bmod n)$, $1 \leq i < j \leq s-1$. The integer b is usually chosen among the primitive roots mod n if n is a prime (section 1.3). Most of the $\boldsymbol{h}$'s in Appendix A.2 have the form (5.2.9), i.e.

$$\boldsymbol{h} = (1, b, \cdots, b^{s-1}). \tag{5.2.10}$$

A table of UD for even n can be obtained by omitting the last row of a table for $n+1$, for instance, the table $U_6(6^6)$ is obtained by omitting the last row of $U_7(7^6)$.

In section 5.4 a table of the best b is given for $n \leq 31$ (Table 5.12) which is enough for most practical problems.

Remark 5.1

By (5.2.8) and (2.1.5), if $\{h_r\}$ are chosen suitably, then

$$\begin{aligned}
&\frac{1}{n}\sum_{k=1}^{n}\prod_{v=1}^{s}\left(1 - \frac{2}{\pi}\ln\left(2\sin\left(\frac{\pi k h_v}{n+1}\right)\right)\right) \\
&\approx \int_0^1 \cdots \int_0^1 \prod_{i=1}^{s}\left(1 - \frac{2}{\pi}\ln(2\sin(\pi x_i))\right)dx_i \\
&= \prod_{i=1}^{s}\int_0^1 \left(1 - \frac{2}{\pi}\ln(2\sin(\pi x_i))\right)dx_i.
\end{aligned} \tag{5.2.11}$$

Since

$$\int_0^1 \left(1 - \frac{2}{\pi}\ln(2\sin(\pi x))\right)dx = 1,$$

the optimization problem seems to find $\boldsymbol{h}$ such that $|D(\boldsymbol{h}) - 1|$ attains minimum. But the integrand on the right hand side of (5.2.11) has singularities 0 and 1 in [0,1], and so (5.2.11) is unreasonable, in particular when n is small.

Remark 5.2

The $D(\boldsymbol{h})$ is only the principal term of D_∞, and we may also use

$$w(\boldsymbol{h}) = \begin{cases} \frac{2}{n}\sum_{k=1}^{\frac{n-1}{2}}\prod_{v=1}^{s}\left(1+2\pi^2 B_2\left(\left(\frac{kh_v}{n}\right)\right)\right), & \text{if } 2 \nmid n, \\ \frac{1}{n}\left(1-\frac{\pi^2}{6}\right)^{\mu}\left(1+\frac{\pi^2}{3}\right)^{s-\mu} & \\ \quad +\frac{2}{n}\sum_{k=1}^{\frac{n}{2}-1}\prod_{v=1}^{s}\left(1+2\pi^2 B_2\left(\left(\frac{kh_v}{n}\right)\right)\right) - 1, & \text{if } 2|n, \end{cases}$$

instead of D_∞, where μ denotes the number of odd integers of h_v $(1 \le v \le s)$ and $B_2(x) = x^2 - x + \frac{1}{6}$ is the Bernoulli polynomial (Hua and Wang (1981) Chapters 4 and 8).

5.3 Data analysis and examples

In this section we use two examples to illustrate how to use UD to arrange the experiments and how to do the data analysis. The work of data analysis in classical experimental designs is often done by analysis of variance or regression analysis. Since the number of experiments in UD is too small compared with the number of levels of factors in experiments, it is difficult to do the analysis of variance in a common way, and so regression analysis is the main tool for data analysis.

Suppose that there are s factors, and each factor has n levels in an experiment. Assume that the table $U_n(n^s)$ is employed to arrange the experiments, and that the respective responses $\{y_k\}$ are obtained. Then we get a set of observations $\{x_{k1}, \cdots, x_{ks}, y_k,\ k = 1, \cdots, n\}$, where $\{x_{ki}\}$ are levels of factors. If the main effects of factors are considered only, then the regression model is linear:

$$EY = \beta_0 + \beta_1 x_1 + \cdots + \beta_s x_s, \qquad (5.3.1)$$

where the coefficients $\beta_0, \beta_1, \cdots, \beta_s$ can be estimated. If the hypothesis testing

$$H_o : \beta_i = 0, \ H_1 : \beta_i \neq 0 \tag{5.3.2}$$

is rejected, then the main effect of i-th factor is significant for Y. Otherwise x_i may be deleted from the model.

If the main effects of second order and interaction effects are also considered, then the simplest model is

$$EY = \beta_0 + \sum_{i=1}^{s} \beta_i x_i + \sum_{i=1}^{s} \sum_{j=i}^{s} \beta_{ij} x_i x_j. \tag{5.3.3}$$

Since the number of terms in (5.3.3) is too large when s is large, the effect for prediction is poor, and so some terms should be deleted from the model which may be done by the various methods for selecting variables in regression analysis, for instance, the stepwise regression method and the best subset regression method (Seber (1977)). If the x_i term appears in the final expression of the model, then the main effect of 1st order of x_i is significant; if the $x_i x_j$ term is in the model, then the interaction effect of 2nd order of x_i and x_j is significant; and if there appears the term x_i^2, then the main effect of 2nd order of x_i is significant.

In some cases we should consider the main effects and interaction effects of high orders, for example $x_i^3, x_i^2 x_j, x_i x_j x_k$, then there are more terms in the right hand side of (5.3.3), and so the work of eliminating variables is still important.

Example 5.2 (continuation)

Table 5.7 shows mortalilies of Reuber H35 rat hepatoma cell after 3 hour treatment of the 17 solutions given by Table 5.4. Three dishes of the hepatoma cell were prepared for each solution, and the assay was repeated for three times. The corresponding results are Y_1, Y_2 and Y_3 and $\bar{Y}$ is the mean of Y_1, Y_2 and Y_3 in Table 5.7.

Table 5.7 *Mortalilies*

Y_1	Y_2	Y_3	$\bar{Y}$
17.95	17.65	18.33	17.9
22.09	22.85	22.62	22.5
31.74	32.79	32.87	32.4
39.37	40.65	37.87	39.3
31.90	31.18	33.75	32.2
31.14	30.66	31.18	31.0
39.81	39.61	40.80	40.0
42.48	41.86	43.79	42.7
24.97	24.65	25.05	24.8
50.29	51.22	20.54	50.6
60.71	60.43	59.69	60.2
67.01	71.99	67.12	68.7
32.77	30.86	33.70	32.4

For illustration of application of the regression analysis in UD. We treat $\bar{Y}$ only and omit the detailed analysis. Since the magnitude of the range of each factor is relative large, it is better to use log Cd, $\cdots$, log Pb instead of Cd, $\cdots$, Pb, respectively. Consider the regression model (5.3.3) and use the stepwise regression to select variables. The final regression equation is

$$\begin{aligned}\hat{\bar{Y}} &= 32.68 + 5.03 \log \mathrm{Cd} + 3.48 \log \mathrm{Cu} + 2.03 \log \mathrm{Ni} \\ &+ 0.55\,(\log \mathrm{Cu})^2 - 0.63\,(\log \mathrm{Zn})^2 \\ &+ 0.94\,(\log \mathrm{Ni})^2 + 0.53\,(\log \mathrm{Cd})(\log \mathrm{Cu}) \\ &- 0.70\,(\log \mathrm{Cd})(\log \mathrm{Cr}) + 0.92\,(\log \mathrm{Cu})(\log \mathrm{Pb}).\end{aligned}$$

The t-ratio of each item in the equation is 11.5, 7.8, 4.9, 2.6, -3.4, 4.1, 2.4, -2.8 and 5.3 respectively and the equation is acceptable. The results show that Cd is the most toxic chemical which, when present, will contribute greatly to cell mortality. The coefficient of $(\log \mathrm{Zn})^2$ is negative. This indicates that Zn may reduce the cytotoxic effect of the other metals. The interactions between Cd and Cu, Cd and Cr, and Cu and Pb are significant also.

The following example gives the comparison of UD with orthogonal design.

Example 5.3

To design a Vinylon (chemical material for making cloth material) product we consider the following factors:

A: temperature (^oC)
B: time (minute)
C: concentration of methanol (g/l)
D: concentration of sulphuric acid (g/l)
E: concentration of mirabilite (g/l).

Factors A,B,C and D have 7 levels and factor E has three levels only. It is recommended to use the **pseudo level method** such that factor E has "7" levels also. For more details see Table 5.8:

Table 5.8 *Factors and levels*

	1	2	3	4	5	6	7
A(^oC)	64	66	68	70	72	74	76
B (m)	14	16	18	20	22	24	26
C(g/l)	18	20	22	24	26	28	30
D(g/l)	206	212	218	224	230	236	242
E(g/l)	70	70	85	85	85	100	100

The engineers made 49 experiments using the orthogonal design (Fang *et al.* (1973)). The interaction effects were not significant and they obtained the following linear regression equation

$$EY = -42.37 + 0.55A + 0.38B + 0.26C + 0.10D - 0.04E, \quad (5.3.4)$$

with multiple correlation coefficient $R = 0.97$ and standard deviation $\hat{\sigma} = 0.83$. Here y stands for the quality of Vinylon. Suppose we use $U_{14}(14^5)$ instead of the orthogonal array, where $U_{14}(14^5)$ is obtained by omitting the last column of $U_{15}(15^5)$. By Appendix A.2, the recommended $\boldsymbol{h}$ is (1, 2, 4, 7, 13) for $s = 5$, and the arrangement of experiments and the respective results are listed in Table 5.9. Since each column of U_{14} has 14 levels and the number of the levels of each factor are less than 14, we use the pseudo levels method to arrange 14 experiments as follows: for factors A,B,C, and D we fuse levels 1 and 2 of U_{14} into new level $1, \cdots$, fuse levels 13 and 14 into new level 7. Factor E is treated similarly. For more details see Table 5.9.

Using linear regression model (5.3.4) with a random error $\varepsilon \sim N(0, 0.83)^2$, by a Monte Carlo simulation the response Y is obtained and its values are listed in Table 5.9, from which we have the regression equation

$$EY = -57.97 + 0.37A + 0.46B + 0.38C + 0.17D + 0.04E \quad (5.3.5)$$

with $R = 0.96$ and $\hat{\sigma} = 1.13$. Moreover, using (5.3.5) to predict Y's values of the 49 experiments made by the engineers, the residual mean value and the standard deviation are

$$\bar{x}_{res} = -0.00816, \ \hat{\sigma}_{res} = 1.734,$$

respectively. The result is not too bad. It is interesting to note that the number of experiments of UD is only $14/49 = 28.57\%$ compared with orthogonal design.

5.4 Measurements of uniformity of design

We have mentioned in section 5.2.3 that there are many methods to measure the uniformity of the design matrix. The methods can be classified as, for example, the NTM, the geometrical method and the statistical method. We have already introduced the NTM in section 5.2, and now we illustrate the other two methods as follows:

5.4.1 Geometry method

(a) Volume distance method

Let $\boldsymbol{A} = (\boldsymbol{a}_{(1)}, \cdots, \boldsymbol{a}_{(n)})' = (a_{ij})$ be a design matrix defined by section 5.2. We call

$$v_{jk}(\boldsymbol{A}) = \prod_{l=1}^{s} |a_{kl} - a_{jl}| \quad (5.4.1)$$

the **volume distance** between experimental points $\boldsymbol{a}_{(j)}$ and $\boldsymbol{a}_{(k)}$ of $\boldsymbol{A}$. Clearly, $v_{jk} = v_{kj}$ and $v_{jj} = 0$. For $1 \leq j < k \leq n$, one can make $2s$ superplanes through $\boldsymbol{a}_{(j)}$ or $\boldsymbol{a}_{(k)}$ such that each superplane is parallel with one of the axes respectively. Then the volume of the rectangle bounded by these superplanes is $v_{jk}(\boldsymbol{A})$. Jiang and Chen (1987) suggested the use of

Table 5.9 *Design and data*

No	A	B	C	D	E	y
1	$A_1(64)$	$B_2(14)$	$C_4(20)$	$D_1(224)$	$E_{13}(100)$	24.08
2	$A_2(64)$	$B_4(16)$	$C_8(24)$	$D_{14}(242)$	$E_{11}(100)$	28.59
3	$A_3(66)$	$B_6(18)$	$C_{12}(28)$	$D_6(218)$	$E_9(85)$	27.88
4	$A_4(66)$	$B_8(20)$	$C_1(18)$	$D_{13}(242)$	$E_7(85)$	27.99
5	$A_5(68)$	$B_{10}(22)$	$C_5(22)$	$D_5(218)$	$E_5(85)$	27.77
6	$A_6(68)$	$B_{12}(24)$	$C_9(26)$	$D_{12}(236)$	$E_3(70)$	31.21
7	$A_7(70)$	$B_{14}(26)$	$C_{13}(30)$	$D_4(212)$	$E_1(70)$	30.83
8	$A_8(70)$	$B_1(14)$	$C_2(18)$	$D_{11}(236)$	$E_{14}(100)$	25.67
9	$A_9(72)$	$B_3(16)$	$C_6(22)$	$D_3(212)$	$E_{12}(100)$	25.31
10	$A_{10}(72)$	$B_5(18)$	$C_{10}(26)$	$D_{10}(230)$	$E_{10}(85)$	31.53
11	$A_{11}(74)$	$B_7(20)$	$C_{14}(30)$	$D_2(206)$	$E_8(85)$	28.03
12	$A_{12}(74)$	$B_9(22)$	$C_3(20)$	$D_9(230)$	$E_6(85)$	31.31
13	$A_{13}(76)$	$B_{11}(24)$	$C_7(24)$	$D_1(206)$	$E_4(70)$	29.16
14	$A_{14}(76)$	$B_{13}(26)$	$C_{11}(28)$	$D_8(224)$	$E_2(70)$	36.39

$$U_2(\boldsymbol{A}) = \min_{1 \le j < k \le n} v_{jk}(\boldsymbol{A}) \tag{5.4.2}$$

as a measurement of the uniformity of $\boldsymbol{A}$. Alternatively Fang and Zhang (1992) suggested the use of

$$U_3(\boldsymbol{A}) = \sum_{1 \le j < k \le n} (v_{jk}(\boldsymbol{A}) - \bar{v}(\boldsymbol{A}))^2, \tag{5.4.3}$$

as a measurement for uniformity, where

$$\bar{v}(\boldsymbol{A}) = \frac{2}{n(n-1)} \sum_{1 \le j < k \le n} v_{jk}(\boldsymbol{A}). \tag{5.4.4}$$

If $U_2(\boldsymbol{A}) \ge U_2(\boldsymbol{B})$ (or $U_3(\boldsymbol{A}) \le U_3(\boldsymbol{B})$), then we say that the set defined by $\boldsymbol{A}$ is scattered more uniformly than that of $\boldsymbol{B}$. This is evident from the view point of geometry: If the experimental points are scattered uniformly on $[1, n]^s$, then the difference of any

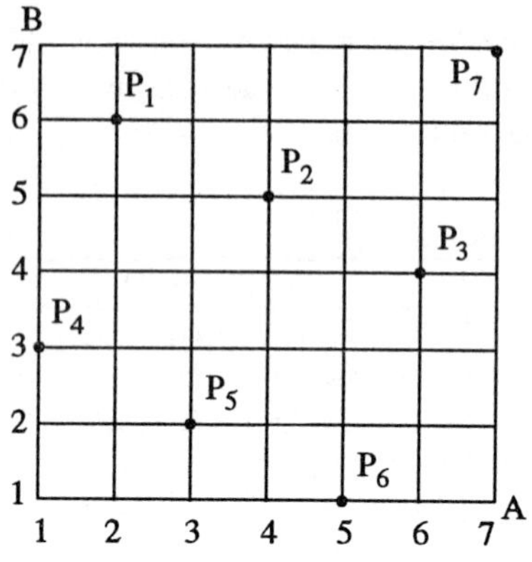

Figure 5.1

two $v_{jk}(\boldsymbol{A})$ should be small. Fang and Zhang (1992) pointed that $U_3(\boldsymbol{A})$ is usually better than $U_2(\boldsymbol{A})$.

(b) Maximum symmetric difference principle

We use Figure 5.1 to illustrate the geometrical meaning of symmetric difference of a design matrix $\boldsymbol{A}$. This is a plot of the experimental points of $U_7(7^2)$ with two factors A and B. From the point of view of A, the sum of absolute value of differences of neighboring levels of B equals to

$$|6-3|+|2-6|+|5-2|+|1-5|+|4-1|+|7-4|=20.$$

On the other hand, from the point of view of B, the respective sum of A is

$$|3-5|+|1-3|+|6-1|+|4-6|+|2-4|+|7-2|=18.$$

The symmetric difference of these two factors is denoted by $SD = 20+18 = 38$. It can be visually seen that the experimental points are scattered uniformly if SD is bigger. Now we give the mathematical definition of **symmetric difference**.

For any natural number n, let

$$\mathcal{A}_n = \{(i_1, \cdots, i_n) : (i_1, \cdots, i_n) \text{ is a permutation of } (1, \cdots, n)\}.$$

Definition 5.3

Suppose that $\boldsymbol{x} = (i_1, \cdots, i_n)$ and $\boldsymbol{y} = (j_1, \cdots, j_n)$ belong to $\mathcal{A}_n$. Then there is a vector $(l_1, \cdots, l_n) \in \mathcal{A}_n$ such that $i_{l_1} = 1, i_{l_2} = 2, \cdots, i_{l_n} = n$. We call

$$ND(\boldsymbol{x}, \boldsymbol{y}) = \sum_{k=1}^{n-1} |j_{l_{k+1}} - j_{l_k}| \tag{5.4.6}$$

the sum of difference of $\boldsymbol{y}$ based on $\boldsymbol{x}$ at lag 1, and

$$SD(\boldsymbol{x}, \boldsymbol{y}) = ND(\boldsymbol{x}, \boldsymbol{y}) + ND(\boldsymbol{y}, \boldsymbol{x}) \tag{5.4.7}$$

is the symmetric difference of $\boldsymbol{x}$ and $\boldsymbol{y}$ at lag 1 or symmetric difference for simplicity.

Example 5.4

For the example of Figure 5.1 we have $\boldsymbol{x} = (2, 4, 6, 1, 3, 5, 7)$ and $\boldsymbol{y} = (6, 5, 4, 3, 2, 1, 7)$. Then $(l_1, \cdots, l_7) = (4, 1, 5, 2, 6, 3, 7)$ because $i_4 = 1, i_1 = 2, \cdots, i_7 = 7$. We have $\boldsymbol{y} \longrightarrow \boldsymbol{y}^* = (3, 6, 2, 5, 1, 4, 7)$ and by (5.4.6)

$$ND(\boldsymbol{x}, \boldsymbol{y}) = |6 - 3| + |2 - 6| + \cdots + |7 - 4| = 20.$$

Similarly $ND(\boldsymbol{y}, \boldsymbol{x}) = 18$, and $SD(\boldsymbol{x}, \boldsymbol{y}) = 38$.

The reason for introducing the vector $(l_1, \cdots, l_n)$ is to order the experimental points according to their x coordinates. Hence $ND(\boldsymbol{x}, \boldsymbol{y})$ equals to the sum of absolute values of the differences of y-coordinates at lag 1, and $ND(\boldsymbol{y}, \boldsymbol{x})$ has the similar meaning. If n points are required to be scattered uniformly, then their symmetric difference should be maximized. This is the so-called **maximum symmetric difference (MSD) principle**. For the high-dimensional case, the symmetric difference is defined as follows:

Definition 5.4
If $\boldsymbol{x}_i \in \mathcal{A}_n$, $i = 1, \cdots, s$, then the symmetric difference of $\boldsymbol{x}_1, \cdots, \boldsymbol{x}_s$ is defined by

$$SD(\boldsymbol{x}_1, \cdots, \boldsymbol{x}_s) = \sum_{1 \leq i < j \leq s} SD(\boldsymbol{x}_i, \boldsymbol{x}_j). \tag{5.4.8}$$

Using MSD, we can find the integer b in (5.2.10) and then the corresponding UD. In order to simplify the computation of SD, we have the following properties of SD:

(i) $SD(\boldsymbol{x}, \boldsymbol{y}) = SD(\boldsymbol{y}, \boldsymbol{x}) = SD(\boldsymbol{x}, n+1-\boldsymbol{y})$ for any $\boldsymbol{x}$, $\boldsymbol{y} \in \mathcal{A}_n$, where $n+1-\boldsymbol{y} = (n+1-j_1, \cdots, n+1-j_n)$ and $\boldsymbol{y} = (j_1, \cdots, j_n)$.

We omit the proof, since it can be derived directly from the definition.

If a is an integer with $0 < a < n$, then by the definition of *glp* set we have

$$j_1 = a, j_2 = 2a, \cdots, j_n = na (\bmod n).$$

Denote $\boldsymbol{y}_a = (j_1, j_2, \cdots, j_n)$. Similarly $\boldsymbol{x}_b = (b, 2b, \cdots, nb) (\bmod n)$, in particular $\boldsymbol{x}_1 = (1, 2, \cdots, n)$. The following properties on symmetric difference are due to Fang and Zhen (1992).

(ii) $SD(\boldsymbol{x}_1, \boldsymbol{y}_a) = SD(\boldsymbol{x}_b, \boldsymbol{y}_1)$, where $b \equiv a^{-1} (\bmod n)$.

(iii) $SD(\boldsymbol{x}_a, \boldsymbol{y}_c) = SD(\boldsymbol{x}_1, \boldsymbol{y}_{a^{-1}c})$.

(iv)

$$SD(\boldsymbol{x}_1, \boldsymbol{y}_a) = (2a-1)(n-a) + (2b-1)(n-b), \tag{5.4.9}$$

where $b \equiv a^{-1} (\bmod n)$.

Formula (5.4.9) provides a computational formula for the symmetric difference. In Example 5.3, we have $\boldsymbol{x} = \boldsymbol{x}_2$, $\boldsymbol{y} = \boldsymbol{y}_6$, $n = 7$ and

$$\begin{aligned} SD(\boldsymbol{x}, \boldsymbol{y}) &= SD(\boldsymbol{x}_2, \boldsymbol{y}_6) = SD(\boldsymbol{x}_1, \boldsymbol{y}_{4\times 6}) \\ &= SD(\boldsymbol{x}_1, \boldsymbol{y}_3) \\ &= (2 \times 3 - 1)(7-3) + (2 \times 5 - 1)(7-5) \\ &= 20 + 18 = 38, \end{aligned}$$

since $2^{-1} = 4(\bmod 7)$ and $3^{-1} = 5(\bmod 7)$. This coincides with the previous conclusion.

(v) If $(a_m, n) = 1$, $m = 1, \cdots, s$, then for any $(j_1, \cdots, j_s) \in \mathcal{A}_n$ we have

$$\begin{aligned} SD(\boldsymbol{x}_{a_1}, \cdots, \boldsymbol{x}_{a_s}) &= SD(\boldsymbol{x}_{a_{j_1}}, \cdots, \boldsymbol{x}_{a_{js}}) \\ &= SD(\boldsymbol{x}_1, \boldsymbol{x}_{a_1^{-1}a_2}, \cdots, \boldsymbol{x}_{a_1^{-1}a_s}). \end{aligned} \tag{5.4.10}$$

Thus we can assume always that $h_1 = 1$ which coincides with the result obtained by discrepancy criterion. The computational results obtained by Fang and Zhen (1992) show that most of 'best' vectors $\boldsymbol{h}$ given by MSD and discrepancy criteria are coincident when $4 \leq n \leq 31$. The MSD is still elementary. However the symmetric difference of a multi-dimensional case is based on the two-dimensional symmetric difference ((5.4.8)) and so the solution obtained by multi-dimensional symmetric difference may be not the best. Table 5.10 shows the differences of the results given by the methods of MSD and discrepancy. We see that the results are all coincident when $n \leq 11$ and the difference increases as n increases.

5.4.2 Statistical method

Suppose that the design matrix $\boldsymbol{Q}$ or the induced matrix $\boldsymbol{X}$ is produced by a *glp* set. If $\boldsymbol{X}$ is regarded as a matrix of s-dimensional observations, then the sample mean and sample covariance matrix are

$$\bar{\boldsymbol{x}} = \frac{1}{2}\mathbf{1}_n \qquad \text{and} \qquad \boldsymbol{S} = \boldsymbol{S}(\boldsymbol{X}) = \frac{1}{n}\boldsymbol{X}'\boldsymbol{D}_n\boldsymbol{X}, \tag{5.4.11}$$

where

$$\boldsymbol{D}_n = \boldsymbol{I}_n - \frac{1}{n}\mathbf{1}_n\mathbf{1}_n' \qquad \text{and} \qquad \mathbf{1}_n = (1, \cdots, 1)' : n \times 1.$$

Table 5.10 *Comparison between the two methods*

		Maximum Symmetric Difference	Discrepancy
n	s	$h_2, h_3, \cdots, h_s$	$h_2, h_3, \cdots, h_s$
13	3	4,5	3,4
	4	3,5,11	6,8,11
	5	3,4,5,11	6,8,9,10
15	3	2,4	3,4
	4	2,4,6	3,4,7
	5	2,3,4,6	2,3,4,7
19	4	7,8,18	6,8,14
	5	7,8,11,18	6,8,14,17
	6	7,8,11,12,18	6,8,10,14,17
21	5	3,6,8,9	3,6,9,11
	7	2,3,6,8,9,11	3,6,8,9,11,12
23	2	13	7
	3	8,13	13,17
25	2	11	11
	4	5,9,13	5,9,17
	8	4,5,8,9,13,16,20	2,7,10,12,15,16,17
27	2	7	6
	3	7,13	6,7
	4	6,7,18	6,14,15
	5	5,9,11,13	6,14,15,17
	6	6,7,12,13,18	6,10,14,15,17
	8	6,7,8,10,14,15,17	3,5,7,9,11,13,17
	9	3,6,7,8,10,14,15,17	3,5,7,9,11,13,15,17
29	3	16,24	9,23
	6	13,15,19,21,24	7,16,23,24,25
	7	13,15,19,21,22,24	7,16,20,23,24,25
	8	12,13,15,19,21,22,24	6,7,8,13,17,19,27
	9	12,13,15,19,21,22,24,25	6,7,8,13,17,19,20,27
	10	11,12,13,15,19,21,22,24,25	6,7,8,13,15,17,19,20,27
	11	6,11,12,13,15,19,21,22,24,25	4,6,7,8,13,15,17,19,20,27
	12	3,5,8,12,13,14,18,19,20,22,23	3,4,6,7,8,13,15,17,19,20,27

Let $\lambda_1 \le \lambda_2 \le \cdots \le \lambda_s$ be the eigenvalues of $\boldsymbol{S}$. It can be shown that (Exercise 5.7)

$$\sum_{i=1}^{s} \lambda_i = \mathrm{tr}(\boldsymbol{S}) = \frac{s}{12}\left(1 - \frac{1}{n^2}\right).$$

All the criteria in optimal regression design are functions of $\{\lambda_i\}$, such as min, max, mean, gmean (geometric mean) and variance. This inspires us to generalize the criteria in optimal regression design as follows:

Let $ID = \{e_i, e_i > 0, i = 1, \cdots, m\}$ be an induced set of $\boldsymbol{Q}$ (or $\boldsymbol{X}$) which expresses the uniformity of $\boldsymbol{Q}$, for example, $ID = \{\lambda_i, i = 1, \cdots, s\}$. Consider the ID-functions:

$$\begin{aligned} ID\mathrm{min}(\boldsymbol{Q}) &= \min_{1\le i\le m} e_i, \\ ID\mathrm{max}(\boldsymbol{Q}) &= \max_{1\le i\le m} e_i, \\ ID\mathrm{mean}(\boldsymbol{Q}) &= \frac{1}{m}\sum_{i=1}^{m} e_i = \bar{e}, \\ ID\mathrm{gmean}(\boldsymbol{Q}) &= \left(\prod_{1}^{m} e_i\right)^{1/m}, \\ ID\mathrm{var}(\boldsymbol{Q}) &= \frac{1}{m}\sum_{i=1}^{m}(e_i - \bar{e})^2. \end{aligned} \tag{5.4.12}$$

The set ID and ID-functions of $\boldsymbol{Q}$ should satisfy the following two conditions:

(a) $ID(\boldsymbol{Q}_1) = ID(\boldsymbol{Q}_2)$ provided $\boldsymbol{Q}_1 \approx \boldsymbol{Q}_2$ (section 5.2). That is, if $\boldsymbol{Q}_1$ and $\boldsymbol{Q}_2$ are equivalent, then they have the same induced set.

(b) Let $ID_f(\boldsymbol{Q})$ be an ID-function on ID set. Then $ID_f(\boldsymbol{Q}_1) = ID_f(\boldsymbol{Q}_2)$ provided $\boldsymbol{Q}_1 \approx \boldsymbol{Q}_2$.

These two properties are clearly necessary for the measurement of uniformity. When $ID = \{\lambda_i\}$ and ID_f is any function in (5.4.12). The properties (a) and (b) are satisfied. Fang and Zhang (1992) proposed several possible induced sets and obtained some of their properties.

The design matrix $\boldsymbol{Q}^*$ is called a uniform design, if it satisfies one of the following conditions:

$$\begin{aligned} ID_{\min}(\boldsymbol{Q}^*) &= \max_{\boldsymbol{Q}} ID_{\min}(\boldsymbol{Q}), \\ ID_{\max}(\boldsymbol{Q}^*) &= \min_{\boldsymbol{Q}} ID_{\max}(\boldsymbol{Q}), \\ ID_{\text{mean}}(\boldsymbol{Q}^*) &= \min_{\boldsymbol{Q}} ID_{\text{mean}}(\boldsymbol{Q}), \\ ID_{\text{gmean}}(\boldsymbol{Q}^*) &= \max_{\boldsymbol{Q}} ID_{\text{gmean}}(\boldsymbol{Q}), \\ ID_{\text{var}}(\boldsymbol{Q}^*) &= \min_{\boldsymbol{Q}} ID_{\text{var}}(\boldsymbol{Q}), \end{aligned} \tag{5.4.13}$$

where ID is not a constant. Using ID_{gmean}, Ding (1986) gives the corresponding UD which is the so-called D-optimal design in the theory of optimal design. The A-optimal criterion is also used by her with the ID-function:

$$ID_A(\boldsymbol{Q}) = \sum_{i=1}^{s} \lambda_i^{-1}. \tag{5.4.14}$$

Her numerical results show that the most uniform designs given by these two methods are coincident with those by approximate discrepancy (5.2.8), but differences still appeared in some cases. Table 5.11 lists only the distinct results for $n \leq 19$. We see that the solutions are distinct as given by A-optimality and approximate discrepancy in many cases, for instance, $(n, s) = (7, 4)$, (11, 6), (13, 7), (15, 5), (17, 9), (19, 10) etc, and that the solutions in many cases are the same by the A-optimality and D-optimality principles but distinct with the results obtained by discrepancy criterion, for example, $(n, s) = (5, 2)$, (11, 2), (13, 2), (13, 3), (17, 2), (19, 2), (19, 3) etc.

We have introduced many measurements for uniformity. How should we to compare them? This is really a difficult problem. Fang and Zhang (1992) give comparisons among 4 induced sets and about 20 measurements of uniformity and discrepancy (exact value) for the cases $s \leq 5$ and $n \leq 31$. They draw the following conclusion: ID_{var} has the best resolving power for uniformity and is the best one among the 5 ID-functions of (5.4.12) while the resolving power of (5.2.8) for uniformity is poor; and $ID = \{\lambda_i\}$ is the best one among the 4 induced sets. Hence they suggest to use

$$ID\text{var}(\boldsymbol{Q}) = \frac{1}{n}\sum_{i=1}^{s}(\lambda_i - \bar{\lambda})^2, \quad \bar{\lambda} = \frac{1}{s}\sum_{i=1}^{s}\lambda_i$$

or

$$U_4(\boldsymbol{Q}) = \sum_{i=1}^{s}(\lambda_i - \bar{\lambda})^2 \tag{5.4.15}$$

as the measurement for uniformity when n is small. If $\boldsymbol{Q}$ is given by the vector of the form (5.2.10), we can find the best b for $n \leq 31$ and $s \leq n/2+1$ by the use of U_4 criterion. Their results are listed in Table 5.12, and we recommend it to the reader who wants to use the UD.

Table 5.11 *Generating Vectors* ($h_1 = 1$)

n	s	A-optimality	D-optimality	Approximate Discrepancy
5	2	4	4	2
7	4	2,4,6	2,3,6 (or 2,4,6)	2,3,6
9	4	2,4,5	2,3,5	2,3,5
11	2	5	5	7
11	6	2,4,5,8,10	2,4,5,8,10 (or 2,3,5,7,10)	2,3,5,7,10
13	2	6	6	5
13	3	6,10	6,10	3,4
13	7	2,3,4,6,8,12	2,3,4,6,8,12 (or 2,6,8,9,10,12)	2,6,8,9,10,12
15	5	2,3,5,7	2,3,5,7 (or 2,3,4,7)	2,3,4,7
17	2	11	11	10
17	9	3,5,9,10, 11,13,15,16	3,5,9,10,11,13,15,16 (or 4,5,6,9,10,14,15,16)	4,5,6,9, 10,14,15,16
19	2	6	6	8
19	3	6,14	6,14	7,8
19	10	3,5,6,10, 11,12,15,17,18	3,5,6,10,11,12,15,17,18 (or 3,4,6,7,8,10,14,17,18)	3,4,6,7,8, 10,14,17,18

Table 5.12 *The best b for UD obtained by using U_4 criterion*

	number of factors														
n	2	3	4	5	6	7	8	9	10	11	12	13	14	15	16
5	2	2													
7	3	3	3												
9	4	4	2												
11	5	7	7	7	7										
13	6	6	6	6	6	6									
15	7	7	7												
17	11	10	10	10	10	10	10								
19	6	14	14	14	14	14	14	14	14						
21	5	5	10	10	10	10	10	10	10	10					
23	9	17	15	17	17	20	20	11	11	11	11				
25	11	11	11	11	8	8	8	8	8	8					
27	16	20	20	20	20	20	20	16							
29	7	13	23	25	25	7	11	11	11	14	14	14	14	19	
31	14	22	12	12	12	12	12	12	12	22	22	22	22	22	22

An alternative measurement for uniformity is the mean square error (MSE) which has been disussed in Chapter 4. With NTLBG algorithm, the generating vectors of uniform design are listed in Appendix A.2.

5.5 Uniform designs of experiments with mixtures

Experiments with mixtures, i.e. there are s factors $x_1, \cdots, x_s$ such that $x_i \geq 0$, $i = 1, \cdots, s$ and $x_1 + \cdots + x_s = 1$, often appear in the design of chemical and metallurgical products, for example food formulations. In the last four decades, a lot of work which appeared in the statistical literature proposed many kinds of designs. Scheffè (1958) introduced the **simplex-lattice designs** and the corresponding polynomial models. Later he (1963) introduced an alternative design, the **simplex-centroid design**, to the general simplex-lattice. Cornell(1975) gave a suggestion of **axial design** and he (1973, 1981) gave a comprehensive review of nearly all the statistical articles on designs of experiments with mixtures and data analysis.

First we shall introduce the simplex-lattice, simplex-centroid and axial designs.

(a) Simplex-lattice design

Suppose that the mixture has s components. Let m be a positive integer and suppose that each component takes $(m + 1)$ equally spaced places from 0 to 1, i.e.

$$x_i = 0, 1/m, 2/m, \cdots, 1, \quad \text{for } i = 1, \cdots, s.$$

For example, when $s = 3$, we have

$m = 1$: 3 design points (1, 0, 0), (0, 1, 0), (0, 0, 1);

$m = 2$: 6 design points (1, 0, 0), (0, 1, 0), (0, 0, 1), (1/2, 1/2, 0), (1/2, 0, 1/2), (0, 1/2, 1/2);

$m = 3$: 10 design points (1, 0, 0), (0, 1, 0), (0, 0, 1), (1/3, 2/3, 0), (1/3, 0, 2/3), (0, 1/3, 2/3), (2/3, 1/3, 0), (2/3, 0, 1/3), (0, 2/3, 1/3) (1/3, 1/3, 1/3).

We would use $\{s, m\}$-simplex-lattice to represent this design which has $\binom{s+m-1}{m}$ design points.

(b) Simplex-centroid design

In an s-component simplex-centroid design, the design points form s pure blends, $\binom{s}{2}$ binary mixtures, $\binom{s}{3}$ ternary mixtures and so on, with the finally overall centroid point $(1/s, \cdots, 1/s)$, the s-ary mixture. So the total number of design points is $2^s - 1$. For example, when $s = 3$, the design points are (1, 0, 0), (0, 1, 0), (0, 0, 1), (1/2, 1/2, 0), (1/2, 0, 1/2), (0, 1/2, 1/2) and (1/3, 1/3, 1/3).

(c) Axial design

The line segment joining a vertex of the simplex T_s with its centroid is called an **axis**. Let d be a positive number such that $0 < d < (s-1)/s$. The experimental points of the axial design are s points on s axes such that each point to the centroid has the same distance d.

Figure 5.2 show the above three designs in $s = 3$. The reader will find that there are at least two problems with these designs:

(i) The experimental points are not scattered very uniformly on the experimental domain T_s.

(ii) Some experimental points are at the boundary of the experimental domain. The experiment is often impossible for many

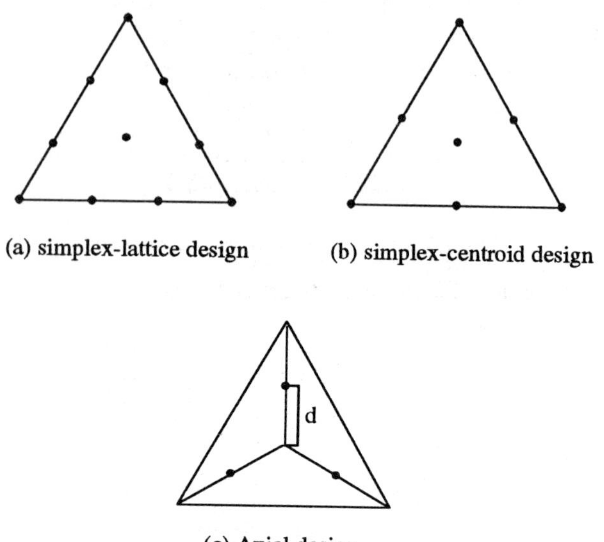

(a) simplex-lattice design (b) simplex-centroid design

(c) Axial design

Figure 5.2

chemical experiments if a component has zero value. So the boundary experimental points are meaningless in that case.

We then need an experimental design, where the experimental points are uniformly scattered on T_s. This is just the idea of UD, and the corresponding design is called to be **uniform design of experiments with mixtures** (UDEM). We shall introduce in this section three kinds of such designs.

5.5.1 Scheffè type design

From Figure 5.2, we see that the distributions of experimental points of the simplex centroid design proposed by Scheffè look like uniformly scattered on T_s, but too many points are located at the boundary of the experimental domain. Hence every experimental point at the boundary of T_s should be removed to an inner point of T_s. A natural way is to keep the pattern of the original design and to contract the boundary points toward the centroid of T_s. We now illustrate the method with an example of simplex-centroid design with $s = 3$. Suppose that the original design is shown by Figure 5.2(b). Let a be a number which will be determined later.

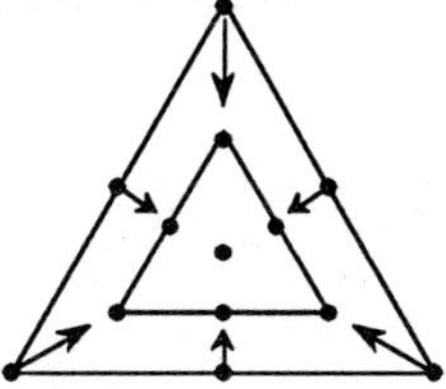

Figure 5.3 Contraction of experimental points

Then we move the three vertices as follows:

$$\begin{aligned}
(1,0,0) &\longrightarrow \left(1-\frac{1}{a},\frac{1}{2a},\frac{1}{2a}\right),\\
(0,1,0) &\longrightarrow \left(\frac{1}{2a},1-\frac{1}{a},\frac{1}{2a}\right),\\
(0,0,1) &\longrightarrow \left(\frac{1}{2a},\frac{1}{2a},1-\frac{1}{a}\right),
\end{aligned} \tag{5.5.1}$$

and the locations of other four points are

$$\begin{aligned}
\left(\frac{1}{2},\frac{1}{2},0\right) &\longrightarrow \left(\frac{1}{2}-\frac{1}{4a},\frac{1}{2}-\frac{1}{4a},\frac{1}{2a}\right),\\
\left(\frac{1}{2},0,\frac{1}{2}\right) &\longrightarrow \left(\frac{1}{2}-\frac{1}{4a},\frac{1}{2a},\frac{1}{2}-\frac{1}{4a}\right),\\
\left(0,\frac{1}{2},\frac{1}{2}\right) &\longrightarrow \left(\frac{1}{2a},\frac{1}{2}-\frac{1}{4a},\frac{1}{2}-\frac{1}{4a}\right),\\
\left(\frac{1}{3},\frac{1}{3},\frac{1}{3}\right) &\longrightarrow \left(\frac{1}{3},\frac{1}{3},\frac{1}{3}\right)
\end{aligned}$$

as shown by Figure 5.3. We wish to find a suitable number a such that these 7 points are scattered uniformly on T_s.

In Chapter 4, we have proposed two measurements, the F-discrepancy method and MSE method to measure the uniformity of a set. Since the amount of computation of the MSE method is comparatively small, we suggest the use of MSE method to measure the uniformity of experimental points. For convenience, we restate the main steps of computing MSE as follows (Chapter 4):

Let $\boldsymbol{x}_1, \cdots, \boldsymbol{x}_n$ be a set of n experimental points on T_s.

(a) Generate an NT-net of N points $\boldsymbol{y}_1, \cdots, \boldsymbol{y}_N$ by (1.5.33) on T_s. These N points are called to be a **training sample**.

(b) The MSE of $\{\boldsymbol{x}_i\}$ is then approximately equal to

$$\begin{aligned}\text{MSE}\{\boldsymbol{x}_i\} &\cong \frac{v(T_s)}{sN}\sum_{i=1}^{N}\min_{1\le j\le n} d^2(\boldsymbol{x}_j,\boldsymbol{y}_i)\frac{1}{v(T_s)} \\ &= \frac{1}{sN}\sum_{i=1}^{N}\min_{1\le j\le n} d^2(\boldsymbol{x}_j,\boldsymbol{y}_i).\end{aligned} \tag{5.5.2}$$

Note that the definition of MSE is slightly different to the previous one by a factor s^{-1} ((1.4.15)) which is used to make a distinction for dimensions.

Suppose that $\boldsymbol{x}_1,\cdots,\boldsymbol{x}_7$ are 7 points obtained by the simplex-centroid design with $s = 3$. Take $N = 610, h_1 = 1$ and $h_2 = 377$. Using the *glp* set generated by $(N; h_1, h_2)$, (1.5.33) and (5.5.2), we obtain that the MSE of $\{\boldsymbol{x}_i\}$ is 0.05553. Let $\boldsymbol{x}_1^*,\cdots,\boldsymbol{x}_7^*$ be the points obtained by contraction of these 7 points. Then the MSE of $\{\boldsymbol{x}^*\}$ depends on the value of a. We find that the minimum of MSE $\{\boldsymbol{x}^*\}$ is 0.02296 with the corresponding $a = 3.761$. This is a great improvement of the original result 0.05553. The modified design is

$$\begin{aligned}
&\boldsymbol{x}_1^* = (0.734, 0.133, 0.133), && \boldsymbol{x}_2^* = (0.133, 0.734, 0.133),\\
&\boldsymbol{x}_3^* = (0.133, 0.133, 0.734), && \boldsymbol{x}_4^* = (0.4335, 0.4335, 0.133),\\
&\boldsymbol{x}_5^* = (0.4335, 0.133, 0.4335), && \boldsymbol{x}_6^* = (0.133, 0.4335, 0.4335),\\
&\boldsymbol{x}_7^* = (1/3, 1/3, 1/3). &&
\end{aligned}$$

The same method can be applied to improve the results obtained by simplex-lattice design. For instance, the MSE of designs {3, 3} is 0.03087 by (5.5.1). After the contraction, the corresponding MSE is 0.01568 with the corresponding $a = 4.836$. For convenience, the above results are listed in Table 5.13, and the distributions of experimental points are shown by Figure 5.4. Without essential difficulty, the reader can obtain the modified design for the general case.

Table 5.13 *Modified Scheffè design*

Design	original MSE	improved MSE	a
Simplex-centroid	0.05553	0.02296	3.761
Simplex lattice {3,3}	0.03087	0.01568	4.836

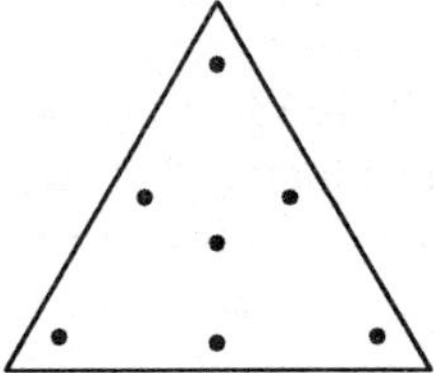
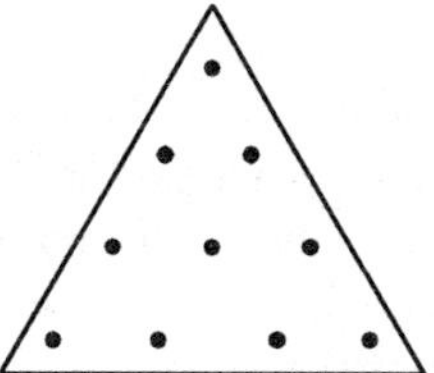

Figure 5.4 Scheffè type design

We may obtain the number d such that the corresponding axial design has the minimum MSE.

5.5.2 *Uniform design - Discrepancy criterion*

The modified Scheffè designs have no boundary point but the distribution of experimental points is often not so uniform as shown by Figure 5.4. The other disadvantage is that the numbers of experimental points in two Scheffè designs are restricted, i.e. the number of experimental points is of the type $\binom{s+m-1}{m}$ in the simplex lattice design, and of the form $2^s - 1$ in the simplex-centroid design. The number of experimental points of the axial design is s. In most chemical or industrial experiments, we meet the requirement that the number of experiments is considerably flexible, and therefore Wang and Fang (1990b) proposed a so-called uniform design for experiments with mixtures (UDEM). The main idea of UDEM is to give the design according to an NT-net on T_s (section 1.5).

Suppose that there are s factors $X_1, \cdots, X_s$ and that the domain for experiments is T_s. If there are n experiments to be arranged, then the design is as follows:

(a) For a given generating vector $(n; h_1, \cdots, h_{s-1})$ from Appendix A we get a *glp* set $\{\boldsymbol{c}_k = (c_{k1}, \cdots, c_{k,s-1})', \ k = 1, \cdots, n\}$ with low discrepancy.

(b) Set

$$\begin{aligned} x_{ki} &= \prod_{j=1}^{i-1} c_{kj}^{\frac{1}{s-j}} (1 - c_{ki}^{\frac{1}{s-i}}), \ i = 1, \cdots, s-1 \\ x_{ks} &= \prod_{j=1}^{s-1} c_{kj}^{\frac{1}{s-j}}, \ k = 1, \cdots, n. \end{aligned} \tag{5.5.3}$$

Then $\{\boldsymbol{x}_k = (x_{k1}, \cdots, x_{ks})', \ k = 1, \cdots, n\}$ is the set of experimental points of UD. For example, we have a generating vector (11; 1, 7) for $n = 11$ and $s = 3$ (Table A.13 in Appendix A). Table 5.14 provides the processes for producing the experimental points, where the columns c_1 and c_2 are the coordinates of $\boldsymbol{c}_k = (c_{k1}, c_{k2})$ and the columns x_1, x_2, x_3 are the coordinates of experimental points defined by (5.5.3). We use $UM_n(n^s)$ to denote the so-obtained UDEM table.

Table 5.14 *Table* $UM_{11}(11^3)$

No	c_1	c_2	x_1	x_2	x_3
1	1/22	13/22	0.787	0.087	0.126
2	3/22	5/22	0.631	0.285	0.084
3	5/22	19/22	0.523	0.065	0.412
4	7/22	11/22	0.436	0.282	0.282
5	9/22	3/22	0.360	0.552	0.087
6	11/22	17/22	0.293	0.161	0.546
7	13/22	9/22	0.231	0.454	0.314
8	15/22	1/22	0.174	0.788	0.038
9	17/22	15/22	0.121	0.280	0.599
10	19/22	7/22	0.071	0.634	0.296
11	21/22	21/22	0.023	0.044	0.993

Example 5.5

Consider a regression model

$$Y = \beta_0 + \sum_{i=1}^{s} \beta_i x_i + \sum_{i=1}^{s} \sum_{j=i}^{s} \beta_{ij} x_i x_j + \epsilon,$$

where ϵ stands for a random error. Since $x_1 + \cdots + x_s = 1$, it can be reduced to the form

$$Y = \beta_0 + \sum_{i=1}^{s-1} \beta_i x_i + \sum_{i=1}^{s-1} \sum_{j=i}^{s-1} x_i x_j + \epsilon,$$

where $\epsilon \sim N(0, \sigma^2)$. Consider the following special model

$$Y = x_1 + x_2 - 3x_1^2 - 3x_2^2 + x_1 x_2 + \epsilon \qquad (5.5.4)$$

and take $\sigma = 0.005$. By simulation and the use of $UM_{17}(17^3)$, we get the data in Table 5.15 for $n = 17$ and $s = 3$. By the least squares method, the regression model is

$$\begin{aligned} EY = & -0.0376 + 1.1162x_1 + 1.1197x_2 \\ & -3.0842x_1^2 - 3.0880x_2^2 + 0.8336x_1x_2 \end{aligned} \tag{5.5.5}$$

which is close to the model (5.5.4). The multiple correlation coefficient of the equation (5.5.5) is $R = 0.9999$ and the residual standard deviation is $\hat{\sigma} = 0.0054$ which is close to $\sigma = 0.005$.

When σ is large, we can not expect such excellent results. For example, consider the model

$$Y = 10 + x_1 - 3x_1^2 - 3x_2^2 + x_1x_2 + \epsilon, \tag{5.5.6}$$

where $\epsilon \sim N(0, \sigma^2)$ with $\sigma = 0.3$. The data listed in Table 5.16 are generated by the use of $UM_{15}(15^3)$ and simulation. The corresponding regression equation now becomes

Table 5.15 *Design* $UM_{17}(17^3)$

No	x_1	x_2	Y
1	0.829	0.076	−1.100
2	0.703	0.253	−0.541
3	0.617	0.102	−0.391
4	0.546	0.307	−0.157
5	0.486	0.045	−0.160
6	0.431	0.284	0.038
7	0.382	0.564	−0.230
8	0.336	0.215	0.146
9	0.293	0.520	−0.103
10	0.252	0.110	0.163
11	0.214	0.439	0.031
12	0.178	0.798	−0.889
13	0.143	0.328	0.134
14	0.109	0.708	−0.644
15	0.076	0.190	0.155
16	0.045	0.590	−0.388
17	0.015	0.029	0.000

Table 5.16 *Design* $UM_{15}(15^3)$

No	x_1	x_2	Y
1	0.817	0.055	8.508
2	0.684	0.179	9.464
3	0.592	0.340	9.935
4	0.517	0.048	9.400
5	0.452	0.201	10.680
6	0.394	0.384	9.748
7	0.342	0.592	9.698
8	0.293	0.118	10.238
9	0.247	0.326	9.809
10	0.204	0.557	9.732
11	0.163	0.809	8.933
12	0.124	0.204	9.971
13	0.087	0.456	9.881
14	0.051	0.727	8.892
15	0.017	0.033	10.139

$$EY = 10.0908 + 0.7972x_1 - 3.4542x_1^2 - 2.6733x_2^2 + 0.8884x_1x_2$$

with $R = 0.9003$ and $\hat{\sigma} = 0.2891$. Note that this equation has some departure from the model (5.5.6), because there are high correlations between x_1 and x_1^2 and the value of σ is large.

5.5.3 *Uniform design - MSE criterion*

The design of experiments with mixtures is in fact the choice of rep-points for $U(T_s)$. There are two methods, the F-discrepancy criterion and the MSE criterion, for finding the rep-points of a distribution. We have illustrated in detail in sections 4.5 and 4.6 the method for finding the rep-points of a multivariate distribution by the MSE principle. The following is the NTLBG algorithm (section 4.5) for finding rep-points for $U(T_s)$ by the MSE criterion.

(a) Generate a training sample $\mathcal{P} = \{\boldsymbol{y}_j,\ j = 1, \cdots, N\}$ for the uniform distribution $U(T_s)$ by the NTM.

(b) Let $k = 0$ and put in initial points $\boldsymbol{X}(k) = \{\boldsymbol{x}_1^{(k)}, \cdots, \boldsymbol{x}_n^{(k)}\}$.

(c) For $j = 1, \cdots, N$, put $\boldsymbol{y}_j$ into the group $t (1 \leq t \leq n)$ and the corresponding index set is denoted by $G_t^{(k)}$, if $\| \boldsymbol{y}_j^{(k)} - \boldsymbol{x}_t^{(k)} \|$ attains its minimum among $\| \boldsymbol{y}_j^{(k)} - \boldsymbol{x}_i^{(k)} \| \ (i = 1, \cdots, n)$.

(d) Calculate the sample mean

$$\boldsymbol{x}_i^{(k+1)} = \frac{1}{N_i} \sum_{j \in G_i^{(k)}} \boldsymbol{y}_j^{(k)}, \ i = 1, \cdots, n,$$

for each group $G_i^{(k)}$, where N_i is the cardinal number of $G_i^{(k)}$.

(e) Set $\boldsymbol{X}(k+1) = (\boldsymbol{x}_1^{(k+1)}, \cdots, \boldsymbol{x}_n^{(k+1)})$.

(f) If $\boldsymbol{X}(k) \neq \boldsymbol{X}(k+1)$, then go to step (c) by using $k+1$ instead of k. Otherwise the process is terminated.

Note that when N is large the termination criterion $\boldsymbol{X}(k) = \boldsymbol{X}(k+1)$ often imposes a heavy computational burden. Therefore we make the following modification: Let ϵ be a preassigned positive number. If

$$\| \boldsymbol{x}_j^{(k)} - \boldsymbol{x}_j^{(k+1)} \| < \epsilon, \ j = 1, \cdots, n,$$

then the iteration process is terminated.

There are several ways of choosing initial points. We only consider two methods. One is to use the experimental points obtained by the modified Scheffè design and the other is to use the experimental points given by UDEM (section 5.5.2). However the latter method may be applied to any integer n, the number of experiments, and the former is restricted to special n only. For illustration we applied the NTLBG algorithm to the two designs: the simplex-centroid and simplex lattice $\{3.3\}$ using the two kinds of initial points and the corresponding results are given in Table 5.15. We conclude that the MSE values of UDEM are much better than those from Scheffè's designs. This means that the uniformities of Scheffè's designs are not very good. After the use of NTLBG algorithm, the uniformities of these two kinds of initial points are all enhanced, and the result related to UDEM is still better.

Table 5.15 *Comparison on MSE values*

Designs	n	MSE of original initial points	MSE of modified Scheffè's designs	MSE of points after NTLBG
Simplex-centroid	7	0.05553	0.02296	0.02266
Simplex lattice {3,3}	10	0.30870	0.01568	0.01457
UDEM	7	0.02720		0.02180
UDEM	10	0.01876		0.01457

When the MSE criterion is applied, we always recommend the UDEM to choose an initial point, and then to find the so-called USM table. For example, Table $USM_n(n^s)$ represents a design of n experimental points, where each point has s components. Tables 5.16 and 5.17 give the experimental design tables $USM_7(7^3)$ and $USM_{10}(10^3)$, and the respective MSEs of experimental points are 0.02180 and 0.01457 (Table 5.15).

Table 5.16 $USM_7(7^3)$

0.7440	0.1434	0.1125
0.5201	0.1323	0.3475
0.4112	0.4381	0.1507
0.2941	0.1803	0.5256
0.1347	0.7403	0.1251
0.1226	0.4648	0.4126
0.1048	0.1344	0.7608

Table 5.17 $USM_{10}(10^3)$

0.7930	0.0926	0.1144
0.5964	0.3019	0.1017
0.5487	0.1051	0.3462
0.3682	0.3250	0.3068
0.3573	0.5360	0.1067
0.3132	0.1169	0.5699
0.1299	0.3633	0.5068
0.1170	0.7890	0.0940
0.0956	0.1234	0.7810
0.1125	0.5875	0.3000

5.6 Design of computer experiments

Numerical methods have been used for years to provide approximate solutions to fluid flow problems that defy analytical solutions because of their complexity. Suppose we have some device or process in which the behavior depends on a random vector $\boldsymbol{x} = (X_1, \cdots, X_s)$. For example, consider an electrical circuit where the performance depends on a number of quantities that vary from circuit to circuit in some random fashion. A mathematical model for the device is developed (e.g. a system of differential equations) from which we can simulate the behavior of the device on a computer. Very often, we want to estimate the expected value of some measure of performance of the device, given by the function $h(\boldsymbol{x})$. Then we want to estimate the expected value $E(h(\boldsymbol{x}))$. If we employ the sample mean method, $E(h(\boldsymbol{x}))$ can be estimated by

$$\bar{h} = \frac{1}{n} \sum_{k=1}^{n} h(\boldsymbol{x}_k), \tag{5.6.1}$$

where $\boldsymbol{x}_1, \cdots, \boldsymbol{x}_n$ are generated by Monte Carlo methods. It is known that the efficiency of (5.6.1) is not high. Therefore, many authors proposed various methods to generate $\boldsymbol{x}_1, \cdots, \boldsymbol{x}_n$ such that the corresponding $\bar{h}$ has better behavior than the simple random sampling (i.e. $\boldsymbol{x}_1, \cdots, \boldsymbol{x}_n$ i.i.d.). McKay, Beckman, and Conover (1979) suggested a method of generating $\boldsymbol{x}_1, \cdots, \boldsymbol{x}_n$ that they call **Latin hypercube sampling** (LHS). Since then many discussions on the LHS and its versions have appeared. Therefore, we can consider their paper as the start of **design of computer experiments** (DCE). A computer experiment consists of prompting computer codes with a series of inputs, executing the codes and collecting the output. An input pair along with its corresponding output is an observation. For example, in the above statement $\{\boldsymbol{x}_1, \cdots, \boldsymbol{x}_n\}$ is the input and $\bar{h}$ is the output. The design of computer experiments is a method for choosing the input such that the output has better behavior. When physical experiments are too complex or too costly to perform, computer models (or codes) based on a mathematical description of the physical system, are adopted as substitute. Therefore, there has been considerable interest in computer experiments in past fifteen years. Many authors, such as Welch (1983), Stein (1987), Ylvisaker (1987), Sacks, Schiller and Welch (1989), Welch *et al.* (1989), Draper and Lin (1990), Morris (1991), Owen (1992), Bernardo *et al.* (1992), Tang

(1992) and Kurker *et al.* (1992) have proposed many efficient methods and given extensive discussions on both designs and applications of computer experiments. The review article by Sacks *et al.* (1989) provides a clear picture in this direction.

In this section we shall not give a thorough introduction to DCE, and our discussion is confined only on the connection between LHS and NTM, because LHS plays an important role in the theory of DCE.

There are various presentations for LHS, and the following one is due to McKay, Beckman and Conover, the creators of LHS. If we wish to ensure that each of the input variables X_k has all portions of its distribution represented by input values, we can divide the range of each X_k into n strata of equal marginal probability $1/n$, and sample once from each stratum. Let this sample be $x_{kj}, j = 1, \cdots, n$. These form that X_k is the component, $k = 1, \cdots, s$, in $\boldsymbol{x}_i, i = 1, \cdots, n$. The components of the various X_k are matched at random. This method of selecting input values is an extension of quota sampling, and can be viewed as an s-dimensional extension of latin square sampling.

The more mathematical presentation of LHS is given by Stein (1987). Suppose that the c.d.f. of the random vector of parameters $\boldsymbol{x} = (X_1, \cdots, X_s)'$ is given by $F(\boldsymbol{x})$, where

$$F(\boldsymbol{x}) = \prod_{k=1}^{s} F_k(x_k), \tag{5.6.2}$$

and F_k is the c.d.f. of X_k, $k = 1, \cdots, s$. The procedure for producing a latin hypercube sample of size n is given as follows:

Step 1 Generate an $n \times s$ matrix $\boldsymbol{P} = (p_{jk})$ where each column of $\boldsymbol{P}$ is an independent random permutation of $\{1, 2, \cdots, n\}$.

Step 2 Generate an $n \times s$ matrix $\boldsymbol{U} = (u_{jk})$ where the u_{jk} are i.i.d. $U(0, 1)$ random variables independent of $\boldsymbol{P}$.

Step 3 Then $\{\boldsymbol{x}_j = (x_{j1}, \cdots, x_{js})', j = 1, \cdots, n\}$, where

$$x_{jk} = F_k^{-1}(n^{-1}(p_{jk} - 1 + u_{jk})), \tag{5.6.3}$$

is a latin hypercube sample of size n of $F(\boldsymbol{x})$.

Example 5.5

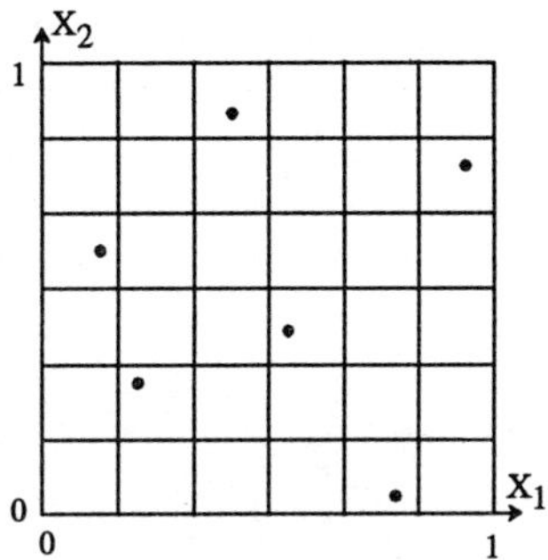

Figure 5.5 A latin hypercube sample

Let $\boldsymbol{x} \sim U(C^s)$. When $s = 2$ and $n = 6$, the LHS is shown by Figure 5.4. We see that $p_{j1}, \cdots, p_{js}$ determine in which "cell" $\boldsymbol{x}_j$ is located, and $u_{j1}, \cdots, u_{js}$ determine where in the cell $\boldsymbol{x}_j$ is located. Note that there is exactly one observation in each row and also in each column in Figure 5.5.

We now compare the variance of $\bar{h}$ in (5.6.1) which depends on whether the simple random sampling or the LHS is used. If simple random sampling is used, then the estimate $\bar{h}$ is unbiased and

$$\mathrm{Var}(\bar{h}) = \mathrm{Var}(h(\boldsymbol{x}))/n. \tag{5.6.4}$$

If LHS is used, then $\bar{h}$ is still unbiased and

$$\mathrm{Var}(\bar{h}) = \mathrm{Var}(h(\boldsymbol{x}))/n + (n-1)\mathrm{Cov}(h(\boldsymbol{x}_1), h(\boldsymbol{x}_2))/n. \tag{5.6.5}$$

Thus LHS lowers the variance if and only if $\mathrm{Cov}(h(\boldsymbol{x}_1), h(\boldsymbol{x}_2)) < 0$. McKay *et al.* (1979) showed that this covariance is negative whenever $h(\boldsymbol{x})$ is monotonic in each variable. Iman and Conover (1980) gave various exact expressions for the variance given in (5.6.5). Stein (1987) pointed out that the covariance term is asymptotically nonpositive under a weak condition as $n \to \infty$. More precisely, let

$$g_k(x_k) = \int_{R^{s-1}} h(\boldsymbol{x}) \prod_{\substack{i=1 \\ i \neq k}} dF_i(x_i) \tag{5.6.6}$$

and let $\{\boldsymbol{x}_{jn}\}$, $j = 1, \cdots, n$, $n = 1, 2, \cdots$ be an infinite triangular array of random vectors such that $\{\boldsymbol{x}_{1n}, \cdots, \boldsymbol{x}_{nn}\}$ is a LHS. Stein (1987) proved the following:

Theorem 5.2
If $Eh^2 < \infty$, then as $n \to \infty$,

$$\begin{aligned} &\text{cov}(h(\boldsymbol{x}_{1n}, h(\boldsymbol{x}_{2n})) \\ &\quad = sn^{-1}(Eh)^2 - n^{-1}\sum_{k=1}^{s}\int g_k(x)^2 dF_k(x) + o(n^{-1}). \end{aligned} \tag{5.6.7}$$

By Jensen's inequality, we have

$$\int g_k(x)^2 dF_k(x) \geq \Big(\int g_k(x) dF_k(x)\Big)^2 = (Eh)^2; \tag{5.6.8}$$

thus

$$\lim_{n\to\infty} n \;\; \text{cov}(h(\boldsymbol{x}_{1n}), h(\boldsymbol{x}_{2n})) \leq 0.$$

We see from Theorem 5.2, for any square integrable $h(\boldsymbol{x})$, LHS does at least as well asymptotically as simple random sampling. We can write Theorem 5.2 in a more interpretable form. Define

$$h_a(\boldsymbol{x}) = \sum_{k=1}^{s} g_k(\boldsymbol{x}_k) - (s-1)Eh \tag{5.6.9}$$

and

$$r(\boldsymbol{x}) = h(\boldsymbol{x}) - h_a(\boldsymbol{x}). \tag{5.6.10}$$

The function $h_a(\boldsymbol{x})$ is the best additive fit to $h(\boldsymbol{x})$; that is

$$\int r^2(\boldsymbol{x}) dF(\boldsymbol{x}) \leq \int \Big[h(\boldsymbol{x}) - \sum_{k=1}^{s} h_k(x_k)\Big]^2 dF(\boldsymbol{x}) \tag{5.6.11}$$

for any set of univariate functions $h_1, \cdots, h_s$. Consequently, as $n \to \infty$ we have

$$\text{Var}\Big(n^{-1}\sum_{j=1}^{n} h(\boldsymbol{x}_{jn})\Big) = n^{-1}\int r(\boldsymbol{x})^2 dF(\boldsymbol{x}) + o(n^{-1}). \quad (5.6.12)$$

Therefore, we essentially filter out the additive components of $h(\boldsymbol{x})$ by using LHS.

Tang (1992) proposed a so-called U-sampling and proved that the U-sampling can improve the LHS in such a way that the U-sample is capable of filtering out all the second-order interactions as well the main effects while the LHS is capable of filtering out only the main effects. Furthermore, Stein (1987) pointed out that if $Eh(\boldsymbol{x})^4 < \infty$, then as $n \to \infty$

$$\sqrt{n}\left(\frac{1}{n}\sum_{j=1}^{n} h(\boldsymbol{x}_{jn}) - Eh(\boldsymbol{x})\right) \xrightarrow{\mathcal{L}} N\Big(0, \int r(\boldsymbol{x})^2 dF(\boldsymbol{x})\Big), \quad (5.6.13)$$

where $\xrightarrow{\mathcal{L}}$ means that the convergence is in sense of distribution.

Owen (1992) suggested an alternative improvement to the LHS by employing orthogonal arrays and we denote it by OALHS. He proved that his technique can improve the LHS in such a way that the OALHS is capable of filtering out all the second-order interactions as well as the main effects, thereby reducing the variance of $\bar{h}$. For comparison we list the variance of $\bar{h}$ for the simple random sampling (SRS), LHS and OALHS as follows:

$$\begin{aligned} \text{SRS}: &\ \frac{1}{n} Var(h(\boldsymbol{x})); \\ \text{LHS}: &\ \frac{1}{n} Var(h(\boldsymbol{x})) - \frac{c}{n} + 0(\frac{1}{n}); \\ \text{OALHS}: &\ \frac{1}{n} Var(h(\boldsymbol{x})) - \frac{d}{n} + 0(n^{-3/2}), \end{aligned}$$

where c and d are positive constants. We can see that all these methods have the same convergence rate of $\bar{h}$ to $E(h(\boldsymbol{x}))$ in the sense $\text{Var}(\bar{h}) = 0(n^{-1})$.

There are certain relationships between LHS and *glp* set on C^s. Each column of the matrix $\boldsymbol{P}$ in LHS is a permutation of $\{1, 2, \cdots, n\}$, and the same conclusion holds for the matrix (q_{kj}) of *glp* set. The columns of $\boldsymbol{P}$ are produced by random choice and those of (q_{jk}) are generated by congruence multiplication. Although there is no independence property between the columns of (q_{kj}), they are approximately uncorrelated. The experimental points in LHS are obtained by a matrix $\boldsymbol{P}$ with an adjugement matrix $\boldsymbol{U}$ of uniform distribution, so that the experimental points (or input) have good randomization properties. The *glp* set is generated without any adjugement. Ylvisaker (1975) showed that the statistical optimal design problem in a two-dimensional random field is equivalent to the problem of finding optimal sampling points for bivariate integral prediction, i.e. the use of a quadrature formula. Stein (1987) has pointed out that the convergence rate of $\bar{h}$ to $E(h(\boldsymbol{x}))$ is $O(n^{-1/2})$ in the sense of probability ((5.6.13)) which is the same order as the application of simple random sample technique, and the only improvement is the variance of limiting distribution. The variances of limiting distribution of LHS and simple random sample are $\int r(\boldsymbol{x})^2 dF(\boldsymbol{x})$ and $\int h(\boldsymbol{x}) dF(\boldsymbol{x})$ respectively. Therefore from the view point of numerical integration, LHS and its improvement OALHS still belong to the variance reduction technique of Monte Carlo method. We have pointed out in Chapter 2 that the convergence rate is $O(n^{-1}\log^s n)$ for the quadrature formula generated by the *glp* set if the integrand is a function with bounded total variation. Therefore we may expect that the number-theoretic method will play an important role in the study of design of computer experiments.

Recently, Fang and Zhu (1993) propose a so-called uniformly random design (URD) with $\mathrm{Var}(\bar{h}) = O(n^{-2}\log^{2s} n)$ which significantly improves that of LHS and OALHS when s is not too large. The procedure of URD is given as follows:

Step 1 Generate a *glp* set $\{\boldsymbol{a}_k, k = 1, \cdots, n\}$ from a generating vector $(n; h_1, \cdots, h_s)$.

Step 2 Generate n i.i.d. random vectors $\boldsymbol{u}_1, \cdots, \boldsymbol{u}_n$ by the Monte Carlo method, each is uniformly distributed on $(-1, 1)^s$. Let

$$\boldsymbol{x}_k = \boldsymbol{a}_k + \boldsymbol{u}_k/2n, \quad k = 1, \cdots, n. \tag{5.6.14}$$

Then $\{\boldsymbol{x}_k, k = 1, \cdots, n\}$ is the required set of URD.

The following results show that $\bar{h}$, in general, is not an unbiased estimator of $Eh(\boldsymbol{x})$, but asymptotically unbiased with convergence rate $O(n^{-1}log^s n)$.

Theorem 5.3
Assume that $h(\boldsymbol{x})$ has bounded partial derivatives with respect to each component of $\boldsymbol{x}$, that is, for some $c > 0$

$$\left|\frac{\partial h(\boldsymbol{x})}{\partial x_i}\right| \leq c, \quad \text{for} \quad i = 1, \cdots, s. \tag{5.6.15}$$

Let $\boldsymbol{N}$ be an infinite subset of natural numbers and $\mathcal{P}_n = (\boldsymbol{a}_1^{(n)}, \cdots, \boldsymbol{a}_n^{(n)}), n \in \boldsymbol{N}$ be a sequence of set of R^s such that $D(n, \mathcal{P}_n) = O(n^{-1}log^s n)$ as $n \to \infty$. Then the estimate

$$\bar{h}_n = \frac{1}{n}\sum_{i=1}^{n} h(\boldsymbol{x}_i^{(n)}) \tag{5.6.16}$$

is asymptotically unbiased, where $\{\boldsymbol{x}_i^{(n)}, i = 1, \cdots, n\}$ is the set of URD based on $\mathcal{P}_n$. More precisely, we have

$$|\bar{h}_n - E(h(\boldsymbol{x}))| = O(n^{-1} \log^s n), \tag{5.6.17}$$

$$|E\bar{h}_n - E(h(\boldsymbol{x}))| = O(n^{-1} \log^s n), \tag{5.6.18}$$

and

$$|\bar{h}_n - E\bar{h}_n| = O(n^{-1} \log^s n). \tag{5.6.19}$$

PROOF. Let $F(\boldsymbol{x})$ be the c.d.f. of $U(c^s)$,

$$h_n^* = \frac{1}{n}\sum_{k=1}^{n} h(\boldsymbol{a}_k^{(n)}) \quad \text{and} \quad F_n(\boldsymbol{y}) = \frac{1}{n}\sum_{k=1}^{n} I\{\boldsymbol{a}_k^{(n)} \leq \boldsymbol{y}\}, \forall \boldsymbol{y} \in R^s.$$

Let $v(h)$ be the total deviation of h in the sense of Hardy and Krause (section 2.1). Condition (5.6.15) implies $v(h) \leq sc$. Now we have

$$\begin{aligned}|h_n^* - E(h(\boldsymbol{x})) &= |\int h(\boldsymbol{y})d_{F_n}(\boldsymbol{y}) - \int h(\boldsymbol{y})d_F(\boldsymbol{y})| \\ &\leq v(h)\sup_{\boldsymbol{y}}|F_n(\boldsymbol{y}) - F(\boldsymbol{y})| \\ &\leq sc \ \sup_{\boldsymbol{y}}|F_n(\boldsymbol{y}) - F(\boldsymbol{y})| \\ &\leq O(n^{-1}\log^s n).\end{aligned}$$

On the other hand, we have

$$\begin{aligned}|\bar{h}_n - h_n^*| &\leq \frac{1}{n}\sum_{k=1}^{n} \sup_{\boldsymbol{u}_k\in(-1,1)^s} |h(\boldsymbol{a}_k^{(n)} - h(\boldsymbol{a}_k^{(n)} + \boldsymbol{u}_k/2n)| \\ &\leq \max_{1\leq k\leq n} \sup_{\boldsymbol{u}_k\in(-1,1)^s} |h(\boldsymbol{a}_k^{(n)}) - h(\boldsymbol{a}_k^{(n)} + \boldsymbol{u}_k/2n)| \\ &\leq sc/n.\end{aligned}$$

The assertion (5.6.17) follows. Consequently, (5.6.18) and (5.6.19) follow. □

Similarly, we can obtain the order of $Var(\bar{h}_n)$ and the mean square error of $\bar{h}_n$ as follows.

Theorem 5.4
Under the same assumption as in Theorem 5.3, we have

$$\mathrm{Var}(\bar{h}_n) = O(n^{-2}\log^{2s} n)$$

and

$$\mathrm{MSE}(h) = E(\bar{h}_n - E(h(\boldsymbol{x}))^2 = O(n^{-2}\log^{2s} n).$$

The reader can find more detailed discussion in Fang and Zhu (1993).

Exercises

5.1 Suppose that there are four factors A, B, C and D and each has four levels $A_1, A_2, A_3, A_4, B_1, \cdots, D_4$. Give a design of experiments by $L_{16}(4^5)$ which is given in Table 5.18.

Table 5.18 $L_{16}(4^5)$

No	1	2	3	4	5
1	1	1	1	1	1
2	1	2	2	2	2
3	1	3	3	3	3
4	1	4	4	4	4
5	2	1	2	3	4
6	2	2	1	4	3
7	2	3	4	1	2
8	2	4	3	2	1
9	3	1	3	4	2
10	3	2	4	3	1
11	3	3	1	2	4
12	3	4	2	1	3
13	4	1	4	2	3
14	4	2	3	1	4
15	4	3	2	4	1
16	4	4	1	3	2

5.2 Transform the experimental points of Exercise 5.1 into C^4 and compute its discrepancy.

5.3 Generate two tables of UD: $U_{13}(13^{12})$ and $U_{15}(15^8)$ by *glp* sets.

5.4 Prove

$$\int_0^1 \left(1 - \frac{2}{\pi}\ln(2\sin(\pi x))\right)dx = 1.$$

5.5 With the data in Table 5.7 and the best subset methods find the regression model for Example 5.2 and compare it with (5.3.6).

5.6 Prove the formula (5.4.9) for symmetric difference between $\boldsymbol{x}_1$ and $\boldsymbol{y}_a$. Can you give a similar formula for $\boldsymbol{x}_a$ and $\boldsymbol{y}_b$?

5.7 Let $\boldsymbol{A} = (a_{ij})$ be an $n \times s$ UD-design matrix and $\boldsymbol{X}$ be its induced matrix (Definition 5.1). Answer the following questions:

(a) Find that the sample means of $\boldsymbol{A}$ and $\boldsymbol{X}$ are $\frac{n+1}{2}\mathbf{1}_n$ and $\frac{1}{2}\mathbf{1}_n$, respectively.

(b) Find that the sample variance of each column of $\boldsymbol{A}$ and $\boldsymbol{X}$ are $\frac{n(n+1)}{12}$ and $\frac{n+1}{12n}$, respectively.

(c) Let $\boldsymbol{S} = \frac{1}{n}\boldsymbol{X}'(\boldsymbol{I}_n - \frac{1}{n}\mathbf{1}_n\mathbf{1}_n')\boldsymbol{X}$. Prove $\mathrm{tr}(\boldsymbol{S}) = \frac{s}{12}(1 - \frac{1}{n^2})$.

5.8 Give the experimental points of the $\{3,4\}$-simplex-lattice design.

5.9 Give the experimental points of simplex-centroid design for the four-dimensional case.

5.10 Find the best d for the axial design in the case $s = 3$ by MSE criterion.

5.11 Generate a latin hypercube sample of size 17 from the standard normal distribution $N_3(\mathbf{0}, \boldsymbol{I}_3)$.

CHAPTER 6

Some applications in statistical inference

Statistical inference consists of mainly estimation theory and testing hypothesis. In this chapter, we shall introduce some applications of the NTM to statistical inference, which include maximum likelihood estimation of parameters, the robust estimation of the mean vector of a multivariate distribution, the test for multinormality and the test for sphericity of a distribution. In the last section we propose a new method of generating an NT-net on the Stiefel manifold which can be applied to projection pursuit.

6.1 Maximum likelihood estimation

Let $\boldsymbol{x}_1, \cdots, \boldsymbol{x}_N$ be a given sample from a population with distribution function $F(\boldsymbol{x}, \boldsymbol{\theta})$, where $\boldsymbol{x} \in R^p$ and the distribution parameter $\boldsymbol{\theta} \in \Theta \subset R^s$. We want to estimate the parameter $\boldsymbol{\theta}$ or a function of $\boldsymbol{\theta}$ from the sample. In section 3.3 we have demonstrated using SNTO to find the MLE of $\boldsymbol{\theta}$ for the case when the sample is univariate. The method may also be applied to multivariate distributions. Here we shall consider the elliptical distributions as an example.

Suppose that $F(\boldsymbol{x}, \boldsymbol{\theta})$ is an elliptical distribution. Then its density $g(\boldsymbol{x}, \boldsymbol{\theta})$, if it exists, has the form

$$g(\boldsymbol{x}, \boldsymbol{\theta}) = |\det \boldsymbol{\Sigma}|^{-\frac{1}{2}} \; h((\boldsymbol{x} - \boldsymbol{\mu})' \boldsymbol{\Sigma}^{-1} (\boldsymbol{x} - \boldsymbol{\mu})), \tag{6.1.1}$$

where $\boldsymbol{\theta} = \{\boldsymbol{\mu}, \boldsymbol{\Sigma}\}, \boldsymbol{\mu} \in R^p$, $\boldsymbol{\Sigma}$ is a $p \times p$ matrix and $\boldsymbol{\Sigma} > 0$ (section 2.2). Without loss of generality, we may assume that $\boldsymbol{\Sigma}$ is the covariance matrix of the distribution (Fang, Kotz and Ng (1990), section 2.5). Let $\boldsymbol{x}_1, \cdots, \boldsymbol{x}_N$ be a sample from distribution (6.1.1) with the likelihood function

$$L(\boldsymbol{\mu}, \boldsymbol{\Sigma}) = |\det \boldsymbol{\Sigma}|^{-\frac{N}{2}} \prod_{i=1}^{N} h((\boldsymbol{x}_i - \boldsymbol{\mu})' \boldsymbol{\Sigma}^{-1} (\boldsymbol{x}_i - \boldsymbol{\mu})). \tag{6.1.2}$$

The maximum likelihood estimates $\hat{\boldsymbol{\mu}}$ and $\hat{\boldsymbol{\Sigma}}$ maximize $L(\boldsymbol{\mu}, \boldsymbol{\Sigma})$ or the logarithm of the likelihood function

$$l(\boldsymbol{\mu}, \boldsymbol{\Sigma}) = -\frac{N}{2} \log |\det \boldsymbol{\Sigma}| + \sum_{i=1}^{N} \log(h((\boldsymbol{x}_i - \boldsymbol{\mu})' \boldsymbol{\Sigma}^{-1} (\boldsymbol{x}_i - \boldsymbol{\mu}))). \tag{6.1.3}$$

When $h(\cdot)$ is an exponential function which corresponds to a normal population, one can find analytic solutions for $\hat{\boldsymbol{\mu}}$ and $\hat{\boldsymbol{\Sigma}}$. In general $\hat{\boldsymbol{\mu}}$ and $\hat{\boldsymbol{\Sigma}}$ have no analytic expressions. Hence we have to find their approximate values by numerical methods.

Since $\boldsymbol{\Sigma}$ is positive definite, it has $p(p+1)/2$ independent elements and is not convenient to find $\hat{\boldsymbol{\Sigma}}$ directly by maximizing $l(\boldsymbol{\mu}, \boldsymbol{\Sigma})$ with respect to $\boldsymbol{\Sigma}$. Therefore, we recommend the Cholesky decomposition

$$\boldsymbol{\Sigma} = \boldsymbol{L}'\boldsymbol{L},$$

where $\boldsymbol{L}$ is the Cholesky root, an upper triangular matrix with positive diagonal elements. It is known that there is a one-to-one correspondence between $\boldsymbol{\Sigma}$ and $\boldsymbol{L}$, and that the number of independent elements of $\boldsymbol{L}$ is $p(p+1)/2$. Therefore, we can denote $L(\boldsymbol{\mu}, \boldsymbol{\Sigma})$ by $L_*(\boldsymbol{\mu}, \boldsymbol{L})$ and consequently $\max_{\boldsymbol{\mu},\boldsymbol{\Sigma}} L(\boldsymbol{\mu}, \boldsymbol{\Sigma}) = \max_{\boldsymbol{\mu},\boldsymbol{L}} L_*(\boldsymbol{\mu}, \boldsymbol{L})$. A similar notation $l_*(\boldsymbol{\mu}, \boldsymbol{L})$ replaces $l(\boldsymbol{\mu}, \boldsymbol{\Sigma})$. Using the moment estimates of $\boldsymbol{\mu}$ and $\boldsymbol{\Sigma}$

$$\begin{aligned} \bar{\boldsymbol{x}} &= \frac{1}{N} \sum_{i=1}^{N} \boldsymbol{x}_i \\ \boldsymbol{S} &= \frac{1}{N-1} \sum_{i=1}^{N} (x_i - \bar{\boldsymbol{x}})(x_i - \bar{\boldsymbol{x}})' \equiv \boldsymbol{L}_1' \boldsymbol{L}_1, \end{aligned} \tag{6.1.4}$$

we may determine an initial domain for finding the MLEs of $\boldsymbol{\mu}$ and $\boldsymbol{\Sigma}$ by means of SNTO (section 3.3). Hence we propose the following algorithm for finding the approximate values of the MLEs of $\boldsymbol{\mu}$ and $\boldsymbol{\Sigma}$.

Step 1 Calculate the sample mean $\bar{\boldsymbol{x}} = (\bar{x}_1, \cdot, \bar{x}_p)'$ and covariance matrix $\boldsymbol{S}$ by (6.1.4).

Step 2 Find the Cholesky root $\boldsymbol{L}_1$ of $\boldsymbol{S}$, i.e.

$$\boldsymbol{S} = \boldsymbol{L}_1' \boldsymbol{L}_1,$$

where $\boldsymbol{L}_1 = (l_{ij})$ is an upper triangular matrix with positive diagonal elements.

Step 3 Set $\boldsymbol{v}^{(0)} = (\bar{x}_1, \cdots, \bar{x}_p, l_{11}, \cdots, l_{1p}, l_{22}, \cdots, l_{2p}, \cdots, l_{pp})'$ and take a positive vector $\boldsymbol{c} \in R^{(p+3)p/2}$ according to the property of h in (6.1.1). Then we may define an initial domain $D^{(0)} = [\boldsymbol{v}^{(0)} - \boldsymbol{c}, \boldsymbol{v}^{(0)} + \boldsymbol{c}]$.

Step 4 Find an approximate maximum $M^*$of $l_*(\boldsymbol{\mu}, \boldsymbol{L})$ or $L_*(\boldsymbol{\mu}, \boldsymbol{L})$ and the respective maximum point $\boldsymbol{v}^*=(v_1^*, \cdots, v_{p(p+3)/2}^*)'$ over $D^{(0)}$ by SNTO. Then the MLEs of $\boldsymbol{\mu}$ and $\boldsymbol{\Sigma}$ can be obtained by $\boldsymbol{v}^*$ as follows: $\hat{\boldsymbol{\mu}} \approx \hat{\boldsymbol{\mu}}^*$ is given by the first p components of $\boldsymbol{v}^*$, and the remaining elements of $\boldsymbol{v}^*$ form an upper triangular matrix $\boldsymbol{L}^*$ by the same relation between $\boldsymbol{v}^{(0)}$ and $\boldsymbol{L}_1$. Then

$$\hat{\boldsymbol{\Sigma}}^* = (\boldsymbol{L}^*)' \boldsymbol{L}^*$$

is the approximate maximum likelihood estimate of $\hat{\boldsymbol{\Sigma}}$.

Example 6.1

By a statistical simulation a sample $\{\boldsymbol{x}_i\}$ of size 500 is generated from a multivariate t-distribution $Mt_3(5, \boldsymbol{\mu}, \boldsymbol{\Sigma})$ where

$$\boldsymbol{\mu} = \mathbf{0} \quad \text{and} \quad \boldsymbol{\Sigma} = \begin{pmatrix} 1.0 & 0.5 & 0.3 \\ 0.5 & 1.0 & 0.2 \\ 0.3 & 0.2 & 1.0 \end{pmatrix}.$$

The likelihood function is

$$L(\boldsymbol{\mu}, \boldsymbol{\Sigma}) = C|\det \boldsymbol{\Sigma}|^{-250} \prod_{i=1}^{500} [1 + \frac{1}{5}(\boldsymbol{x}_i - \boldsymbol{\mu})' \boldsymbol{\Sigma}^{-1} (\boldsymbol{x}_i - \boldsymbol{\mu})]^{-4},$$

where C is the normalizing constant, and the logarithm of the likelihood function after deleting the constant C is

$$l(\boldsymbol{\mu}, \boldsymbol{\Sigma}) = -250|\det \boldsymbol{\Sigma}| - 4\sum_{i=1}^{500}[1 + \frac{1}{5}(\boldsymbol{x}_i - \boldsymbol{\mu})'\boldsymbol{\Sigma}^{-1}(\boldsymbol{x}_i - \boldsymbol{\mu})].$$

The sample mean vector and sample covariance matrix are calculated as follows:

$$\bar{\boldsymbol{X}} = \begin{pmatrix} -0.0661 \\ 0.0626 \\ -0.0083 \end{pmatrix}, \qquad \boldsymbol{S} = \begin{pmatrix} 1.7162 & 0.8318 & 0.4567 \\ 0.8318 & 1.5923 & 0.2680 \\ 0.4567 & 0.2680 & 1.6053 \end{pmatrix}.$$

The corresponding value of $-l(\bar{\boldsymbol{x}}, \boldsymbol{S})$ is 1042.1. After applying the above algorithm with $n_1 = 159,053$, $n_2 = 100,063$ and $\boldsymbol{c} = (1, \cdots, 1)'$ in SNTO we obtain the MLEs of $\boldsymbol{\mu}$ and $\boldsymbol{\Sigma}$ are:

$$\hat{\boldsymbol{\mu}} = \begin{pmatrix} -0.0926 \\ 0.0765 \\ -0.0324 \end{pmatrix} \quad \text{and} \quad \hat{\boldsymbol{\Sigma}} = \begin{pmatrix} 1.1681 & 0.5712 & 0.3451 \\ 0.5712 & 1.0356 & 0.2411 \\ 0.3451 & 0.2411 & 1.0042 \end{pmatrix}.$$

With $-l(\hat{\boldsymbol{\mu}}, \hat{\boldsymbol{\Sigma}}) = 1040.4 < 1042.1 = -l(\bar{\boldsymbol{x}}, \boldsymbol{S})$. The numerical results obtained by SNTO coincide with those obtained by using other optimization methods.

Let us now consider hypotheses testing for the parameters $\boldsymbol{\mu}$ and $\boldsymbol{\Sigma}$. Let Ω be the parameter space $(\boldsymbol{\mu}, \boldsymbol{\Sigma})$ and consider the following hypotheses

$$\begin{aligned} H_0 &:(\boldsymbol{\mu}, \boldsymbol{\Sigma}) \in \omega \subset \Omega, \\ H_1 &:(\boldsymbol{\mu}, \boldsymbol{\Sigma}) \in \Omega/\omega. \end{aligned} \tag{6.1.5}$$

Then the likelihood ratio criterion of the hypothesis H_0 is as follows

$$\lambda = \frac{\max_\omega L(\boldsymbol{\mu}, \boldsymbol{\Sigma})}{\max_\Omega L(\boldsymbol{\mu}, \boldsymbol{\Sigma})}, \tag{6.1.6}$$

where $L(\boldsymbol{\mu}, \boldsymbol{\Sigma})$ is the likelihood function in (6.1.2). The numerator and denominator can be obtained by using the SNTO separately. Then the λ value can be compared with the critical value of the test.

If the population distribution does not belong to the class of elliptical distributions, the maximum likelihood estimate and likelihood ratio statistic might be obtained in a similar way

6.2 Robust estimation of the mean vector

Robust parameter estimation usually has the good property of being stable under change of disturbance distribution. For example, suppose that $X_1, \cdots, X_N$ is a sample from the normal population $N(\mu, \sigma^2)$. Then the usual estimation of μ is given by the sample mean

$$\bar{X} = \frac{1}{N}\sum_{i=1}^{n} X_i.$$

This is not a robust estimate because if there is a large deviation in an observation, then it may lead also to a large deviation in $\bar{X}$. Let $X_{(1)} \leq X_{(2)} \leq \cdots \leq X_{(N)}$ be order statistics of the sample. The truncated estimate

$$\bar{X}_{(k)} = \frac{1}{N-2k}\sum_{i=k+1}^{N-k} X_{(i)}$$

($k = 1, 2, \cdots,$ and $k < N/10$) is an unbiased estimator of μ which has the good property of not being affected by extreme values in the sample. Hence $\bar{X}_{(k)}$ is a robust estimate. We may also use the sample median

$$X_M = \begin{cases} X_{(\frac{(N+1)}{2})}, & \text{if } N \text{ is odd,} \\ (X_{(\frac{N}{2})} + X_{(\frac{N}{2}+1)})/2, & \text{otherwise} \end{cases}$$

as a robust estimate of μ.

Robust estimation in univariate situations has been studied by a number of authors, e.g. Tukey (1960), Huber (1964, 1972, 1973) and Andrews *et al.* (1972). The α-trimmed mean and median are popular robust estimates of the location parameter. The idea can be generalized to multivariate case in different ways, such as

Bickel (1964), Gentleman (1965), Gnanadesikan (1977), Hampel *et al.* (1986) and Rousseeuw and Leroy (1987). We will introduce a popular method emphasized by Rousseeuw and Zomeren (1990). The method is called the **minimum volume ellipsoid estimator** (MVE). Let $\boldsymbol{x}_1, \cdots, \boldsymbol{x}_N$ be a sample from a s-dimensional population and suppose we want to estimate its "center" and "scatter" by means of a column vector $\hat{\boldsymbol{\mu}}$ and a $s \times s$ matrix $\boldsymbol{V}$. The MVE is defined as the pair $(\hat{\boldsymbol{\mu}}, \boldsymbol{V})$ such that the determinant of $\boldsymbol{V}$ is minimized subject to

$$\# \quad \{i : (\boldsymbol{x}_i - \hat{\boldsymbol{\mu}})'\boldsymbol{V}^{-1}(\boldsymbol{x}_i - \hat{\boldsymbol{\mu}}) \leq a^2\} \geq h,$$

where $h = [(N + S + 1)/2]$ and $[x]$ is the integer part of x. The number a^2 is a fixed constant, which can be chosen from $\chi^2_{s,.50}$ when we expect the majority of the data to come from a normal distribution. The robust distances are defined relative to the MVE:

$$RD_i = [(\boldsymbol{x}_i - \hat{\boldsymbol{\mu}})'\boldsymbol{V}^{-1}(\boldsymbol{x}_i - \hat{\boldsymbol{\mu}})]^{1/2}. \tag{6.2.1}$$

One can then compute a weighted mean

$$\boldsymbol{\mu}_1 = (\sum_{i=1}^{N} w_i)^{-1} \sum_{i=1}^{N} w_i \boldsymbol{x}_i \tag{6.2.2}$$

and a weighted covariance matrix

$$\boldsymbol{V}_1 = (\sum_{i=1}^{N} w_i - 1)^{-1} \sum_{i=1}^{N} (\boldsymbol{x}_i - \hat{\boldsymbol{\mu}}_1)(\boldsymbol{x}_i - \hat{\boldsymbol{\mu}}_1)' \tag{6.2.3}$$

where the weights $w_i = w(RD_i)$ depend on the robust distances.

There are two approximate algorithms for the MVE, one of which is called the **projection algorithm** which is a variant of an algorithm of Gasko and Donoho (1982). For each $\boldsymbol{x}_i$ we consider

$$u_i = \max_{\boldsymbol{a} \in U_s} |\boldsymbol{x}_i'\boldsymbol{a} - L(\boldsymbol{x}_1'\boldsymbol{a}, \cdots, \boldsymbol{x}_N'\boldsymbol{a})| / S(\boldsymbol{x}_1'\boldsymbol{a}, \cdots, \boldsymbol{x}_N'\boldsymbol{a}) \tag{6.2.4}$$

where L and S are the MVE estimates in one dimension, which we compute as follows. For any set of numbers $z_1 \le z_2 \le \cdots \le z_N$ one can determine its shortest half by taking the smallest of the differences : $z_h - z_1, z_{h+1} - z_2, \cdots, z_N - z_{N-h+1}$. If the smallest difference is $z_j - z_{j-h+1}$ we put L equal to the midpoint of the corresponding half, $L(z_1, \cdots, z_N) = (z_j + z_{j-h+1})/2$ and S as its length $S(z_1, \cdots, z_N) = c(N)(z_j - z_{j-h+1})$ up to a correction factor $c(N)$, which depends on the sample size. Note that (6.2.4) is exactly the one-dimensional version of RD_i in (6.2.1). In the past one had the difficulty of calculating u_i for all directions $\boldsymbol{a}$, and it has been suggested to take all $\boldsymbol{a}$ of the form $\boldsymbol{x}_i - \boldsymbol{m}, i = 1, \cdots, N$, where $\boldsymbol{m} = (\text{median}_{1 \le j \le N}\ x_{j1}, \cdots, \text{median}_{1 \le j \le N}\ x_{js})'$. Obviously, it is not enough to take only these directions, and the NTM can be applied to calculate u_i by the use of an NT-net on U_s.

Furthermore, the idea of projection and the α-trimmed mean can motivate a new robust estimator of location by using NT-nets on a sphere. Some examples are given for the **ϵ-contaminated model**.

Consider the following ϵ-contaminated model

$$(1 - \epsilon)N_p(\boldsymbol{\mu}_1, \boldsymbol{\Sigma}) + \epsilon N_p(\boldsymbol{\mu}_2, \boldsymbol{\Sigma}), \quad \boldsymbol{\mu}_1 \ne k\boldsymbol{\mu}_2, \tag{6.2.5}$$

where ϵ is a small positive number. Suppose that a sample $\boldsymbol{x}_1, \cdots, \boldsymbol{x}_N$ is drawn from the distribution (6.2.5). This means that some observations in the sample are contaminated. We want to estimate the original mean vector $\boldsymbol{\mu}_1$. Set $f_i(\boldsymbol{c}) = \boldsymbol{c}'\boldsymbol{x}_i/\|\boldsymbol{c}'\boldsymbol{x}_i\|$, $i = 1, \cdots, N$ for any $\boldsymbol{c} \in U_p$. It is more likely that $f_i(\boldsymbol{c})$ will attain its maximum at $\boldsymbol{\mu}_1/\|\boldsymbol{\mu}_1\|$ than at any other direction. With this in mind we propose the following procedure.

Let $\{\boldsymbol{c}_k, k = 1, \cdots, n\}$ be an NT-net on U_p. Set $e_{ki} = f_i(\boldsymbol{c}_k)$, $i = 1, \cdots, N$, $k = 1, \cdots, n$. We then sort $\{e_{ki}\}$ for each k

$$e_{k(k_1)} \le e_{k(k_2)} \le \cdots \le e_{k(k_N)}, \quad k = 1, \cdots, n. \tag{6.2.2}$$

Discard the first $n_1 = [N\epsilon]$ smaller order statistics in each k, and obtain the sum of remaining ones

$$e_{k\cdot} = \sum_{j=n_1+1}^{N} e_{k(k_j)}, \quad k = 1, \cdots, n.$$

We propose the following estimate of $\boldsymbol{\mu}_1$:

$$\bar{x}_T = \frac{1}{N - n_1} \sum_{j=n_1+1}^{N} x_{k_0(k_{0j})},$$

where k_0 satisfies $e_{k_0 \cdot} = \max_k e_{k \cdot}$.

The above method of estimation proposed by Fang, Yuan and Bentler (1992) is somewhat like the α-trimmed mean in the univariate situation, and contains the maximum likelihood idea. It is obvious that the method can be used in location estimation of other continuous multivariate distribution families. In the following example the reader can see that the above method does improve the usual estimate $\bar{\boldsymbol{x}}$.

Example 6.2

A sample of size 30 from the model (6.2.5) with $\epsilon = 0.1$, $\boldsymbol{\mu}_1 = (-1.584, 7.350)'$ and $\boldsymbol{\mu}_2 = (4.496, 1.509)'$ is generated by statistical simulation as follows:

X_1	X_2	X_1	X_2	X_1	X_2
−1.901	6.358	−3.474	8.331	−3.394	7.867
−0.114	7.783	−1.116	6.246	−2.605	6.056
−0.685	6.058	−1.666	7.237	−2.601	9.775
−1.283	7.748	−0.057	6.966	−2.179	7.016
−1.915	7.697	−0.910	6.954	−2.655	7.506
0.141	6.220	0.391	8.110	−1.571	7.043
−0.702	7.930	−1.640	6.108	−2.446	6.348
−0.169	7.087	−0.998	7.053	4.123	1.254
−1.466	5.547	−0.563	6.639	4.675	0.168
−1.256	6.684	−1.875	5.828	3.598	1.900

The usual sample mean is

$$\bar{\boldsymbol{x}} = (-0.877, 6.451)',$$

and with 180 projection directions the new estimate is

$$\bar{\boldsymbol{x}}_T = (-1.434, 7.045)'$$

and the distances $d(\boldsymbol{\mu}_1, \bar{\boldsymbol{x}}) = 1.309, d(\boldsymbol{\mu}_1, \bar{\boldsymbol{x}}_T) = 0.116$. The latter is better than the usual estimate $\bar{\boldsymbol{x}}$.

To evaluate the reliability of the conclusion we repeat the above process twenty times by letting $\boldsymbol{\mu}_1$ and $\boldsymbol{\mu}_2$ be randomly generated from a normal population. The corresponding $D_0 = d(\boldsymbol{\mu}_1, \bar{\boldsymbol{x}})$ and $D_1 = d(\boldsymbol{\mu}_1, \bar{\boldsymbol{x}}_T)$ are listed in Table 6.1, which shows that in most cases D_0 is much bigger than D_1.

Applying this method to samples of size 40 and 50, we obtain the same conclusion. Our simulation indicates that when $D = d(\boldsymbol{\mu}_1/\|\boldsymbol{\mu}_1\|, \boldsymbol{\mu}_2/\|\boldsymbol{\mu}_2\|)$ in the last column of Table 6.1 is small, D_0 and D_1 are close each other. Hence the method is particularly efficient when D is large, for example $\boldsymbol{\mu}_1 \perp \boldsymbol{\mu}_2$, i.e. $\boldsymbol{\mu}_1$ and $\boldsymbol{\mu}_2$ are orthogonal.

Table 6.1 D_0, D_1, *and* D

D_0	D_1	D
0.101	0.063	0.078
1.477	0.007	2.757
0.853	0.227	2.771
0.693	0.072	1.087
0.693	0.015	1.835
3.901	0.034	3.328
0.497	0.278	0.007
1.901	0.014	3.143
0.047	0.016	0.032
1.930	0.147	3.145
0.128	0.058	3.062
0.331	0.002	2.906
1.469	0.051	3.998
0.247	0.003	0.815
0.170	0.104	0.514
0.534	0.072	3.241
0.308	0.288	0.583
0.255	0.043	0.460
1.431	2.001	0.009
1.389	0.000	3.993

6.3 Tests for multinormality (I)

Testing multinormality has been receiving considerable attentions in past few decades. Mardia (1970, 1971, 1974, 1975), Malkovich and Afifi (1973), Hensler, Mehrotra and Michalek (1977), Cox and Small (1978), Gnanadesikan (1977), Bera and John (1983), Baringhaus and Henze (1988), Csörgö (1989) and Horswell and Looney (1992), for example, constitute a large literature. But so far there is no method that has been universally accepted. Recently, following Malkovich and Afifi (1973) and Machado (1983) Fang, Yuan and Bentler (1992) proposed another approach using the union-intersection principle and the NTM. It is well known that a p-dimensional random vector $\boldsymbol{x}$ is distributed according to a multinormal distribution if and only if for each $\boldsymbol{c} \in R^p$, $\boldsymbol{c}'\boldsymbol{x}$ is univariate normal. We can assume $\|\boldsymbol{c}\| = 1$ i.e., $\boldsymbol{c} \in U_p$, without loss of generality. If we can find the "worst" projection direction $\boldsymbol{c}_0$, then the test of multinormality is equivalent to testing normality of $\boldsymbol{c}_0'\boldsymbol{x}$. Suppose that $\{\boldsymbol{c}_k, k = 1, \cdots, n\}$ is a large NT-net on U_p. Then the worst direction $\boldsymbol{c}^*$ among $\{\boldsymbol{c}_k\}$ should be close to $\boldsymbol{c}_0$. Hence the test of multinormality is approximately equivalent to testing the normality of $\boldsymbol{c}^{*\prime}\boldsymbol{x}$, i.e.

$$
\begin{aligned}
&H_0 : \boldsymbol{x} \sim \text{a multinormal distribution,} \\
&H_1 : \boldsymbol{x} \sim \text{a non-normal distribution} \qquad (6.3.1)
\end{aligned}
$$

$\Longleftrightarrow$

$$
\begin{aligned}
&H_0 : \boldsymbol{c}'\boldsymbol{x} \sim \text{a normal distribution for each } \boldsymbol{c} \in U_p, \\
&H_1 : \boldsymbol{c}'\boldsymbol{x} \sim \text{a non-normal distribution for some } \boldsymbol{c} \in U_p \qquad (6.3.2)
\end{aligned}
$$

$\Longleftrightarrow\!\!\!\!\!\approx$ (approximately equivalent)

$$
\begin{aligned}
&H_0 : \boldsymbol{c}_k'\boldsymbol{x} \sim \text{a normal distribution for } 1 \le k \le n, \\
&H_1 : \boldsymbol{c}_k'\boldsymbol{x} \sim \text{a non-normal distribution for some k} \qquad (6.3.3)
\end{aligned}
$$

$\Longleftrightarrow$

$$
\begin{aligned}
&H_0 : \boldsymbol{c}^{*\prime}\boldsymbol{x} \sim \text{a normal distribution,} \\
&H_1 : \boldsymbol{c}^{*\prime}\boldsymbol{x} \sim \text{a non-normal distribution.}
\end{aligned}
$$

There are many methods for testing normality. It is popular to use the **sample skewness and kurtosis** to test for univariate normality even though the method is not necessarily the best one. Let $X_1, \cdots, X_N$ be a sample. Then the sample skewness and kurtosis are defined respectively by

$$Sk = \frac{1}{N}\sum_{i=1}^{N}(X_i - \bar{X})^3 \Big/ \Big[\frac{1}{N}\sum_{i=1}^{N}(X_i - \bar{X})^2\Big]^{3/2}$$
$$= \sqrt{N}\sum_{i=1}^{N}(X_i - \bar{X})^3 \Big/ \Big[\sum_{i=1}^{N}(X_i - \bar{X})^2\Big]^{3/2}$$

and

$$Ku = \frac{1}{N}\sum_{i=1}^{N}(X_i - \bar{X})^4 \Big/ \Big[\frac{1}{N}\sum_{i=1}^{N}(X_i - \bar{X})^2\Big]^{2} - 3$$
$$= N\sum_{i=1}^{N}(X_i - \bar{X})^4 \Big/ \Big[\sum_{i=1}^{N}(X_i - \bar{X})^2\Big]^{2} - 3,$$

where $\bar{X}$ is the sample mean. The critical points can be found in Pearson and Hartley (1956).

Let $Sk(\boldsymbol{b})$ and $Ku(\boldsymbol{b})$ be the sample skewness and kurtosis of $\{\boldsymbol{b}'\boldsymbol{x}_i, i = 1, \cdots, N\}$, respectively, i.e.

$$Sk(\boldsymbol{b}) = \frac{\sqrt{N}\sum_{j=1}^{N}(\boldsymbol{b}'\boldsymbol{x}_j - \boldsymbol{b}'\bar{\boldsymbol{x}})^3}{[\sum_{j=1}^{N}(\boldsymbol{b}'\boldsymbol{x}_j - \boldsymbol{b}'\bar{\boldsymbol{x}})^2]^{3/2}} \tag{6.3.4}$$

and

$$Ku(\boldsymbol{b}) = \frac{N\sum_{j=1}^{N}(\boldsymbol{b}'\boldsymbol{x}_j - \boldsymbol{b}'\bar{\boldsymbol{x}})^4}{[\sum_{j=1}^{N}(\boldsymbol{b}'\boldsymbol{x}_j - \boldsymbol{b}'\bar{\boldsymbol{x}})^2]^{2}} - 3, \tag{6.3.5}$$

where $\bar{\boldsymbol{x}}$ is the sample mean. The worst direction $\boldsymbol{b}_0$ can be considered as

$$Sk(\boldsymbol{b}_0) = \max_{\boldsymbol{b}\in U_p}|Sk(\boldsymbol{b})| \cong \max_{k}|Sk(\boldsymbol{b}_k)|$$

or

$$Ku(\boldsymbol{b}_0) = \max_{\boldsymbol{b}\in U_p}|Ku(\boldsymbol{b})| \cong \max_{k}|Ku(\boldsymbol{b}_k)|.$$

Let $\{\boldsymbol{b}_k, k = 1, \cdots, n\}$ be an NT-net on U_s. Hence, the statistics

$$\begin{aligned} Sk_{\max} &= \max_{k}|Sk(\boldsymbol{b}_k)| \\ Ku_{\max} &= \max_{k}|Ku(\boldsymbol{b}_k)| \end{aligned} \tag{6.3.6}$$

are close to $Sk(\boldsymbol{b}_0)$ and $Ku(\boldsymbol{b}_0)$ respectively if n is large and can be used for testing multinormality. For a given significance level α, the rejection region is

$$Sk_{\max} > Sk(\alpha) \qquad \text{or} \qquad Ku_{\max} > Ku(\alpha).$$

A statistical table of $Sk(\alpha)$ and $Ku(\alpha)$ for $\alpha = 1\%$ and $\alpha = 5\%$ with $p = 2, 3, 4, 5$ for various sample sizes is given in Tables 6.2 and 6.3 by simulation calculation with sample size 2000 for each p (Fang, Yuan and Bentler 1992).

The following examples are helpful for the illustration of above method.

Example 6.3

This example, which arose from the problem of standardizing the sizes for men's clothes in China in 1976, involves the data of 12 measurements of the body(cm):

X_1 : height from the waist up,
X_2 : arm length,
X_3 : bust,
X_4 : neck,
X_5 : shoulder length,
X_6 : width of the front part of chest,
X_7 : width of the back part of chest,
X_8 : height,
X_9 : height without head and neck,
X_{10} : height from the waist down,
X_{11} : waist circumference,
X_{12} : buttocks.

Table 6.2 *Mean, variance and critical points of* $Sk_{\max}$, $P(Sk_{\max} > Sk(\alpha)) = \alpha$

sample size	5%	1%	mean
			$p = 2$
30	1.136	1.450	0.630
40	1.022	1.307	0.570
50	0.924	1.181	0.519
60	0.842	1.074	0.475
70	0.776	0.985	0.440
80	0.727	0.915	0.412
90	0.693	0.863	0.393
100	0.676	0.829	0.382
			$p = 3$
30	1.407	1.717	0.883
40	1.266	1.540	0.798
50	1.142	1.387	0.724
60	1.037	1.258	0.662
70	0.949	1.152	0.610
80	0.880	1.071	0.570
90	0.829	1.013	0.541
100	0.796	0.979	0.523
			$p = 4$
40	1.428	1.702	0.960
50	1.303	1.550	0.882
60	1.191	1.416	0.813
70	1.094	1.299	0.752
80	1.011	1.199	0.701
90	0.943	1.117	0.658
100	0.888	1.053	0.623
110	0.848	1.005	0.598
120	0.822	0.975	0.581
			$p = 5$
40	1.540	1.843	1.097
50	1.407	1.675	1.005
60	1.288	1.527	0.924
70	1.185	1.398	0.853
80	1.098	1.288	0.793
90	1.025	1.197	0.743
100	0.968	1.126	0.703
110	0.926	1.074	0.674
120	0.899	1.042	0.655

Table 6.3 *Mean, variance and critical points of* $Ku_{\max}$, $P(Ku_{\max} > Ku(\alpha)) = \alpha$

sample size	5%	1%	mean
			$p = 2$
30	2.281	3.684	1.150
40	2.096	3.371	1.058
50	1.932	3.081	0.977
60	1.788	2.816	0.907
70	1.663	2.574	0.848
80	1.559	2.357	0.801
90	1.475	2.163	0.764
100	1.411	1.993	0.739
			$p = 3$
30	3.342	4.759	1.650
40	3.051	4.354	1.523
50	2.794	3.991	1.410
60	2.571	3.671	1.313
70	2.384	3.394	1.231
80	2.231	3.159	1.164
90	2.113	2.967	1.112
100	2.029	2.819	1.076
			$p = 4$
40	3.901	5.455	1.995
50	3.582	4.996	1.855
60	3.295	4.586	1.729
70	3.039	4.227	1.617
80	2.816	3.919	1.520
90	2.624	3.661	1.438
100	2.465	3.453	1.369
110	2.337	3.296	1.316
120	2.241	3.189	1.277
			$p = 5$
40	4.491	6.183	2.479
50	4.166	5.678	2.296
60	3.866	5.224	2.132
70	3.591	4.821	1.984
80	3.342	4.468	1.855
90	3.118	4.165	1.743
100	2.919	3.913	1.649
110	2.745	3.712	1.572
120	2.598	3.561	1.513

To demonstrate the method we only consider a subsample of the data. A sample of size 100, which can be found in Fang, Yuan and Bentler (1992), was drawn. The sample marginal skewness and kurtosis of each measurement is given in Table 6.4, which shows that each measurement can be individually considered to be from a normal distribution except X_{11}. If we are interested in testing multinormality of a subset of measurements, for example, testing multinormality of $\boldsymbol{x} = (X_1, X_3, X_8, X_{10}, X_{12})'$, we denote its observations by $\{\boldsymbol{x}_i, i = 1, \cdots, 100\}$ and do the following steps:

Table 6.4 *Sample Skewness and Kurtosis*

measurement	Skewness	Kurtosis
X_1	−0.13959	−0.18798
X_2	0.36731	0.09630
X_3	0.47527	0.10808
X_4	0.37394	−0.26673
X_5	0.30296	−0.52081
X_6	0.25955	0.15814
X_7	0.38965	0.80882
X_8	0.03461	−0.45418
X_9	0.03273	−0.54070
X_{10}	0.00703	−0.52704
X_{11}	0.67409	0.10105
X_{12}	0.47567	−0.31017

Step 1 Choose the number $n = 932$ of projection directions. We here use the *glp* set with generating vector (932; 1, 116, 288, 314) to generate 932 directions $\boldsymbol{c}_k$ on U_5.

Step 2 Calculate the 932 pairs of sample skewness and kurtosis of $\{\boldsymbol{c}_k'\boldsymbol{x}_i, i = 1, \cdots, 100\}, k = 1, \cdots, 932$, and find the respective maximums of their absolute values

$$Sk_{max} = 0.7293, \qquad Ku_{max} = 1.0911.$$

Since both $Sk_{\max}$ and $Ku_{\max}$ are less than the respective 5%-quantities in Tables 6.2 and 6.3, we may conclude that $\boldsymbol{x} = (X_1, X_3, X_8, X_{10}, X_{12})'$ has a multivariate normal distribution.

The following numerical results show that the first three subsets of measurements are all multivariate normal while the last two

The following numerical results show that the first three subsets of measurements are all multivariate normal while the last two subsets are all non-normal:

$$\{X_1, X_3, X_8, X_{10}\}: \ Sk_{\max} = 0.6913, \ Ku_{\max} = 0.9395,$$
$$\{X_1, X_8, X_{10}, X_{12}\}: \ Sk_{\max} = 0.7042, \ Ku_{\max} = 0.7886,$$
$$\{X_3, X_8, X_{10}, X_{12}\}: \ Sk_{\max} = 0.7411, \ Ku_{\max} = 0.8843,$$
$$\{X_4, X_5, X_6, X_{11}\}: \ Sk_{\max} = 0.7776, \ Ku_{\max} = 1.1484,$$
$$\{X_2, X_4, X_6, X_{11}\}: \ Sk_{\max} = 0.2798, \ Ku_{\max} = 1.6158.$$

Example 6.4

Let $\boldsymbol{x} = (X_1, X_2, X_3, X_4)'$ have independent components with $X_i \sim N(0,1)$, $i = 1,2,3$ and $\log X_4 \sim N(0,1)$, i.e. X_4 has the standard lognormal distribution. Ten samples of size 80 are generated by standard simulation techniques. Their $Sk_{\max}$'s and $Ku_{\max}$'s are listed in Table 6.5, in which the first part gives ten pairs of $Sk_{\max}$'s and $Ku_{\max}$'s based on 377 projection directions for the first three variables and the second part presents another ten pairs based on 597 projection directions for the all four variables. The results in Table 6.5 show that the hypothesis of multinormality of (X_1, X_2, X_3) is always accepted, and the hypothesis that (X_1, X_2, X_3, X_4) is from a multinormal distribution is rejected at all times.

Table 6.5 *Tests for Multinormality*

$p = 3$			$p = 4$		
$Sk_{\max}$	$Ku_{\max}$	Conclusion	$Sk_{\max}$	$Ku_{\max}$	Conclusion
0.4429	0.6930	A	6.020	41.391	R
0.6938	0.8305	A	4.992	30.364	R
0.4736	0.8955	A	2.411	7.391	R
0.4943	0.9091	A	1.895	3.234	R
0.5875	0.9139	A	1.638	3.754	R
0.8823	1.1531	A	2.188	5.137	R
0.3676	0.8682	A	1.066	0.967	R
0.6105	1.0520	A	1.786	2.805	R
0.1775	0.7545	A	2.635	8.292	R
0.2989	1.0035	A	2.387	6.885	R

The above method where the multivariate test is reduced to the univariate test is called the union-intersection principle in the classical theory of multivariate analysis. In recent years, the idea has developed into an active topic called projection pursuit (PP) (Huber (1985)). We shall illustrate it in detail in section 6.6.

Li and Zha (1991) used the PP method to assess the multivariate goodness of fit with the one-dimensional Neyman test statistic (Neyman (1937)). Suppose that $Y_1, \cdots, Y_N$ is a sample from a population. Their underlying distribution function G is unknown. We want to test

$$\begin{aligned} H_0: &\quad G = F, \\ H_1: &\quad G \neq F, \end{aligned} \tag{6.3.7}$$

where $F(x)$ is a known c.d.f. Neyman suggested the use of statistic

$$K_N = \frac{1}{N}\sum_{i=1}^{m}\Big[\sum_{j=1}^{N}\pi_i(F(Y_j))\Big]^2 \tag{6.3.8}$$

to test the hypothesis (6.3.4), where $\pi_0 \equiv 1$, $\pi_1, \cdots, \pi_m$ is a system of orthonormal polynomials on $(0,1)$, i.e.

$$\int_0^1 \pi_i(y)\pi_j(y)dy = \begin{cases} 0, & i \neq j, \\ 1, & i = j. \end{cases}$$

K_N ((6.3.5)) is called the **Neyman statistic**. It can be proved that under the null hypothesis, the limiting distribution of K_N as $N \to \infty$ is χ^2_m, the chi-square distribution with m degrees of freedom.

Now, let $\boldsymbol{x}_1, \cdots, \boldsymbol{x}_N$ be an s-dimensional sample from an unknown underlying distribution function $G(\boldsymbol{x})$. We want to test

$$\begin{aligned} H_0: &\quad G = F, \\ H_1: &\quad G \neq F, \end{aligned} \tag{6.3.9}$$

where $F(\boldsymbol{x})$ is a known and continuous multivariate c.d.f. Let

$$F_{\boldsymbol{a}}(\boldsymbol{x}) = F(\boldsymbol{a}'\boldsymbol{x}) \text{ and } G_{\boldsymbol{a}}(\boldsymbol{x}) = G(\boldsymbol{a}'\boldsymbol{x})$$

for any $\boldsymbol{a} \in U_s$. Then the Neyman statistic can be used for the null hypothesis

$$H_0^{\boldsymbol{a}}: \ G_{\boldsymbol{a}} = F_{\boldsymbol{a}}.$$

The corresponding Neyman statistic is

$$K_N^{\boldsymbol{a}} = \frac{1}{N}\sum_{i=1}^{m}\Big[\sum_{j=1}^{N}\pi_i(F_{\boldsymbol{a}}(\boldsymbol{x}_j))\Big]^2.$$

The test statistic for (6.3.9) by using the PP idea is

$$K_N = \sup_{\|\boldsymbol{a}\|=1} K_N^{\boldsymbol{a}},$$

which is called the **PP Neyman statistic.** Li and Zha (1991) have found the limiting distribution of K_N and the value of K_N can be approximated by the NTM.

6.4 Tests for multinormality (II)

In the last section we have discussed tests for multinormality. We continue to study this problem in this section and introduce some new tests for multinormality which are based on entropy, density estimation, the projection pursuit method and the NTM, and was proposed by Zhu, Wong and Fang (1993).

It is well-known that the **entropy** of a s-dimensional distribution $F(\boldsymbol{x})$ with a p.d.f. $f(\boldsymbol{x})$ is defined by

$$H(f) = -\int_D \log f(\boldsymbol{x})dF(\boldsymbol{x}) \qquad (6.4.1)$$

where $D = \{\boldsymbol{x} : f(\boldsymbol{x}) > 0\}$ is the support of f. Let $\boldsymbol{x}_1, \cdots, \boldsymbol{x}_N$ be a sample from the population F and F_N be its empirical distribution. Let $\hat{f}$ be an estimate of f by some nonparametric method. Then the **sample entropy** would have the form

$$H_N(f) = -\frac{1}{N}\sum_{j=1}^{N}\log \hat{f}(\boldsymbol{x}_j). \qquad (6.4.2)$$

When $s = 1$ Vasicek (1976) suggested using a nearest neighbour estimate as $\hat{f}$ with the form

$$\hat{f}(x_{(j)}) = \frac{N}{2m}(X_{(j+m)} - X_{(j-m)}), \tag{6.4.3}$$

where $X_{(1)}, \cdots, X_{(N)}$ are the order statistics of $X_1, \cdots, X_N$, and m is a positive integer that acts as a smoothing parameter. It is known that the entropy of the standard normal distribution $N(0,1)$ is

$$H(\phi) = -\int_{-\infty}^{\infty} \log \phi(x) d\Phi(x) = \frac{1}{2}\log(2\pi e),$$

where ϕ and Φ are the p.d.f. and c.d.f. of $N(0,1)$, respectively. Therefore, when the mean and variance of F is equal to 0 and 1 respectively, we might use

$$T = |H_n(f) - \frac{1}{2}\log(2\pi e)|$$

or

$$\begin{aligned} T^* &= |\exp(H_n(f)) - \exp(\frac{1}{2}\log(2\pi e))| \\ &= |\exp(H_n(f)) - \sqrt{2\pi e}| \end{aligned}$$

for testing normality, because $H(f) \leq \frac{1}{2}\log(2\pi e)$ for any f and the equality holds iff $f \equiv \phi$. The idea can be similarly extended to the case $s > 1$.

Suppose that we want to test the hypothesis (6.3.1). Equivalently we may test the hypothesis (6.3.2). Let $\boldsymbol{x} \sim F(\boldsymbol{x})$ and have the first two order moments. Now for any $\boldsymbol{a} \in R^s$

$$Z(\boldsymbol{a}) = \frac{\boldsymbol{a}'\boldsymbol{x} - \boldsymbol{a}'\boldsymbol{\mu}}{\boldsymbol{a}'\Sigma\boldsymbol{a}} \sim N(0,1),$$

where $\boldsymbol{\mu} = E(\boldsymbol{x})$ and $\Sigma = \mathrm{Cov}(\boldsymbol{x})$. Given a sample $\boldsymbol{x}_1, \cdots, \boldsymbol{x}_N$ of $F(\boldsymbol{x})$ we can find the sample mean vector $\bar{\boldsymbol{x}}$ and the sample covariance matrix $\boldsymbol{V}$ and an estimate of $Z(\boldsymbol{a})$ by

$$\hat{Z}(\boldsymbol{a}) = \frac{\boldsymbol{a}'\boldsymbol{x} - \boldsymbol{a}'\bar{\boldsymbol{x}}}{\boldsymbol{a}'\boldsymbol{V}\boldsymbol{a}}.$$

Let $(\boldsymbol{a}'\boldsymbol{x})_{(1)}, \cdots, (\boldsymbol{a}'\boldsymbol{x})_{(N)}$ be the order statistics of $\boldsymbol{a}'\boldsymbol{x}_1, \cdots, \boldsymbol{a}'\boldsymbol{x}_N$. Then the corresponding order variables $\hat{Z}_{(1)}(\boldsymbol{a}) \leq \hat{Z}_{(2)}(\boldsymbol{a}) \leq \cdots \leq \hat{Z}_{(N)}(\boldsymbol{a})$ of $\hat{Z}_1(\boldsymbol{a}), \cdots, \hat{Z}_N(\boldsymbol{a})$ have the form

$$\hat{Z}_{(j)}(\boldsymbol{a}) = \frac{(\boldsymbol{a}'\boldsymbol{x})_{(j)} - \boldsymbol{a}'\bar{\boldsymbol{x}}}{\boldsymbol{a}'\boldsymbol{V}\boldsymbol{a}}, \quad j = 1, \cdots, N. \tag{6.4.4}$$

Similar to Vasicek's idea and considering robustness, Zhu, Wong and Fang (1993) suggested using

$$H_{np}(\boldsymbol{a}, f) = \frac{-1}{(1-2p)N} \sum_{j=pN}^{(1-p)N} \{\log \frac{N}{2m}[\hat{Z}_{(j+m)}(\boldsymbol{a}) - \hat{Z}_{(j-m)}(\boldsymbol{a})]\} \tag{6.4.5}$$

as an estimate of $(2p)$-trimmed projected entropy $H_p(\boldsymbol{a}, f)$, where

$$H_p(\boldsymbol{a}, f) = \frac{1}{1-2p} \int_{b_p}^{b_{1-p}} \log f_{\boldsymbol{a}}(z) dF_{\boldsymbol{a}}(z), \tag{6.4.6}$$

$f_{\boldsymbol{a}}$ and $F_{\boldsymbol{a}}$ are the p.d.f. and c.d.f. of $Z(\boldsymbol{a})$, and b_p and b_{1-p} of p- and $(1-p)$- quantiles of $Z(\boldsymbol{a})$, respectively. The p value should be small, for example, $p = 0.01$. The $(2p)$-trimmed entropy of $N(0.1)$ is

$$H_p = \frac{1}{1-2p} \int_{Z_p}^{Z_{1-p}} \log \phi(x) dF(x), \tag{6.4.7}$$

where Z_p and Z_{1-p} are p- and $(1-p)$- quantiles of $N(0,1)$, respectively. In terms of $Z(\boldsymbol{a})$, the hypothesis (6.3.2) can be expressed as

$$H_0 : Z(\boldsymbol{a}) \sim N(0,1) \quad \text{for each} \quad \boldsymbol{a} \in U_s. \tag{6.4.8}$$

Hence a Kolmogorov-Smirnov type statistic

$$T_1(N, p) = \sup_{\boldsymbol{a} \in U_s} |\exp(H_{Np}(\boldsymbol{a}, f)) - \exp(H_p)| \tag{6.4.9}$$

and a Cramer-von Mises type statistic

$$T_2(N, p) = \int_{\boldsymbol{a} \in U_s} \{\exp(H_{Np}(\boldsymbol{a}, f) - \exp(H_p)\}^2 d\nu, \tag{6.4.10}$$

where $d\nu$ is the volume element of U_s, are recomended to test (6.4.8). Some properties of large sample of $T_1(n,p)$ and $T_2(n,p)$ are given by Zhu, Wong and Fang (1993). These properties show that the two statistics $T_1(n,p)$ and $T_2(n,p)$ are reasonable.

Let $\{\boldsymbol{a}_k, k = 1, \cdots, n\}$ be an NT-net on U_s. If n is large, the hypothesis (6.4.8) can be approximated by

$$H_o : Z(\boldsymbol{a}_k) \sim N(0,1), \quad k = 1, \cdots, n, \tag{6.4.11}$$

and the corresponding approximations of $T_1(N,P)$ and $T_2(N,P)$ are

$$E_1(N,p) = \max_{1\le i\le n} |\exp(H_{np}(\boldsymbol{a}_i, f) - \exp(H_p)| \tag{6.4.12}$$

and

$$E_2(N,p) = \frac{1}{n}\sum_{i=1}^{n}\{\exp(H_{np}(\boldsymbol{a}_i, f) - \exp(H_p)\}^2, \tag{6.4.13}$$

respectively. The critical values $T_1(N,p)$ and $T_2(N,p)$, for $n = 30 \sim 100, s = 2 \sim 5$ and $\alpha = 1\%, 5\%$ and 10% are given in Tables 6.7 and 6.8, respectively. These values are calculated by Monte Carlo simulation.

Table 6.6 *Powers of $Sk_{\max}$, $Ku_{\max}$, T_1 and T_2*

No	*Alternatives*	$Sk_{\max}$	$Ku_{\max}$	T_1	T_2
1.	$\boldsymbol{x} \sim N_4(\boldsymbol{0}, \boldsymbol{I}_4)$	0.00	0.00	0.06	0.06
2.	$\boldsymbol{x} = \{E_1, E_2, E_3, E_4\}$	1.00	0.94	1.00	1.00
3.	$\boldsymbol{x} = \{\chi_1, \chi_2, \chi_3, \chi_4\}$	1.00	0.96	1.00	1.00
4.	$\boldsymbol{x} \sim Mt_4(50, \boldsymbol{0}, \boldsymbol{I}_4)$	0.00	0.02	0.06	0.06
5.	$\boldsymbol{x} \sim U(B_4)$	0.00	0.00	1.00	0.88
6.	$\boldsymbol{x} = \{G_1, G_2, G_3, G_4\}$	0.90	0.68	1.00	1.00
7.	$\boldsymbol{x} = \{B_1, B_2, B_3, B_4\}$	1.00	0.90	1.00	1.00
8.	$\boldsymbol{x} = \{N_3(\boldsymbol{0}, \boldsymbol{I}_3), E_4\}$	0.76	0.46	0.98	1.00
9.	$\boldsymbol{x} = \{N_2(\boldsymbol{0}, \boldsymbol{I}_2), E_3, E_4\}$	0.96	0.80	1.00	1.00
10.	$\boldsymbol{x} = \{N_3(\boldsymbol{0}, \boldsymbol{I}_3), \chi_4)\}$	0.92	0.52	1.00	1.00
11.	$\boldsymbol{x} = \{N_2(\boldsymbol{0}, \boldsymbol{I}_2), \chi_3, \chi_4\}$	0.98	0.70	1.00	1.00

Table 6.7 *Critical values of* $T_1(n,p), P(T_1(n,p) > T_1(n,p,\alpha)) = \alpha, p = 1\%$

size	α	10%	5%	1%
$d = 2$	$N = 180$			
30		1.312	1.395	1.575
40		1.087	1.164	1.327
50		0.806	0.866	0.997
60		0.696	0.751	0.849
70		0.642	0.691	0.776
80		0.595	0.633	0.729
90		0.562	0.598	0.669
100		0.534	0.568	0.640
$d = 3$	$N = 233$			
30		1.461	1.539	1.710
40		1.229	1.285	1.451
50		0.934	0.984	1.092
60		0.791	0.836	0.938
70		0.717	0.761	0.850
80		0.667	0.702	0.767
90		0.631	0.668	0.734
100		0.595	0.620	0.679
$d = 4$	$N = 266$			
30		1.503	1.572	1.680
40		1.238	1.300	1.407
50		0.967	1.017	1.143
60		0.821	0.856	0.949
70		0.736	0.781	0.855
80		0.688	0.723	0.802
90		0.645	0.675	0.748
100		0.614	0.645	0.705
$d = 5$	$N = 440$			
30		1.556	1.611	1.736
40		1.296	1.347	1.480
50		1.002	1.048	1.136
60		0.857	0.901	0.982
70		0.771	0.820	0.902
80		0.710	0.743	0.818
90		0.668	0.697	0.762
100		0.631	0.657	0.731

Table 6.8 *Critical values of* $T_2(n,p)$, $P(T_2(n,p) > T_2(n,p,\alpha)) = \alpha, p = 1\%$

size	α	10%	5%	1%
$d = 2$	$n = 180$			
30		0.569	0.653	0.833
40		0.369	0.419	0.543
50		0.231	0.254	0.319
60		0.155	0.172	0.212
70		0.128	0.141	0.176
80		0.109	0.121	0.146
90		0.097	0.107	0.127
100		0.087	0.095	0.117
$d = 3$	$N = 233$			
30		0.502	0.547	0.642
40		0.325	0.352	0.409
50		0.196	0.209	0.244
60		0.136	0.144	0.165
70		0.112	0.119	0.138
80		0.097	0.103	0.119
90		0.086	0.091	0.099
100		0.078	0.083	0.094
$d = 4$	$N = 266$			
30		0.482	0.509	0.552
40		0.304	0.325	0.361
50		0.184	0.195	0.214
60		0.127	0.134	0.146
70		0.105	0.110	0.118
80		0.091	0.095	0.103
90		0.081	0.084	0.091
100		0.074	0.077	0.085
$d = 5$	$N = 440$			
30		0.456	0.477	0.513
40		0.293	0.305	0.328
50		0.175	0.182	0.197
60		0.123	0.129	0.138
70		0.101	0.105	0.111
80		0.087	0.090	0.095
90		0.078	0.080	0.085
100		0.071	0.074	0.079

To study power of T_1 and T_2 several alternatives are considered as listed in Table 6.6, where $\{E_1, E_2, E_3, E_4\}$ denotes that $E_1 \sim E_4$ are independent each having the standard exponential distribution; $\{N_3(\mathbf{0}, \boldsymbol{I}_3), E_4\}$ denotes that the first three components of $\boldsymbol{x}$ have the distribution $N_3(\mathbf{0}, \boldsymbol{I}_3)$ and is independent of E_4 which has the standard exponential distribution. A similar explanation holds for the others where χ stands for the chi-square distribution with 2 degrees of freedom, G the gamma distribution $Ga(2,1)$, and B the beta distribution $beta(3,2)$. Table 6.6 lists powers of statistics $Sk_{\max}, ku_{\max}, T_1(N,p)$ and $T_2(N,p)$ for comparisons, where $\alpha = 5\%, N = 100$ and $s = 4$. We can see that T_1 and T_2 have better powers that of $Sk_{\max}, Ku_{\max}$. The reader can find more conclusions from Table 6.6.

Example 6.5

The *Iris Setosa, Iris versicolor and Iris virginica* data sets are well-known in multivariate literatures. The last two data sets are often used as examples in classification and algorithm of clustering. The first one had been used by Rincon-Gallardo *et al.* (1979), Koziol (1982) and Csörgö (1989) in tests for multivariate normality. The *Iris setosa* data set was collected originally by Anderson (1935) and analyzed by Fisher (1936). Small (1980) re-examined the data based on the marginal skewness and kurtosis, and concluded that it significantly departs from the quadrivariate normality. Royston (1983) also obtained the same conclusion by using the generalized Shapiro-Wilk W test. We use this data set to illustrate the approach in this section.

The *Iris Setosa* data consists of 50 quadrivariate observations. The four variables are as follows:

$$\begin{aligned} X_1 &: \text{ sepal length,} \\ X_2 &: \text{ sepal width,} \\ X_3 &: \text{ petal length,} \\ X_4 &: \text{ petal width.} \end{aligned}$$

It has been shown that the petal length and width are not normal while the sepal length and width are normal by univariate normality tests.

First we test the quadrivariate normality of $\boldsymbol{x} = (X_1, X_2, X_3, X_4)'$. We find that $T_1(50, 1\%) = 2.396 > 0.991 = T_1(50, 1\%, 5\%)$ and $T_2(50, 1\%) = 0.7439 > 0.194 = T_2(50, 1\%, 5\%)$. Therefore, we conclude that $\boldsymbol{x}$ is not normal, which is consistent with the result of Small (1980). Secondly, we consider the tests for some subsets of $\boldsymbol{x}$ as follows:

$$\begin{aligned}
\{X_1, X_2\}:\ & T_1 = 0.8524,\ T_2 = 0.2299,\ \text{Accepted;}\\
\{X_1, X_3\}:\ & T_1 = 1.7858,\ T_2 = 0.4322,\ \text{Rejected;}\\
\{X_1, X_4\}:\ & T_1 = 3.4154,\ T_2 = 2.6100,\ \text{Rejected;}\\
\{X_2, X_3\}:\ & T_1 = 1.4948,\ T_2 = 0.2794,\ \text{Rejected;}\\
\{X_2, X_4\}:\ & T_1 = 3.3753,\ T_2 = 2.3215,\ \text{Rejected;}\\
\{X_3, X_4\}:\ & T_1 = 3.3299,\ T_2 = 2.0559,\ \text{Rejected.}
\end{aligned}$$

The values of the T_1 and T_2 for all paired subsets involving X_3 or X_4 indicate non-normality. The reader is encouraged to use T_1 and T_2 on all tripled subsets of $\boldsymbol{x}$.

6.5 Tests for sphericity

"Test for sphericity" can have various senses. Given a sample $\boldsymbol{x}_1, \cdots, \boldsymbol{x}_N$ from a multivariate normal distribution $N_s(\mathbf{0}, \boldsymbol{\Sigma})$ we want to test for the null distribution being spherical (Definition 2.1), which is equivalent to testing $H_0 : \boldsymbol{\Sigma} = \boldsymbol{I}_s$. This test is often called test for sphericity in multivariate analysis. Alternatively, consider two classes of elliptically symmetric distributions: $\mathcal{F}_o = \{F(\boldsymbol{x}) : F(\boldsymbol{x}) \text{ is spherical and has a p.d.f.}\}$ and $\mathcal{F}_1 = \{F(\boldsymbol{x}) : F(\boldsymbol{x}) \text{ is elliptical } EC_s(\mathbf{0}, \boldsymbol{\Sigma}, g) \text{ and has a p.d.f.}\}$. We want to test for spherical symmetry, i.e., $H_o : F(\boldsymbol{x}) \in \mathcal{F}_o$ agains $H_1 : F(\boldsymbol{x}) \in \mathcal{F}_1$ which is another kind of test for sphericity (Fang and Zhang (1990), Section 5.5). In this section we consider a test for sphericity in the more general sense as follows. Given an s-dimensional distribution function $G(\boldsymbol{x})$, we often require to test the following hypothesis of sphericity

$$\begin{aligned}
H_0 :\ & G(\boldsymbol{x}) \text{ is spherical,}\\
H_1 :\ & G(\boldsymbol{x}) \text{ is not spherical.}
\end{aligned} \tag{6.5.1}$$

The following fact is pointed out by Fang, Kotz and Ng (1990, Theorem 2.5).

Lemma 6.1
Let $\boldsymbol{x} \in R^s$ be a random vector. Then $\boldsymbol{x}$ has a spherical distribution if and only if for each $\boldsymbol{a} \in U_s$

$$\boldsymbol{a}'\boldsymbol{x} \stackrel{\mathrm{d}}{=} X_1,$$

where X_1 is the first component of $\boldsymbol{x}$.

In view of Lemma 6.1 the hypothesis (6.5.1) can be changed into

$$\begin{aligned} H_0 &: \text{all } \boldsymbol{a}'\boldsymbol{x},\ \boldsymbol{a} \in U_s, \text{ have the same distribution} \\ H_1 &: \boldsymbol{a}'\boldsymbol{x} \text{ and } \boldsymbol{b}'\boldsymbol{x} \text{ have different distributions} \\ &\quad \text{for some } \boldsymbol{a}, \boldsymbol{b} \in U_s. \end{aligned} \tag{6.5.2}$$

Let $\{\boldsymbol{a}_1, \cdots, \boldsymbol{a}_m\}$ be an NT-net on U_p. If m is large enough, the hypothesis (6.5.2) can be approximated by

$$H_0^* : \boldsymbol{a}_i'\boldsymbol{x},\ i = 1, \cdots, m, \text{ have the same distribution.} \tag{6.5.3}$$

Fang, Zhu and Bentler (1993) suggested some **Wilcoxon Type statistics** for testing (6.5.3). For give $0 < k < l \leq m$, consider a two sample problem:

$$\begin{aligned} \text{I}: &\ \boldsymbol{a}_k'\boldsymbol{x}_1, \cdots, \boldsymbol{a}_k'\boldsymbol{x}_N, \\ \text{II}: &\ \boldsymbol{a}_l'\boldsymbol{x}_1, \cdots, \boldsymbol{a}_l'\boldsymbol{x}_N, \end{aligned}$$

where $\boldsymbol{x}_1, \ldots, \boldsymbol{x}_N$ is a sample from the population. The Wilcoxon Type statistics for this two-sample problem is (Serfling, (1980), p.175)

$$V_N(\boldsymbol{a}_k, \boldsymbol{a}_l) = \frac{1}{N^2} \sum_{i=1}^{N} \sum_{j=1}^{N} I\{\boldsymbol{a}_k'\boldsymbol{x}_i < \boldsymbol{a}_l'\boldsymbol{x}_j\}, \tag{6.5.4}$$

where $I\{A\}$ is the indicator function of A. The following statistic

$$T_N = \min_{\substack{1 \leq k, l \leq m \\ k \neq l}} \{V_N(\boldsymbol{a}_k, \boldsymbol{a}_l)\} \tag{6.5.5}$$

is recommended to test the hypothesis (6.5.3).

The remaining problem is to find the distribution of T_N. When N and m are small statistical simulation may help us to obtain an approximation to the distribution. When N is large, we may use the limiting distribution of T_N. Note that $V_N(\boldsymbol{a}_k, \boldsymbol{a}_l)$ is constructed by two dependent samples, not two independent samples, and so differs from the traditional two-sample Wilcoxon statistics. The following result due to Fang, Zhu and Bentler (1993) gives the limiting distribution of T_N only in the case of the $\boldsymbol{a}'_k$ being orthogonal.

Theorem 6.1

Assume that G is spherically symmetric and has zero probability at the origin. Then the distribution of the row random vector $\{\sqrt{N}(V_N(\boldsymbol{a}_k, \boldsymbol{a}_l) - \frac{1}{2})\colon \boldsymbol{a}_k \perp \boldsymbol{a}_l,\ 1 \leq k < l \leq s\}$ converges to an $s(s-1)/2$ variate normal distribution $N(\boldsymbol{0}, \frac{1}{12}\boldsymbol{V})$, where the elements of $\boldsymbol{V}$ are of the form

$$V(k,l,k_1,l_1) = \begin{cases} 0, & \text{for } k,l,l_1,k_1 \text{ are different each other,} \\ 1, & \text{for } k = k_1, l \neq l_1 \text{ or } k \neq k_1, l = l_1, \\ 2, & \text{for } k = k_1 \text{ and } l = l_1, \\ -1, & \text{for } k = l_1 \text{ or } k_1 = l. \end{cases} \tag{6.5.6}$$

We see that the limiting distribution of $\{\sqrt{N}(V_N(\boldsymbol{a}_k, \boldsymbol{a}_l) - \frac{1}{2})\colon 1 \leq k < l \leq s\}$ is independent of G, and so is $(T_N - \frac{1}{2})$.

Theorem 6.2

Under the assumptions in Theorem 6.1, we have for any real number λ

$$\begin{aligned} &\lim_{n\to\infty} P\Big\{\sqrt{N}\Big(T_N - \frac{1}{2}\Big) \leq -\lambda\Big\} \\ &= P\{\min_{\substack{1\leq k,l\leq s \\ k\neq l}} \{Y_l - Y_k\} \leq -\lambda\} \\ &= 1 - \int_{-\infty}^{\infty} s(\Phi(\sqrt{12}\lambda + z) - \Phi(z))^{s-1}\phi(z)dz, \end{aligned} \tag{6.5.7}$$

where $Y_1, \cdots, Y_s$ are independent random variables from $N(0, 1/12)$, and $\Phi(\cdot)$ and $\phi(\cdot)$ are the standard normal distribution function and its density function respectively.

In terms of Theorem 6.2, we can make a necessary test for sphericity of G, because under H_0, G must have, at any s orthogonal directions $\boldsymbol{a}_1, \cdots, \boldsymbol{a}_s$, the same marginal distribution. Some critical values of λ_α that

$$P\{\min_{\substack{1\leq k,l\leq s \\ k\neq l}} \{Y_l - Y_k\} < -\lambda_\alpha\} = \alpha \tag{6.5.8}$$

can be found in Table 6.9. The proofs of the above theorems are given in Fang, Zhu and Bentler (1993). Why we call the test a necessary test? Because it is the necessary condition that under H_0, G has the same marginal distribution at any s orthogonal directions. Therefore, if $\min_{\substack{1\leq k,l\leq s \\ k\neq l}}\{Y_l - Y_k\} > -\lambda_\alpha$, the hypothesis H_0 is rejected at significance level α; on the other hand if $\min_{\substack{1\leq k,l\leq s \\ k\neq l}}\{Y_l - Y_k\} < -\lambda_\alpha$, the hypothesis is not really accepted.

We now pay attention to the properties of $\boldsymbol{V}$. It is clear that $\boldsymbol{V}$ is singular, and we shall look for the eigenvalues and rank of $\boldsymbol{V}$. In the following, we first give an example to illustrate the properties of $\boldsymbol{V}$, then present the general result.

Example 6.6

When $s = 4$, the covariance matrix $\boldsymbol{V}$ can be calculated by (6.5.6) and is given by

$$\boldsymbol{V} = \begin{array}{c} \begin{matrix} (1,2) & (1,3) & (1,4) & (2,3) & (2,4) & (3,4) \end{matrix} \\ \begin{pmatrix} 2 & 1 & 1 & -1 & -1 & 0 \\ 1 & 2 & 1 & 1 & 0 & -1 \\ 1 & 1 & 2 & 0 & 1 & 1 \\ -1 & 1 & 0 & 2 & 1 & -1 \\ -1 & 0 & 1 & 1 & 2 & 1 \\ 0 & -1 & 1 & -1 & 1 & 2 \end{pmatrix} \end{array}. \tag{6.5.9}$$

We are going to show that the rank $(\boldsymbol{V}) = s - 1 = 3$ and the eigenvalues of $\boldsymbol{V}$ are 0 and 4. Note

$$|\boldsymbol{V} - \lambda \boldsymbol{I}| = \begin{array}{c} \begin{matrix} (1,2) & (1,3) & (1,4) & (2,3) & (2,4) & (3,4) \end{matrix} \\ \begin{vmatrix} 2-\lambda & 1 & 1 & -1 & -1 & 0 \\ 1 & 2-\lambda & 1 & 1 & 0 & -1 \\ 1 & 1 & 2-\lambda & 0 & 1 & 1 \\ -1 & 1 & 0 & 2-\lambda & 1 & -1 \\ -1 & 0 & 1 & 1 & 2-\lambda & 1 \\ 0 & -1 & 1 & -1 & 1 & 2-\lambda \end{vmatrix} \end{array}.$$

We make column transformations $(2,3) \Longrightarrow (2,3) + (1,2) - (1,3)$, $(2,4) \Longrightarrow (2,4) + (1,2) - (1,4)$ and $(3,4) \Longrightarrow (3,4) + (1,2) - (1,4)$ to derive

$$|\boldsymbol{V} - \lambda \boldsymbol{I}| = \lambda^3 \begin{array}{c} \begin{matrix} (1,2) & (1,3) & (1,4) & (2,3) & (2,4) & (3,4) \end{matrix} \\ \begin{vmatrix} 2-\lambda & 1 & 1 & -1 & -1 & 0 \\ 1 & 2-\lambda & 1 & 1 & 0 & -1 \\ 1 & 1 & 2-\lambda & 0 & 1 & 1 \\ -1 & 1 & 0 & -1 & 1 & -1 \\ -1 & 0 & 1 & 1 & -1 & 1 \\ 0 & -1 & 1 & -1 & 1 & -1 \end{vmatrix} \end{array}.$$

We then make column transformations $(1,2) \Longrightarrow (1,2) - (2,3) - (2,4)$, $(1,3) \Longrightarrow (1,3) + (2,3) - (3,4)$ and $(1,4) \Longrightarrow (1,4) + (2,4) + (3,4)$ to obtain

$$\begin{aligned} |\boldsymbol{V} - \lambda \boldsymbol{I}| &= \lambda^3 \begin{array}{c} \begin{matrix} (1,2) & (1,3) & (1,4) & (2,3) & (2,4) & (3,4) \end{matrix} \\ \begin{vmatrix} 4-\lambda & 0 & 4-\lambda & 1 & -1 & 0 \\ 0 & 4-\lambda & 4-\lambda & 1 & 0 & -1 \\ 0 & 0 & 4-\lambda & 0 & 1 & 1 \\ 0 & 0 & 0 & -1 & 0 & 0 \\ 0 & 0 & 0 & 0 & -1 & 0 \\ 0 & 0 & 0 & 0 & 0 & -1 \end{vmatrix} \end{array}. \\ &= (-\lambda)^3 (4-\lambda)^3. \end{aligned}$$

This fact implies that there exists an orthogonal matrix $\boldsymbol{Q}$ such that

$$\boldsymbol{Q}'\boldsymbol{V}\boldsymbol{Q} = s \begin{pmatrix} \boldsymbol{I}_{s-1} & \boldsymbol{O} \\ \boldsymbol{O} & \boldsymbol{O} \end{pmatrix},$$

By the same technique used in Example 6.4, we have the following general result:

Lemma 6.2
For $s \geq 2$ there exists a $\boldsymbol{Q} \in O(s(s-1)/2)$, the set of orthogonal matrices of order $\frac{s(s-1)}{2}$, such that

$$\boldsymbol{Q}'\boldsymbol{V}\boldsymbol{Q} = s\begin{pmatrix} \boldsymbol{I}_{s-1} & \boldsymbol{O} \\ \boldsymbol{O} & \boldsymbol{O} \end{pmatrix}. \tag{6.5.10}$$

PROOF Consider the determinant $|\boldsymbol{V} - \lambda\boldsymbol{I}| = |\{\boldsymbol{u}(k,l) : 1 \leq k < l \leq s\}| = 0$, where the $\boldsymbol{u}(k,l)$'s are the column vectors. First we make the column transformations (we omit $\boldsymbol{u}$ for simplicity)

$$(k,l) \Longrightarrow (k,l) + (1,k) - (1,l)$$

for all $1 < k < l \leq s$. Then take a factor $\lambda^{(s-1)(s-2)/2}$ out of the determinant. Finally we use the following column transformations

$$\begin{aligned}(1,l) \Longrightarrow &(1,l) + (2,l) + \cdots (l-1,l) - (l,l+1) - \cdots - (l,p) \\ &\text{for } l = 2, \cdots, p,\end{aligned}$$

and find

$$|\boldsymbol{V} - \lambda\boldsymbol{I}| = (-\lambda)^{(s-1)(s-2)/2}(s-\lambda)^{s-1}. \qquad \square$$

We now proceed to discuss how to apply the above necessary test for sphericity. First we need to choose s projection directions $\boldsymbol{a}_1, \cdots, \boldsymbol{a}_s$ randomly which are orthogonal each other. Equivalently we generate an orthogonal matrix $\boldsymbol{G}$ from the uniform distribution on $O(s)$, i.e. $U(O(s))$. An efficient method is proposed by Anderson, Olkin and Underhill (1987). In next section we shall give a detailed discussion on this problem.

After the projection directions are chosen, we calculate the value of $\sqrt{N}(T_N - \frac{1}{2})$, and then compare it with the critical value λ_α at $\alpha = 1\%$ or $\alpha = 5\%$, which are listed in Table 6.9 for $3 \leq s \leq 10$.

Table 6.9 *The critical values of* λ_α

s	$\alpha = 1\%$	$\alpha = 5\%$	s	$\alpha = 1\%$	$\alpha = 5\%$
3	1.19	0.96	7	1.41	1.21
4	1.27	1.05	8	1.44	1.24
5	1.33	1.12	9	1.47	1.27
6	1.38	1.17	10	1.49	1.49

In the following, we carry out a simulation by using a sample from 4-dimensional multivariate distribution with density function

$$\frac{\Gamma(10)}{(10\pi)^2\Gamma(5)}(1+\boldsymbol{x}'\boldsymbol{x}/10)^{-7}.$$

The sample of size 200 is generated by the Monte Carlo method. Generate an orthogonal random matrix $\boldsymbol{G} = (\boldsymbol{a}_1, \cdots, \boldsymbol{a}_4)$, and calculate $V_N(\boldsymbol{a}_i, \boldsymbol{a}_j)$ to derive $\sqrt{N}(T_N - 1/2)$. In Table 6.10 we present the values of $\boldsymbol{a}_1, \cdots, \boldsymbol{a}_4$, $V(i,j) \equiv V_N(\boldsymbol{a}_i, \boldsymbol{a}_j)$, $1 \leq i, j \leq 4$, and $\sqrt{200}(T_{200} - 1/2)$. Then compare $\sqrt{200}(T_{200} - 1/2) = -0.4968$ with $\lambda_{0.01}$ or $\lambda_{0.05}$ for $s = 4$ which are listed in Table 6.9. We can accept the hypothesis of the sample coming from a spherical distribution.

Furthermore, if the practical problem needs strict sphericity for the distribution, it does not suffice to illustrate its sphericity by means of the above test for a group of the orthogonal projections. In this case, we shall adopt the idea of Bonferroni multiple comparison, that is, generate randomly p orthogonal matrix

Table 6.10 *The Values of* $\boldsymbol{a}_i$*'s and* $V(i,j)$*'s*

$\boldsymbol{a}_1$	0.81480	−0.19538	−0.01916	−0.54549
$\boldsymbol{a}_2$	0.23949	−0.00261	−0.88917	0.38990
$\boldsymbol{a}_3$	0.03868	0.95250	−0.11488	−0.27936
$\boldsymbol{a}_4$	0.52654	0.23357	0.44251	0.68730
$V(1,j)$	0.50000	0.46651	0.49595	0.46487
$V(2,j)$	0.63349	0.50000	0.53314	0.49648
$V(3,j)$	0.50405	0.46056	0.50000	0.47196
$V(4,j)$	0.53513	0.50352	0.52804	0.50000
T_{200}		0.46487		

$\boldsymbol{G}_i$, $i = 1, \cdots, p$, such as $p = 5$. Then let α/p and $\lambda_{\alpha/p}$ be the respective new significance level and the critical value; for example,

$\alpha = 0.05$, $p = 5$, $\alpha/p = 0.01$. The statistic is $\min_{1\le i\le p} \sqrt{N}(T_N^{(i)} - \frac{1}{2})$, where $T_N^{(i)}$ corresponds to G_i. Of course, it is an interesting matter to study the limit distribution of $\min_{1\le i\le p} \sqrt{N}(T_N^{(i)} - \frac{1}{2})$.

6.6 Projection Pursuit

6.6.1 Projection Pursuit Methods

The term "**projection pursuit**" (PP) was first used by Friedman and Tukey in 1974 and was comprehensively discussed by Huber (1985). The PP method reveals structure in the original data by offering selected low-dimensional orthogonal projections of it for inspection. We cannot directly appreciate patterns of variation in more than three dimensions. In practice, the projection will be onto one- or two-dimensional space.

Let $\boldsymbol{X}$ be an $N \times p$ matrix of observations with p variables and N observations. For any $\boldsymbol{a} \in R^p$, $\boldsymbol{X}\boldsymbol{a}$ is a column N-vector which is the orthogonal projection of the sample onto direction $\boldsymbol{a}$. Without loss of generality we always assume $\boldsymbol{a}'\boldsymbol{a} = 1$, or $\boldsymbol{a} \in U_p$. If H is a function that measures the interest of a one-dimensional sample, then $I(\boldsymbol{a}) \equiv H(\boldsymbol{X}\boldsymbol{a})$ is called a **projection index**. For example the skewness and kurtosis of $\boldsymbol{a}'\boldsymbol{x}_1, \cdots, \boldsymbol{a}'\boldsymbol{x}_N$ in section 6.3 are two projection indexes. PP attempts to find the projection direction $\boldsymbol{a}$ which is a global optimum of $I(\boldsymbol{a})$, or at least local optimum of $I(\boldsymbol{a})$.

Principal component analysis (PCA), canonical correlation analysis, and **correspondence analysis** are based on the same idea as PP. For example, $I(\boldsymbol{a})$ is the total sample variance of $\boldsymbol{X}\boldsymbol{a}$ in PCA. Fortunately we can find the analytic solution of the global optimum of $I(\boldsymbol{a})$ in these methods. In most cases, we have to use numerical optimization methods to find approximations to the optimum of $I(\boldsymbol{a})$. For example, if $I(\boldsymbol{a})$ is the sample skewness or the sample Kurtosis (section 6.3). Assume that $I(\boldsymbol{a})$ is continuous on U_p. Therefore, we may use SNTO to find approximation of $\max_{\boldsymbol{a}\in U_p} I(\boldsymbol{a})$ and the corresponding maximum direction $\boldsymbol{a}^*$. We used this idea in section 6.3.

For orthogonal projection onto a space of dimension $q > 1$, $\boldsymbol{a}$ is replaced by a $p \times q$ matrix $\boldsymbol{A}$, so $\boldsymbol{X}\boldsymbol{A}$ is an $N \times q$ matrix, and the corresponding projection index becomes $I(\boldsymbol{A}) = H(\boldsymbol{X}\boldsymbol{A})$. For most cases we can assume $\boldsymbol{A} \in O(p, q)$, where

$$O(p,q) = \{\boldsymbol{U} : \boldsymbol{U} : p \times q,\ \boldsymbol{U}'\boldsymbol{U} = \boldsymbol{I}_q\} \tag{6.6.1}$$

is the **Stiefel manifold**. Obviously, $O(p,q) = O(p)$ when $p = q$. If we attempt to find the optimum $\boldsymbol{A}^*$ over $O(p,q)$ by SNTO we need a method to generate a set of matrices which are uniformly scattered on $O(p,q)$, i.e., an NT-net on $O(p,q)$.

A simple way to generate an NT-net on $O(p,q)$ is based on the following theorem which is an extension of (2.2.23).

Theorem 6.3

Let random matrix $\boldsymbol{X}$ be uniformly distributed on $O(p,q)$. Then $\boldsymbol{X}$ has a stochastic representation

$$\boldsymbol{X} \stackrel{d}{=} \boldsymbol{Y}(\boldsymbol{Y}'\boldsymbol{Y})^{-1/2}, \tag{6.6.2}$$

where $\boldsymbol{Y}$ is an $p \times q$ random matrix with i.i.d. elements each having the standard normal distribution $N(0,1)$, i.e., $\boldsymbol{Y}$ has a matrix normal distribution $N_{p\times q}(\boldsymbol{0}, \boldsymbol{I}_p \otimes \boldsymbol{I}_q)$. Here $\boldsymbol{A}^{-1/2}$ denotes the non-negative definite square root of $\boldsymbol{A}$.

The proof of Theorem 6.3 can be found in Fang and Zhang (1990, p.101). Let $\boldsymbol{A} = (\boldsymbol{a}_1, \cdots, \boldsymbol{a}_q)$ be an $p \times q$ matrix. Define an pq-dimensional vector

$$\text{vec}\,(\boldsymbol{A}) = \begin{pmatrix} \boldsymbol{a}_1 \\ \vdots \\ \boldsymbol{a}_q \end{pmatrix}, \tag{6.6.3}$$

where "vec" can be realized as an operator. Obviously, $\boldsymbol{Y} \sim N_{p\times q}(\boldsymbol{0}, \boldsymbol{I}_p \otimes \boldsymbol{I}_q)$ if $\text{vec}(\boldsymbol{Y}) \sim N_{pq}(\boldsymbol{0}, \boldsymbol{I}_{pq})$. With the fact (6.6.2) we propose the following algorithm:

Step 1 Generate an NT-net $\{\boldsymbol{c}_k, k = 1, \cdots, n\}$ on C^{pq} with which we can obtain a set of rep-points of $N_{pq}(\boldsymbol{0}, \boldsymbol{I}_{pq})$, $\{\boldsymbol{y}_k, k = 1, \cdots, n\}$, by some method introduced before (section 4.2) and consequently we have a set of rep-points of $N_{p\times q}(\boldsymbol{0}, \boldsymbol{I}_p \otimes \boldsymbol{I}_q)$, $\{\boldsymbol{Y}_k, k = 1, \cdots, n\}$, such that $\text{vec}(\boldsymbol{Y}_k) = \boldsymbol{y}_k, k = 1, \cdots, n$.

Step 2 Then $\mathcal{P}_F = \{\boldsymbol{X}_k = \boldsymbol{Y}_k(\boldsymbol{Y}_k'\boldsymbol{Y}_k)^{-1/2}, k = 1, \cdots, n\}$ is an NT-net on $O(p,q)$.

This algorithm is simple, but the uniformity of $\mathcal{P}_F$ on $O(p,q)$ is not good when n is small. The dimensionality of $O(p,q)$ is $d_{pq} =$

$pq - q(q+1)/2$. We use pq degrees of freedom to generate an d_{pq}-dimensional manifold, and waste $q(q+1)/2$ degrees of freedom. This fact causes poor uniformly of $\mathcal{P}_F$.

Another method which uses only d_{pq} variables was proposed by Fang and Li (1993).

In fact, q mutually perpendicular projection directions may be determined successively as follows: Take $\boldsymbol{a}_1 \in U_p$ arbitrarily. Consider the $(p-1)$-dimensional subspace perpendicular to $\boldsymbol{a}_1$ and take any $\boldsymbol{a}_2^*$ in it which induces a vector $\boldsymbol{a}_2 \in U_p$ in R^p. Then consider the $(p-2)$-dimensional subspace perpendicular to $\boldsymbol{a}_1$ and $\boldsymbol{a}_2$ and take an $\boldsymbol{a}_3^*$ in it which induces $\boldsymbol{a}_3 \in U_p$ in R^p and so on, until q directions $\boldsymbol{a}_1, \cdots, \boldsymbol{a}_q$ are determined successively. For example, when $p = 3, q = 2$, we choose any $\boldsymbol{a}_1$ on U_3. Now the subspace perpendicular to $\boldsymbol{a}_1$ is a two-dimensional plane. Take any direction $\boldsymbol{a}_2^*$ in this plane which induces $\boldsymbol{a}_2 \in U_3$ in R^3. Now $\{\boldsymbol{a}_1, \boldsymbol{a}_2\}$ is a pair of mutually perpendicular projection directions of R^3. It follows from (1.5.27) that $\boldsymbol{a}_1$ is determined by two angles ϕ_1 and ϕ_2, and $\boldsymbol{a}_2$ by only one angle ϕ_3. Hence if we take an NT-net on C^3, then we may get a uniformly scattered set of matrices on $O(3,2)$ by (1.5.27) and the above algorithm. To obtain an effective computation algorithm we need to solve the following two problem: (a) For any given t mutually perpendicular directions $\{\boldsymbol{a}_1, \cdots, \boldsymbol{a}_t\}$, how to find their orthogonal subspace $R_{\perp}^{p-t}$? (b) For any $(p-t)$-dimensional vector $\boldsymbol{a}^*$ in $R_{\perp}^{p-t}$, how to represent it as a vector in R^p? However these two problems can be solved satisfactorily by the so-called Givens transform in matrix algebra. Let us first introduce the Givens transform.

6.6.2 The Givens transform

Definition 6.1

A **Givens matrix** of order p is an orthogonal matrix having the following form

$$\boldsymbol{G}_{ij} \equiv \boldsymbol{G}_{ij}(\phi_{ij}) = \begin{bmatrix} \boldsymbol{I}_{i-1} & \boldsymbol{0} & \boldsymbol{0} & \boldsymbol{0} & \boldsymbol{0} \\ \boldsymbol{0} & \cos\phi_{ij} & \boldsymbol{0} & -\sin\phi_{ij} & \boldsymbol{0} \\ \boldsymbol{0} & \boldsymbol{0} & \boldsymbol{I}_{j-i+1} & \boldsymbol{0} & \boldsymbol{0} \\ \boldsymbol{0} & \sin\phi_{ij} & \boldsymbol{0} & \cos\phi_{ij} & \boldsymbol{0} \\ \boldsymbol{0} & \boldsymbol{0} & \boldsymbol{0} & \boldsymbol{0} & \boldsymbol{I}_{p-j} \end{bmatrix}, \tag{6.6.4}$$

where $i < j$ and $\phi_{ij} \in (-\frac{\pi}{2}, \frac{\pi}{2}]$.

Let $\boldsymbol{U} \equiv (u_{ij})$ be any $p \times q$ matrix. Then $\boldsymbol{G}'_{ij}\boldsymbol{U} \equiv (u^*_{ij})$ has the same rows as $\boldsymbol{U}$ except the ith and jth rows which become

$$\begin{aligned} u^*_{ik} &= u_{ik}\cos\phi_{ij} + u_{jk}\sin\phi_{ij} \\ u^*_{jk} &= -u_{ik}\sin\phi_{ij} + u_{jk}\cos\phi_{ij} \end{aligned}, \quad k = 1, \cdots, q.$$

If we want $u^*_{ji} = 0$, then we have $u_{ii}\sin\phi_{ij} = u_{ji}\cos\phi_{ij}$, and so

$$\phi_{ij} = \begin{cases} \pi/2, & \text{if } u_{ii} = 0 \\ \arctan(u_{ji}/u_{ii}), & \text{otherwise.} \end{cases} \tag{6.6.5}$$

We take $\boldsymbol{G}_{ij}$ $(i < j)$ in the following such that ϕ_{ij} is defined by (6.6.5), i.e. $u^*_{ji} = 0$. Hence $\boldsymbol{G}_{ij}$ can be regarded as an matrix operator.

Operating Givens matrices to a matrix $\boldsymbol{A}$ we transform it such that a specific column of $\boldsymbol{A}$ has all components zero except one. For example, when $\boldsymbol{U} \in U(p,q)$, $(\boldsymbol{G}'_{1p}\boldsymbol{G}'_{1,p-1}\cdots\boldsymbol{G}'_{12})\boldsymbol{U}$ has 0's in the first column besides the first element which is ± 1 since $\boldsymbol{U}'\boldsymbol{U} = \boldsymbol{I}_q$. Since $(\boldsymbol{G}'_{1p}\cdots\boldsymbol{G}'_{12})\boldsymbol{U} \in O(p,q)$, we have

$$\boldsymbol{U}^{(2)} = \boldsymbol{G}'_{1p}\cdots\boldsymbol{G}'_{12}\boldsymbol{U}^{(1)} = \begin{pmatrix} \epsilon_1 & \boldsymbol{0} \\ \boldsymbol{0} & \boldsymbol{U}_2 \end{pmatrix}, \tag{6.6.6}$$

where $\epsilon_1 = \pm 1, \boldsymbol{U}^{(1)} = \boldsymbol{U}$ and $\boldsymbol{U}_2 \in O(p-1, q-1)$. Similarly, for $t = 2, 3, \cdots, q$

$$\boldsymbol{G}'_{t,p}\cdots\boldsymbol{G}'_{t,t+1}\boldsymbol{U}^{(t)} = \begin{pmatrix} \boldsymbol{\varepsilon}_t & \boldsymbol{0} \\ \boldsymbol{0} & \boldsymbol{U}_{t+1} \end{pmatrix} \equiv \boldsymbol{U}^{(t+1)},$$

where $\boldsymbol{U}_{t+1} \in O(p-t, q-t)$, $\boldsymbol{\varepsilon}_t = \text{diag}(\varepsilon_1, \cdots, \varepsilon_t)$ and $\varepsilon_j = \pm 1, j = 1, \cdots, t$, and finally

$$\boldsymbol{G}'_{qp}\cdots\boldsymbol{G}'_{q,q+1}\boldsymbol{U}^{(q)} = \begin{pmatrix} \boldsymbol{\varepsilon}_q \\ \boldsymbol{0} \end{pmatrix}. \tag{6.6.7}$$

Therefore we have the following theorems which are essentially due to Anderson, Olkin and Underhill (1987).

Theorem 6.4
Any $\boldsymbol{U} \in O(p,q)$ $(q \leq p)$ can be expressed as

$$\boldsymbol{U} = (\boldsymbol{G}_{12} \cdots \boldsymbol{G}_{1p})(\boldsymbol{G}_{23} \cdots \boldsymbol{G}_{2p}) \cdots (\boldsymbol{G}_{q,q+1} \cdots \boldsymbol{G}_{q,p}) \begin{pmatrix} \boldsymbol{\varepsilon}_q \\ \boldsymbol{0} \end{pmatrix}, \tag{6.6.8}$$

where $\boldsymbol{\varepsilon}_q = \text{diag}(\varepsilon_1, \cdots, \varepsilon_q)$, $\varepsilon_j = \pm 1, j = 1, \cdots, q$; and $\boldsymbol{G}_{ij}$ are Givens matrices, $-\pi/2 < \phi_{ij} \leq \pi/2$, $i = 1, \cdots, q$, $j = i+1, \cdots, p$. In particular, when $p = q$, $\boldsymbol{U}$ has the parameter representation

$$\boldsymbol{U} = (\boldsymbol{G}_{12} \cdots \boldsymbol{G}_{1p})(\boldsymbol{G}_{23} \cdots \boldsymbol{G}_{2p}) \cdots (\boldsymbol{G}_{p-1,p})\boldsymbol{\varepsilon}_p. \tag{6.6.9}$$

Theorem 6.5
Suppose that random matrix $\boldsymbol{U}$ is uniformly distributed on $O(p,q)(q \leq p)$. Let $\boldsymbol{U}$ express as (6.6.8) which is a stochastic representation of $\boldsymbol{U}$. Then

(i) $\boldsymbol{G}_{ij}(i = 1, \cdots, q, j = 1, \cdots, p; i < j)$ and $\varepsilon_j(j = 1, \cdots, q)$ are mutually independent; or $\phi_{ij}(i = 1, \cdots, q, j = 1, \cdots, p, i < j)$ and $\varepsilon_j(j = 1, \cdots, q)$ are mutually independent.
(ii) $P(\varepsilon_j = 1) = P(\varepsilon_j = -1) = 0.5, j = 1, \cdots, q$.
(iii) The joint density of ϕ_{ij}'s is

$$c\Big(\prod_{j=2}^{p} \cos^{j-2}\theta_{1j}\Big)\Big(\prod_{j=3}^{p} \cos^{j-3}\theta_{2j}\Big) \cdots \Big(\prod_{j=q+1}^{p} \cos^{j-q-1}\theta_{qj}\Big), \tag{6.6.10}$$

where c is the normalizing constant.

If we directly use Theorems 6.4 and 6.5 and the inverse transformation method to generate an NT-net on $O(p,q)$, it takes a long time to find the inverse (or its approximation) of F_j where the c.d.f. F_j has a p.d.f.

$$p_j(\phi) = \begin{cases} \frac{1}{B(\frac{1}{2}, \frac{j-2+1}{2})} \cos^{j-2}(\phi), & \text{if } -\pi/2 \leq \phi < \pi/2, \\ 0, & \text{otherwise.} \end{cases}$$

We have discussed this difficulty in section 4.3 and recommended the TFWW algorithm to generate an NT-net on U_s which does not involve the calculation of inverse of F_j's. In fact, the TFWW algorithm can be applied in the current case. It is known that each column of $\boldsymbol{U} \sim U(O(p,q))$ has a uniform distribution on U_p (Fang and Zhang (1990)). Denote $\boldsymbol{U} = (\boldsymbol{u}_1, \cdots, \boldsymbol{u}_q)$. It is easy to see that the first column $\boldsymbol{v}_1$ of $\boldsymbol{V}_1 = (\boldsymbol{G}_{12} \cdots \boldsymbol{G}_{1p}) \begin{pmatrix} \varepsilon_1 & \boldsymbol{O} \\ \boldsymbol{O} & \boldsymbol{I}_{p-1} \end{pmatrix}$ is just $\boldsymbol{u}_1$ and so that $\boldsymbol{v}_1 \sim U(U_p)$. Moreover $\boldsymbol{V}_1$ is completely determined by $\boldsymbol{v}_1$. Thus, if we can generate $\boldsymbol{v}_1 \sim U(U_p)$, then we obtain $\boldsymbol{V}_1$. By the definition of $\boldsymbol{V}_1$ we find

$$\begin{cases} v_{11} & = \varepsilon_1 \cos(\theta_{1p})\cos(\theta_{1,p-1}) \cdots \cos(\theta_{12}) \\ & \vdots \\ v_{1,p-1} & = \varepsilon_1 \cos(\theta_{1p})\sin(\theta_{1,p-1}) \\ v_{1p} & = \varepsilon_1 \sin(\theta_{1p}) \end{cases} \tag{6.6.11}$$

where $\boldsymbol{v}_1 = (v_{11}, \cdots, v_{1p})'$. Since $-\pi/2 < \theta_{ij} \leq \pi/2$, we have $\varepsilon_1 = \text{sign}(v_{11})$ where sign(.) is the signum function. If we generate $\boldsymbol{v}_1$ by the TFWW algorithm, we can find $\boldsymbol{V}_1$ by (6.6.11). More precisely, let

$$\begin{cases} y_1 & = v_{11}/v_{12} = \text{ctg}(\theta_{12}) \\ y_2 & = v_{12}/v_{13} = \sin(\theta_{12})\text{ctg}(\theta_{13}) \\ & \vdots \\ y_{p-1} & = v_{1,p-1}/v_{1p} = \sin(\theta_{1,p-1})\text{ctg}(\theta_{1p}) \\ y_p & = v_{1p} = \varepsilon_1 \sin(\theta_{1p}) \end{cases} \tag{6.6.12}$$

and $z_1 = 1, z_i = \sin(\theta_{1i}), i = 2, \cdots, p$. We can recursively compute

$$\begin{aligned} &z_p = \sin(\theta_{1p}) = \varepsilon_1 y_p \quad \text{and} \\ &z_i = y_i z_{i+1} / \sqrt{1 - z_{i+1}^2}\,\text{sign}(z_{i+1}), \quad i = p-1, \cdots, 2. \end{aligned} \tag{6.6.13}$$

Consequently, we can express $\cos(\theta_{ij})$ by

$$\cos(\theta_{ij}) = y_{j-1} z_j / z_{j-1} \quad , \quad j = 2, \cdots, p. \tag{6.6.14}$$

Substituting $\cos(\theta_{ij}), \sin(\theta_{ij})$ and ε_1 into $\boldsymbol{V}_1$ we can express the elements of $\boldsymbol{V}_1$ as follows:

$$\begin{aligned}
&\boldsymbol{V}_1(i,1) = v_{1i}, \quad i = 1, \cdots, p,\\
&\boldsymbol{V}_1(k,k) = y_{k-1}z_k/z_{k-1} \quad , \quad k = 2, \cdots, p,\\
&\boldsymbol{V}_1(l,k) = 0, \quad 3 \le k < l \le p-1,\\
&\boldsymbol{V}_1(l,k) = -z_k z_{k-1}(\prod_{j=l}^{k-2} y_j), \quad 1 \le l < k \le p.
\end{aligned} \tag{6.6.15}$$

Similarly, we denote

$$\boldsymbol{V}_k = \boldsymbol{G}_{k,k+1} \cdots \boldsymbol{G}_{kp} \begin{pmatrix} \boldsymbol{I}_{k-1} & \boldsymbol{0} & \boldsymbol{0} \\ \boldsymbol{0} & \varepsilon_k & 0 \\ 0 & 0 & I_{p-k} \end{pmatrix}, k = 2, \cdots, q,$$

and express elements of $\boldsymbol{V}_k$ in terms of a vector which is uniformly distributed on U_{p-k+1}. Finally, we get

$$\boldsymbol{U} = \boldsymbol{V}_1 \cdots \boldsymbol{V}_q \begin{pmatrix} \boldsymbol{I}_q \\ \boldsymbol{0} \end{pmatrix}.$$

Based on the above discussion and the TFWW algorithm, Fang and Li (1993) proposed an algorithm to generate an NT-net on $O(p,q)$. Their numerical results show that the algorithm is satisfactory in both uniformity and computing time.

Exercise

6.1 Let $X \sim Ga(\beta, \lambda)$, be the gamma distribution defined in (2.6.1). The following data is a sample from the population $Ga(\beta, \lambda)$.
   (a) Find moment estimates $\hat{\beta}^*$ and $\hat{\lambda}^*$ of β and λ.
   (b) In terms of $\hat{\beta}^*$, $\hat{\lambda}^*$ and SNTO, find the MLEs of β and λ.

1.5128	17.6464	22.7419	8.9765	5.5807
9.6622	6.8015	2.9317	6.5333	4.3420
12.8354	12.1527	7.5309	22.2337	2.2453
4.4472	17.0835	20.2310	5.7031	11.6051
7.5139	3.6220	11.4734	5.2327	1.9192
11.5441	15.7105	9.8498	9.0798	10.8729
3.8403	3.3202	2.8350	6.0044	9.4557
9.3823	7.1123	19.4210	3.7043	13.5312
12.2513	18.0835	10.0428	8.6268	11.6838
10.8290	10.2125	23.0184	5.2033	11.8362

6.2 Let

$$\Sigma = \begin{pmatrix} 984 & 562 & 476 & 1208 & 870 \\ 562 & 391 & 324 & 734 & 517 \\ 476 & 324 & 299 & 542 & 437 \\ 1208 & 734 & 642 & 1535 & 1079 \\ 870 & 517 & 437 & 1079 & 777 \end{pmatrix}$$

Find the Cholesky root of $\boldsymbol{\Sigma}$.

6.3 With the data set in Example 6.3 answer the following questions:

(a) Test multinormality of $\boldsymbol{x} = (X_1, X_3, X_8, X_{10})'$ by the statistics $Sk_{\max}$ and $Ku_{\max}$.

(b) If the hypothesis in (a) is accepted, $\boldsymbol{x}$ can be considered from a normal distribution $N_4(\boldsymbol{\mu}, \boldsymbol{\Sigma})$. Give the moment estimates $\hat{\boldsymbol{\mu}}^*$ and $\hat{\boldsymbol{\Sigma}}^*$ of $\boldsymbol{\mu}$ and $\boldsymbol{\Sigma}$.

(c) With SNTO and $(\hat{\boldsymbol{\mu}}^*, \hat{\boldsymbol{\Sigma}}^*)$ find the MLEs $\hat{\boldsymbol{\mu}}$ and $\hat{\boldsymbol{\Sigma}}$ of $\boldsymbol{\mu}$ and $\boldsymbol{\Sigma}$.

6.4 With the data set discussed in Example 6.5 test the multivariate normality for the following tripled subsets: $\{X_1, X_2, X_3\}, \{X_1, X_2, X_4\}, \{X_1, X_3, X_4\}$ and $\{X_2, X_3, X_4\}$.

6.5 Write a computer program for the minimum volume ellipsoid estimator with the projection algorithm by means of an NT-net on U_s.

6.6 Write a computer program to generate an NT-net on $O(p, q)$ with the stochastic representation (6.6.2).

APPENDIX A

Tables of *glp* set

A.1 The *glp* sets with large n

The data in the following 12 tables are essentially taken from Hua and Wang (1981). *The Applications of Number Theory to Numerical Analysis,* Springer-Verlag and Science Press. The authors would like to thank Springer-Verlag and Science Press for granting permission to reprint them in this volume. In these tables, the generating vectors $(n; h_1, \cdots, h_s)$ with $h_1 = 1$ and so the corresponding *glp* sets

$$\left(\frac{k-0.5}{n}, \left\{\frac{h_2 k - 0.5}{n}\right\}, \cdots, \left\{\frac{h_s k - 0.5}{n}\right\}\right) \quad (1 \leq k \leq n)$$

are given, and the values of $W_2(n, \boldsymbol{h})$, $W_4(n, \boldsymbol{h})$ and $\rho(n, \boldsymbol{h})$ in the original tables are neglected. These sets are mainly used to evaluate the definite multiple integral, in particular, the probabilities and moments of multivariate distributions, as well as optimization.

Table A.1 $s = 2$ $(n = F_m, h_1 = 1, h_2 = F_{m-1})$

n	8	13	21	34	55	89	144	233	377	610
h_2	5	8	13	21	34	55	89	144	233	377
n	987	1597	2584	4181	6765	10,946	17,711	28,657	46,368	75,025
h_2	610	987	1597	2584	4181	6765	10,946	17,711	28,657	46,368

Table A.2 $s = 3, h_1 = 1$

n	35	101	135	185	266	418	597	828	1010
h_2	11	40	29	26	27	90	63	285	140
h_3	16	85	42	64	69	130	169	358	237
n	1220	1459	1626	1958	2440	3237	4044	5037	6066
h_2	319	256	572	202	638	456	400	580	600
h_3	510	373	712	696	1002	1107	1054	1997	1581
n	8191	10,007	20,039	28,117	39,029	57,091	82,001	140,052	314,694
h_2	739	544	5704	19,449	10,607	48,188	21,252	34,590	77,723
h_3	5515	5733	12,319	5600	26,871	21,101	67,997	112,313	252,365

Table A.3 $s = 4, h_1 = 1$

n	307	562	701	1019	2129	3001	4001	5003	6007
h_2	42	53	82	71	766	174	113	792	1351
h_3	229	89	415	765	1281	266	766	1889	5080
h_4	101	221	382	865	1906	1269	2537	191	3086
n	8191	10,007	20,039	28,117	39,029	57,091	82,001	100,063	147,312
h_2	2488	1206	19,668	17,549	30,699	52,590	57,270	92,313	136,641
h_3	5939	3421	17,407	1900	34,367	48,787	58,903	24,700	116,072
h_4	7859	2842	14,600	24,455	605	38,790	17,672	95,582	76,424

Table A.4 $s = 5, h_1 = 1$

n	1069	1543	2129	3001	4001	5003	6007	8191
h_2	63	58	618	408	1534	840	509	1386
h_3	762	278	833	1409	568	117	780	4302
h_4	970	694	1705	1681	3095	3593	558	7715
h_5	177	134	1964	1620	2544	1311	1693	3735
n	10,007	15,019	20,039	33,139	51,097	71,053	100,063	374,181
h_2	198	10,641	11,327	32,133	44,672	33,755	90,036	343,867
h_3	9183	2640	11,251	17,866	45,346	65,170	77,477	255,381
h_4	6967	6710	12,076	21,281	7044	12,740	27,253	310,881
h_5	8507	784	18,677	32,247	14,242	6878	6222	115,892

Table A.5 $s = 6, h_1 = 1$

n	2129	3001	4001	5003	6007	8191	10,007	15,019
h_2	41	233	1751	2037	312	1632	2240	8743
h_3	1681	271	1235	1882	1232	1349	4093	8358
h_4	793	122	1945	1336	5943	6380	1908	6559
h_5	578	1417	844	4803	4060	1399	931	2795
h_6	279	51	1475	2846	5250	6070	3984	772

n	20,039	33,139	51,097	71,053	100,063	114,174	302,686
h_2	5557	18,236	9931	18,010	43,307	107,538	285,095
h_3	150	1831	7551	3155	15,440	88,018	233,344
h_4	11,951	19,143	29,683	50,203	39,114	15,543	41,204
h_5	2461	5522	44,446	6065	43,534	80,974	214,668
h_6	9179	22,910	17,340	13,328	29,955	56,747	150,441

Table A.6 $s = 7, h_1 = 1$

n	3997	11,215	15,019	24,041	33,139	46,213	57,091	71,053
h_2	3888	10,909	12,439	1833	7642	37,900	35,571	31,874
h_3	3564	10,000	2983	18,190	9246	17,534	45,299	36,082
h_4	3034	8512	8607	21,444	5584	41,873	51,436	13,810
h_5	2311	6485	7041	23,858	23,035	32,280	34,679	6605
h_6	1417	3976	7210	1135	32,241	15,251	1472	68,784
h_7	375	1053	6741	12,929	30,396	26,909	8065	9848

n	84,523	100,063	172,155	234,646	462,891	769,518	957,838
h_2	82,217	39,040	167,459	228,245	450,265	748,528	931,711
h_3	75,364	62,047	153,499	209,218	412,730	686,129	854,041
h_4	64,149	89,839	130,657	178,084	351,310	584,024	726,949
h_5	48,878	6347	99,554	135,691	267,681	444,998	553,900
h_6	29,969	30,892	61,040	83,197	164,124	272,843	339,614
h_7	7936	64,404	18,165	22,032	43,464	72,255	89,937

Table A.7 $s = 8, h_1 = 1$

n	3997	11,215	24,041	28,832	33,139	46,213	57,091	71,053
h_2	3888	10,909	17,441	27,850	3520	5347	17,411	60,759
h_3	3564	10,000	21,749	24,938	29,553	30,775	46,802	26,413
h_4	3034	8512	5411	20,195	3239	35,645	9779	24,409
h_5	2311	6485	12,326	13,782	1464	11,403	16,807	48,215
h_6	1417	3976	3144	5918	16,735	16,894	35,302	51,048
h_7	375	1053	21,024	25,703	19,197	32,016	1416	19,876
h_8	3211	9010	6252	15,781	3019	16,600	47,755	29,096

n	84,523	100,063	172,155	234,646	462,891	769,518	957,838
h_2	82,217	4344	167,459	228,245	450,265	748,528	931,711
h_3	75,364	58,492	153,499	209,218	412,730	686,129	854,041
h_4	64,149	29,291	130,657	178,084	351,310	584,024	726,949
h_5	48,878	60,031	99,554	135,691	267,681	444,998	553,900
h_6	29,969	10,486	61,040	83,197	164,124	272,843	339,614
h_7	7936	22,519	18,165	22,032	43,464	72,255	89,937
h_8	67,905	60,985	138,308	188,512	371,882	618,224	769,518

Table A.8 $s = 9, h_1 = 1$

n	3997	11,215	33,139	42,570	46,213	57,091	71,053
h_2	3888	10,909	68	41,409	8871	20,176	26,454
h_3	3564	10,000	4624	37,957	40,115	12,146	13,119
h_4	3034	8512	16,181	32,308	20,065	23,124	27,174
h_5	2311	6485	6721	24,617	30,352	2172	17,795
h_6	1417	3976	26,221	15,094	15,654	33,475	22,805
h_7	375	1053	26,661	3997	42,782	5070	43,500
h_8	3211	9010	23,442	34,200	17,966	42,339	45,665
h_9	1962	5506	3384	20,901	33,962	36,122	49,857

n	100,063	159,053	172,155	234,646	462,891	769,518	957,838
h_2	70,893	60,128	167,459	228,245	450,265	748,528	931,711
h_3	53,211	101,694	153,499	209,218	412,730	686,129	854,041
h_4	12,386	23,300	130,657	178,084	351,310	584,024	726,949
h_5	27,873	43,576	99,554	135,691	267,681	444,998	553,900
h_6	56,528	57,659	61,040	83,197	164,124	272,843	339,614
h_7	16,417	42,111	18,165	22,032	43,464	72,255	89,937
h_8	17,628	85,501	138,308	188,512	371,882	618,224	769,518
h_9	14,997	93,062	84,523	115,204	227,266	377,811	470,271

Table A.9 $s = 10, h_1 = 1$

n	4661	13,587	24,076	58,358	85,633
h_2	4574	13,334	23,628	57,271	37,677
h_3	4315	12,579	22,290	54,030	35,345
h_4	3889	11,337	20,090	48,695	3864
h_5	3304	9631	17,066	41,366	54,821
h_6	2570	7492	13,276	32,180	74,078
h_7	1702	4961	8790	21,307	30,354
h_8	715	2084	3692	8950	57,935
h_9	4289	12,502	22,153	53,697	51,906
h_{10}	3122	9100	16,125	39,086	56,279
n	103,661	115,069	130,703	155,093	805,098
h_2	45,681	65,470	64,709	90,485	790,101
h_3	57,831	650	53,373	20,662	745,388
h_4	80,987	95,039	17,385	110,048	671,792
h_5	9718	77,293	5244	102,308	570,685
h_6	51,556	98,366	29,008	148,396	443,949
h_7	55,377	70,366	52,889	125,399	293,946
h_8	37,354	74,605	66,949	124,635	123,470
h_9	4353	55,507	51,906	10,480	740,795
h_{10}	27,595	49,201	110,363	44,198	539,222

Table A.10 $s = 11, h_1 = 1$

n	4661	13,587	24,076	58,358	297,974
h_2	4574	13,334	23,628	57,271	294,481
h_3	4315	12,579	22,290	54,030	284,041
h_4	3889	11,337	20,090	48,695	266,778
h_5	3304	9631	17,066	41,366	242,894
h_6	2570	7492	13,276	32,180	212,668
h_7	1702	4961	8790	21,307	176,456
h_8	715	2084	3692	8950	134,682
h_9	4289	12,502	22,153	53,697	87,835
h_{10}	3122	9100	16,125	39,086	36,464
h_{11}	1897	5529	9797	23,747	279,147

n	698,047	1,243,423	2,226,963	7,494,007
h_2	685,041	1,228,845	2,200,854	7,354,408
h_3	646,274	1,185,282	2,122,833	6,838,211
h_4	582,461	1,113,244	1,993,814	6,253,169
h_5	494,796	1,013,577	1,815,311	5,312,043
h_6	384,914	887,449	1,589,415	4,132,365
h_7	254,860	736,338	1,318,777	2,736,109
h_8	107,051	562,016	1,006,567	1,149,286
h_9	642,292	366,527	656,448	6,895,461
h_{10}	467,527	152,163	272,523	5,019,180
h_{11}	284,044	1,164,860	2,086,257	3,049,402

Table A.11 $s = 12, 13, 14, h_1 = 1$

n	18,984	53,328	77,431	297,974	1,243,423
h_2	18,761	52,703	76,523	294,481	1,228,845
h_3	18,096	50,834	73,810	284,041	1,185,282
h_4	16,996	47,745	69,324	266,778	1,113,244
h_5	15,475	43,470	63,118	242,894	1,013,577
h_6	13,549	38,061	55,264	212,668	887,449
h_7	11,242	31,580	45,854	176,456	736,338
h_8	8581	24,104	34,998	134,682	562,016
h_9	5596	15,720	22,825	87,835	366,527
h_{10}	2323	6526	9476	36,464	152,163
h_{11}	17,785	49,959	72,539	279,147	1,164,860
h_{12}	14,053	39,477	57,320	220,583	920,477
h_{13}	10,158	28,534	41,430	159,433	665,302
h_{14}	6,143	17,255	25,054	96,414	402,327

n	2,428,705	14,753,436	19,984,698	34,248,063
h_2	2,400,231	14,580,465	19,750,396	33,846,536
h_3	2,315,141	14,063,582	19,050,236	32,646,662
h_4	2,174,435	13,208,845	17,892,427	30,662,508
h_5	1,979,761	12,026,276	16,290,543	27,917,337
h_6	1,733,402	10,529,739	14,263,366	24,443,334
h_7	1,438,245	8,736,780	11,834,661	20,281,228
h_8	1,097,753	6,668,420	9,032,903	15,479,816
h_9	715,916	4,348,908	5,890,941	10,095,390
h_{10}	297,211	1,805,439	2,445,610	4,191,077
h_{11}	2,275,252	13,821,268	18,722,002	32,084,164
h_{12}	1,797,913	10,921,619	14,794,199	25,353,030
h_{13}	1,299,495	7,893,924	10,692,946	18,324,655
h_{14}	785,841	4,773,681	6,466,329	11,081,440

Table A.12 $s = 15, 16, 17, 18, h_1 = 1$

n	70,864	139,489	1,139,691	2,422,957
h_2	70,353	138,484	1,131,480	2,398,094
h_3	68,825	135,476	1,106,904	2,323,761
h_4	66,291	130,487	1,066,142	2,200,720
h_5	62,768	123,553	1,009,487	2,030,234
h_6	58,283	114,724	937,347	1,814,052
h_7	52,867	104,063	850,242	1,554,392
h_8	46,559	91,647	748,799	1,253,920
h_9	39,405	77,566	633,750	915,717
h_{10}	31,457	61,921	505,923	543,256
h_{11}	22,772	44,825	366,239	140,357
h_{12}	13,412	26,401	215,705	2,134,112
h_{13}	3445	6781	55,406	1,683,011
h_{14}	63,806	125,597	1,026,186	1,214,641
h_{15}	52,844	104,019	849,882	733,806
h_{16}	41,501	81,691	667,455	
h_{17}	29,859	58,775	480,219	
h_{18}	18,002	35,435	289,522	

n	4,395,774	14,271,038	55,879,244
h_2	4,364,102	14,168,215	55,476,633
h_3	4,269,316	13,860,486	54,271,700
h_4	4,112,097	13,350,069	52,273,127
h_5	3,893,578	12,640,642	49,495,314
h_6	3,615,335	11,737,315	45,958,274
h_7	3,279,371	10,646,597	41,687,493
h_8	2,888,108	9,376,347	36,713,742
h_9	2,444,365	7,935,718	31,072,856
h_{10}	1,951,338	6,335,088	24,805,477
h_{11}	1,412,580	4,585,990	17,956,764
h_{12}	831,972	2,701,027	10,576,061
h_{13}	213,699	693,780	50,314,090
h_{14}	3,957,988	12,849,750	41,669,876
h_{15}	3,277,986	10,642,098	32,725,430
h_{16}	2,574,365	8,357,770	23,545,197
h_{17}	1,852,197	6,013,224	14,195,319
h_{18}	1,116,683	3,625,352	2,716,545

A.2 The *glp* sets with small n

The following tables give the generating vectors $(n; h_1, \cdots, h_s)$ with $h_1 = 1$ for small n. The criterion for choosing the generating vectors is the mean square error (MSE). Therefore, these tables are, in particular, recommended for experimental design and have been computed by Mr. Ke-Hai Yuan. Some tables of the generating vectors by using other criteria can be found in Chapter 5.

Table A.13 $s = 2, h_1 = 1$

n	5	7	9	11	13	15	17	19	21	23	25	27	29	31
h_2	2	3	4	7	5	11	5	14	13	9	7	16	23	21

Table A.14 $s = 3, h_1 = 1$

n	5	7	9	11	13	15	17	19	21	23	25	27	29	31
h_2	2	2	2	3	3	2	3	3	4	15	8	20	16	11
h_3	4	4	4	5	9	7	9	9	10	18	14	22	24	28

Table A.15 $s = 4, h_1 = 1$

n	7	9	11	13	15	17	19	21	23	25	27	29	31
h_2	2	2	2	6	2	2	5	2	2	4	5	4	15
h_3	3	4	5	8	4	4	7	10	5	6	17	6	19
h_4	6	7	7	10	8	8	9	17	10	9	25	16	22

Table A.16 $s = 5, h_1 = 1$

n	11	13	15	17	19	21	23	25	27	29	31
h_2	3	6	2	3	4	2	6	4	2	14	2
h_3	4	8	4	9	6	10	11	6	4	18	4
h_4	5	9	7	10	7	13	13	9	8	20	8
h_5	9	10	13	13	17	16	20	11	16	22	16

Table A.17 $s = 6, h_1 = 1$

n	11	13	17	19	21	23	25	27	29	31
h_2	2	2	3	4	2	5	8	2	14	6
h_3	4	3	5	5	4	6	12	4	18	9
h_4	5	4	9	6	8	11	14	5	19	11
h_5	8	6	10	7	11	13	18	8	20	28
h_6	10	8	13	17	16	20	21	16	22	29

Table A.18 $s = 7, h_1 = 1$

n	13	17	19	21	23	25	27	29	31
h_2	2	3	2	2	2	8	2	5	5
h_3	3	5	4	4	4	12	4	14	7
h_4	4	9	7	5	8	14	5	18	8
h_5	6	10	8	8	9	18	8	19	10
h_6	8	13	13	10	16	19	10	20	14
h_7	12	15	16	19	18	21	16	22	16

Table A.19 $s = 8, h_1 = 1$

n	17	19	23	25	27	29	31
h_2	3	2	2	2	2	2	5
h_3	5	4	4	3	4	3	7
h_4	9	7	8	4	5	4	8
h_5	10	8	9	6	8	6	10
h_6	11	13	13	9	10	8	14
h_7	13	14	16	12	16	12	16
h_8	15	16	18	18	20	16	19

Table A.20 $s = 9, h_1 = 1$

n	17	19	23	25	27	29	31
h_2	3	2	2	2	4	2	5
h_3	5	4	3	3	7	3	7
h_4	9	7	4	4	10	4	8
h_5	10	8	8	6	13	6	10
h_6	11	9	9	9	16	8	14
h_7	13	13	13	11	19	12	16
h_8	15	14	16	12	22	16	18
h_9	16	16	18	18	25	24	19

Table A.21 $s = 10, 11, 12, h_1 = 1$

n	h_2	h_3	h_4	h_5	h_6	h_7	h_8	h_9	h_{10}	h_{11}	h_{12}
19	2	4	7	8	9	13	14	16	18		
23	2	3	4	6	8	9	12	13	16		
25	2	3	4	6	8	9	11	12	18		
27	2	4	5	8	10	13	16	20	26		
29	2	3	4	6	8	12	16	19	24		
31	4	5	7	8	10	14	16	18	19		
23	2	3	4	6	8	9	12	13	16	18	
25	2	3	4	6	8	9	11	12	18	24	
29	2	3	4	6	8	9	12	16	19	24	
31	4	5	6	9	11	13	19	23	28	29	
23	2	4	5	8	9	10	11	16	17	20	22
29	6	11	12	13	15	19	21	22	24	25	27
31	4	5	7	8	9	10	14	16	18	19	25

Table A.22 $s = 13, 14, 15, h_1 = 1$

n	h_2	h_3	h_4	h_5	h_6	h_7	h_8	h_9	h_{10}	h_{11}	h_{12}	h_{13}	h_{14}	h_{15}
29	2	3	4	6	7	8	9	12	16	18	19	24		
31	2	4	5	7	8	10	14	16	18	19	20	25		
29	2	3	4	6	7	8	9	12	14	16	18	19	24	
31	2	4	5	7	8	10	14	16	18	19	20	25	28	
29	2	3	4	6	7	8	9	12	14	16	18	19	24	28
31	2	4	5	7	8	9	10	14	16	18	19	20	25	28

APPENDIX B

Integrations and uniform distributions on D

Consider a multiple integral over a closed and bounded domain D of an s-dimensional space

$$I = \int_D g(x_1, \cdots, x_s)\, dx_1 \cdots dx_s. \tag{B.1}$$

Let $x_1, \cdots, x_s$ be one-to-one transformed to new independent variables $y_1, \cdots, y_s$ through

$$\begin{cases} x_1 = f_1(y_1, \cdots, y_s), \\ \quad\cdots \\ x_s = f_s(y_1, \cdots, y_s), \end{cases} \tag{B.2}$$

where the f_i are continuously differentiable. The transformation will be written as $\boldsymbol{x} = \boldsymbol{f}(\boldsymbol{y})$ or $\boldsymbol{y} = \boldsymbol{f}^{-1}(\boldsymbol{x})$. The absolute value of the determinant of $\partial \boldsymbol{x}'/\partial \boldsymbol{y}$, denoted by

$$\begin{aligned} J(\boldsymbol{x} \to \boldsymbol{y}) &= \det \left(\frac{\partial(x_1, \cdots, x_s)}{\partial(y_1, \cdots, y_s)} \right)_+ \\ &= \det \begin{bmatrix} \frac{\partial x_1}{\partial y_1} & \cdots & \frac{\partial x_s}{\partial y_1} \\ \vdots & & \vdots \\ \frac{\partial x_1}{\partial y_s} & \cdots & \frac{\partial x_s}{\partial y_s} \end{bmatrix}_+, \end{aligned} \tag{B.3}$$

will be called the Jacobian of the transformation from $\boldsymbol{x}$ to $\boldsymbol{y}$. Now (B.1) can be expressed as

$$I = \int_{D^*} g(\boldsymbol{f}(\boldsymbol{y})) J(\boldsymbol{x} \to \boldsymbol{y})\, d\boldsymbol{y}, \tag{B.4}$$

where

$$D^* = \{\boldsymbol{y} : \boldsymbol{y} = \boldsymbol{f}^{-1}(\boldsymbol{x}), \boldsymbol{x} \in D\}. \tag{B.5}$$

More generally, if the number of y_i is less than s, say t, i.e.

$$\begin{cases} x_1 = f_1(y_1, \cdots, y_t), \\ \quad \cdots \\ x_s = f_s(y_1, \cdots, y_t), \end{cases} \tag{B.6}$$

the formula (B.4) still holds (cf. Hua (1984) Ch. 8), but

$$J(\boldsymbol{x} \to \boldsymbol{y}) = \det(\boldsymbol{T}\boldsymbol{T}')^{1/2}, \tag{B.7}$$

where

$$\boldsymbol{T} = \begin{pmatrix} \frac{\partial x_1}{\partial y_1} & \cdots & \frac{\partial x_s}{\partial y_1} \\ \vdots & & \vdots \\ \frac{\partial x_1}{\partial y_t} & \cdots & \frac{\partial x_s}{\partial y_t} \end{pmatrix}. \tag{B.8}$$

Let the random vector $\boldsymbol{x} = (X_1, \cdots, X_s)'$ be uniformly distributed on D. Then the p.d.f. of $\boldsymbol{x}$ is

$$p(\boldsymbol{x}) = \begin{cases} 1/v(D), & \text{if } \boldsymbol{x} \in D, \\ 0, & \text{otherwise,} \end{cases} \tag{B.9}$$

where

$$v(D) = \int_{D^*} J(\boldsymbol{x} \to \boldsymbol{y}) d\boldsymbol{y} \tag{B.10}$$

is the volume of D. Consider the transformation

$$X_i = f_i(Y_1, \cdots, Y_t), \quad i = 1, \cdots, s, \ t \leq s.$$

Then the p.d.f. of $Y_1, \cdots, Y_t$ is

$$\frac{1}{v(D)} J(\boldsymbol{x} \to \boldsymbol{y}).$$

If the latter can be expressed as product of t univariate density functions, i.e.

$$\frac{1}{v(D)} J(\boldsymbol{x} \to \boldsymbol{y}) = \prod_{j=1}^{t} p_j(y_j), \tag{B.11}$$

then we can indicate that

(a) $Y_1, \cdots, Y_t$ are independent,
(b) the p.d.f. of Y_j is $p_j(y_j)$, $j = 1, \cdots, t$.

We now apply the above general theory to the cases of $D = A_s, B_s, U_s, V_s$ and T_s.

B.1 The domain A_s

Consider an integral

$$I_1 = \int_{A_s} f(\boldsymbol{x}) d\boldsymbol{x}, \tag{B.12}$$

where

$$A_s = \{(x_1, \cdots, x_s : 0 \leq x_1 \leq x_2 \leq \cdots \leq x_s \leq 1\} \tag{B.13}$$

defined in (1.4.5). Let $\boldsymbol{x} = \boldsymbol{x}_1(\boldsymbol{\phi})$ be the transformation as follows

$$\begin{cases} x_1 = \phi_1 \phi_2 \cdots \phi_s, \\ x_2 = \phi_2 \cdots \phi_s, \\ \quad \cdots \\ x_{s-1} = \phi_{s-1} \phi_s, \\ x_s = \phi_s, \end{cases} \tag{B.14}$$

where $\boldsymbol{\phi} = (\phi_1, \cdots, \phi_s) \in C^s$. This maps C^s into A_s. Let

$$\boldsymbol{T}_1 = \left(\frac{\partial x_j}{\partial \phi_i} \right) = (t_{ij}), \quad i, j = 1, \cdots, s.$$

Obviously, $\boldsymbol{T}_1$ is a lower triangular matrix, and the Jacobian

$$
\begin{aligned}
J_1(\boldsymbol{\phi}_1) = \det(\boldsymbol{T}_1)_+ = t_{11}t_{22}\cdots t_{ss} &= \prod_{i=1}^{s} \frac{\partial x_i}{\partial \phi_i} \\
&= (\phi_2 \cdots \phi_s)(\phi_3 \cdots \phi_s) \cdots (\phi_s)(1) \\
&= \prod_{i=2}^{s} \phi_i^{i-1}.
\end{aligned}
\tag{B.15}
$$

Therefore, we have

$$
I_1 = \int_{C^s} f(\boldsymbol{x}_1(\boldsymbol{\phi})) \prod_{i=2}^{s} \phi_i^{i-1} \, d\boldsymbol{\phi}. \tag{B.16}
$$

In particular, when $f \equiv 1$ we have

$$
\begin{aligned}
v(A_s) = I_1 = \int_{A_s} dv &= \int_{C_s} \prod_{i=2}^{s} \phi_i^{i-1} \, d\boldsymbol{\phi} \\
&= \prod_{i=1}^{s} \int_0^1 \phi_i^{i-1} \, d\phi_i = 1/s!.
\end{aligned}
\tag{B.17}
$$

Let $\boldsymbol{x} = (X_1, \cdots, X_s)$ be uniformly distributed on A_s. Then the density function of $\boldsymbol{x}$ is

$$
p_1(\boldsymbol{x}) = \begin{cases} s!, & \text{if } \boldsymbol{x} \in A_s, \\ 0, & \text{otherwise.} \end{cases}
$$

Consider the transformation (B.14). The joint density function of $\phi_1, \cdots, \phi_s$ is

$$
p_1(\boldsymbol{x}_1(\boldsymbol{\phi})) J_1(\boldsymbol{\phi}) = s! \prod_{i=2}^{s} \phi_i^{i-1} = \prod_{i=1}^{s} (i\phi_i^{i-1})
$$

which implies that

(a) $\phi_1, \cdots, \phi_s$ are independent,
(b) for $1 \leq j \leq s$, ϕ_j has p.d.f. $j\phi^{j-1}$ and c.d.f.

$$F_j(\phi) = \begin{cases} 0, & \phi < 0, \\ \phi^j, & 0 \leq \phi \leq 1, \\ 1, & \phi > 1. \end{cases} \tag{B.18}$$

B.2 The domains B_s and U_s

Let B_s and U_s be the unit ball and the unit sphere of R^s, i.e.

$$B_s = \{(x_1, \cdots, x_s) : x_1^2 + \cdots + x_s^2 \leq 1\} \tag{B.19}$$

and

$$U_s = \{(x_1, \cdots, x_s) : x_1^2 + \cdots + x_s^2 = 1\}. \tag{B.20}$$

Consider the integral

$$I_2 = \int_{B_s} f(\boldsymbol{x})\, d\boldsymbol{x}$$

and the spherical coordinate transformation $\boldsymbol{x} = \boldsymbol{x}_2(\boldsymbol{\phi})$ as follows

$$\begin{cases} x_1 = \phi_1 \cos(\pi\phi_2), \\ x_2 = \phi_1 \sin(\pi\phi_2)\cos(\pi\phi_3), \\ \quad \cdots \\ x_{s-1} = \phi_1 \sin(\pi\phi_2)\cdots\sin(\pi\phi_{s-1})\cos(2\pi\phi_s), \\ x_s = \phi_1 \sin(\pi\phi_2)\cdots\sin(\pi\phi_{s-1})\sin(2\pi\phi_s), \end{cases} \tag{B.21}$$

where $\boldsymbol{\phi} = (\phi_1, \cdots, \phi_s) \in C^s$. This maps C^s into B_s. We want to find the Jacobian of the transformation. There are several ways to calculate the Jacobian. It seems that the following method is a simple one which is based on the following

Lemma B.1
If

$$\begin{aligned} y_1 &= f_1(x_1, \cdots, x_n), \\ y_2 &= f_2(y_1, x_2, \cdots, x_n), \\ &\cdots \\ y_n &= f_n(y_1, y_2, \cdots, y_{n-1}, x_n), \end{aligned}$$

then the Jacobian of $(y_1, \cdots, y_n) \to (x_1, \cdots, x_n)$ is

$$\begin{aligned} &J((y_1, \cdots, y_n) \to (x_1, \cdots, x_n)) \\ &= \prod_{i=1}^{n} J(y_i \to x_i) = \prod_{i=1}^{n} \left| \frac{\partial f_i}{\partial x_i} \right|. \end{aligned} \tag{B.22}$$

PROOF Let

$$t_{ij} = \begin{cases} \dfrac{\partial f_i}{\partial x_j}, & i \le j, \\ 0, & i > j, \end{cases}$$

$$g_{ij} = \begin{cases} -\dfrac{\partial f_i}{\partial y_j}, & j < i, \\ 1, & i = j, \\ 0, & j > i, \end{cases}$$

$\boldsymbol{T} = (t_{ij})$, $\boldsymbol{G} = (g_{ij})$, $\boldsymbol{dx} = (dx_1, \cdots, dx_n)'$ and $\boldsymbol{dy} = (dy_1, \cdots, dy_n)'$. As $y_i = f_i(y_1, \cdots, y_{i-1}, x_i, \cdots, x_n)$ the differential of y_i is

$$\begin{aligned} dy_i &= \sum_{j=1}^{i-1} \frac{\partial f_i}{\partial y_j} dy_j + \sum_{j=i}^{n} \frac{\partial f_i}{\partial x_j} dx_j \\ &= -\sum_{j=1}^{i-1} g_{ij} dy_j + \sum_{j=i}^{n} t_{ij} dx_j, \end{aligned}$$

i.e.

$$\boldsymbol{G dy} = \boldsymbol{T dx} \quad \text{or} \quad \boldsymbol{dy} = \boldsymbol{G}^{-1}\boldsymbol{T dx}.$$

Therefore, the Jacobian

$$J(\boldsymbol{y} \to \boldsymbol{x}) = |\det(\boldsymbol{G}^{-1}\boldsymbol{T})|_+ = |\det(\boldsymbol{G}^{-1})|_+ \, |\det(\boldsymbol{T})|_+.$$

Since $\boldsymbol{G}$ is a upper triangular matrix with diagonal elements 1 the inverse matrix $\boldsymbol{G}^{-1}$ is an upper triangular matrix with diagonal elements 1 also. We have $\det(\boldsymbol{G}^{-1}) = 1$. Note that $\boldsymbol{T}$ is a lower triangular matrix, and the assertion follows by

$$J(\boldsymbol{y} \to \boldsymbol{x}) = |\det(\boldsymbol{T})|_+ = \prod_{i=1}^{n} \left| \frac{\partial f_i}{\partial x_i} \right|. \qquad \square$$

We now apply Lemma B.1 to (B.21) and show that

$$J_2(\boldsymbol{\phi}) = J(\boldsymbol{x} \to \boldsymbol{\phi}) = 2\pi^{s-1} \phi_1^{s-1} \prod_{i=2}^{s-1} \sin^{s-i}(\pi \phi_i). \tag{B.23}$$

In fact, we have

$$\begin{aligned} x_s^2 &= \phi_1^2 \Big(\prod_{i=2}^{s-1} \sin^2(\pi\phi_i) \Big) \sin^2(2\pi\phi_s), \\ x_{s-1}^2 + x_s^2 &= \phi_1^2 \prod_{i=2}^{s-1} \sin^2(\pi\phi_i), \\ &\cdots \\ x_2^2 + \cdots + x_s^2 &= \phi_1^2 \sin^2(\pi\phi_2), \\ x_1^2 + \cdots + x_s^2 &= \phi_1^2. \end{aligned}$$

By Lemma B.1 it follows that

$$\begin{aligned} J(x_s \to \phi_s) &= 2\pi(\phi_1^2/x_s) \prod_{i=2}^{s-1} \sin^2(\pi\phi_i) \sin(2\pi\phi_s) \cos(2\pi\phi_s), \\ J(x_{s-1} \to \phi_{s-1}) &= \pi(\phi_1^2/x_{s-1}) \prod_{i=2}^{s-2} \sin^2(\pi\phi_i) \\ &\qquad \sin(\pi\phi_{s-1}) \cos(\pi\phi_{s-1}), \\ &\cdots \\ J(x_2 \to \phi_2) &= \pi(\phi_1^2/x_2) \sin(\pi\phi_2) \cos(\pi\phi_2), \\ J(x_1 \to \phi_1) &= \phi_1/x_1, \end{aligned}$$

and therefore

$$J_2(\boldsymbol{\phi}) = \prod_{i=1}^{s} J(x_i \to \phi_i) = 2\pi^{s-1}\phi_1^{s-1} \prod_{i=2}^{s-1} \sin^{s-i}(\pi\phi_i)$$

which completes the proof of (B.23).

By (B.23) we have

$$\begin{aligned} I_2 &= \int_{B_s} f(\boldsymbol{x})\, d\boldsymbol{x} \\ &= 2\pi^{s-1} \int_{C^s} f(\boldsymbol{x}_2(\boldsymbol{\phi}))\phi_1^{s-1} \prod_{i=2}^{s-1} \sin^{s-i}(\pi\phi_i)\, d\phi_i. \end{aligned} \tag{B.24}$$

In particular, the volume of B_s is given by

$$\begin{aligned} v(B_s) &= 2\pi^{s-1} \int_0^1 \phi_1^{s-1}\, d\phi_1 \prod_{i=2}^{s-1} \int_0^1 \sin^{s-i}(\pi\phi_i)\, d\phi_i \\ &= \frac{2}{s} \prod_{i=2}^{s} B\Big(\frac{1}{2}, \frac{s-i+1}{2}\Big), \end{aligned}$$

because for any positive integer m

$$\int_0^1 \sin^m(\pi x)\, dx = \frac{1}{\pi} B\Big(\frac{1}{2}, \frac{m+1}{2}\Big).$$

Since

$$B\Big(\frac{1}{2}, \frac{m+1}{2}\Big) = \frac{\Gamma(\frac{1}{2})\Gamma(\frac{m+1}{2})}{\Gamma(\frac{m+2}{2})}$$

and

$$\Gamma\Big(\frac{1}{2}\Big) = \sqrt{\pi},$$

we have

$$v(B_s) = \frac{2\pi^{s/2}}{s\Gamma(s/2)}. \tag{B.25}$$

Let $\boldsymbol{x} = (X_1, \cdots, X_s)$ be uniformly distributed on B_s. Then the p.d.f. of $\boldsymbol{x}$ is

$$p_2(\boldsymbol{x}) = \begin{cases} \dfrac{s\Gamma(s/2)}{2\pi^{s/2}}, & \text{if } \boldsymbol{x} \in B_s, \\ 0, & \text{otherwise,} \end{cases}$$

and the p.d.f. of $\phi_1, \cdots, \phi_s$ in (B.23) is

$$\begin{aligned} p_2(\boldsymbol{x}_2(\boldsymbol{\phi}))J_2(\boldsymbol{\phi}) &= \frac{s\Gamma(s/2)}{2\pi^{s/2}} 2\pi^{s-1}\phi_1^{s-1}\prod_{i=2}^{s-1}\sin^{s-i}(\pi\phi_i) \\ &= (s\phi_1^{s-1})\prod_{i=2}^{s-1}\frac{\pi\sin^{s-i}(\pi\phi_i)}{B(\frac{1}{2},\frac{s-i+1}{2})} \end{aligned}$$

which implies that

(a) $\phi_1, \cdots, \phi_s$ are independent;

(b) ϕ_1 has a p.d.f.

$$p_1(\phi) = \begin{cases} s\phi^{s-1}, & 0 \le \phi \le 1, \\ 0, & \text{otherwise;} \end{cases} \tag{B.26}$$

$\phi_i (2 \le i \le s-1)$ has a p.d.f.

$$p_i(\phi) = \begin{cases} \dfrac{\pi\sin^{s-i}(\pi\phi)}{B(\frac{1}{2},\frac{s-i+1}{2})}, & \text{if } 0 \le \phi \le 1, \\ 0, & \text{otherwise,} \end{cases} \tag{B.27}$$

and $\phi_s \sim U(0,1)$.

Now we are going to treat the integral

$$I_3 = \int_{U_s} f(\boldsymbol{x})\,dv,$$

where dv is the volume element of U_s. Consider the spherical coordinate transformation $\boldsymbol{x} = \boldsymbol{x}_3(\boldsymbol{\phi})$ as follows

$$\begin{cases} x_1 = \cos(\pi\phi_1), \\ x_2 = \sin(\pi\phi_1)\cos(\pi\phi_2), \\ \quad \cdots \\ x_{s-1} = \sin(\pi\phi_1)\cdots\sin(\pi\phi_{s-2})\cos(2\pi\phi_{s-1}), \\ x_s = \sin(\pi\phi_1)\cdots\sin(\pi\phi_{s-2})\sin(2\pi\phi_{s-1}). \end{cases} \tag{B.28}$$

We are going to show that the Jacobian of $\boldsymbol{x}$ respect to $\boldsymbol{\phi}$ is

$$J_3(\boldsymbol{\phi}) = \det(\boldsymbol{T}\boldsymbol{T}')^{1/2} = 2\pi^{s-1}\prod_{i=1}^{s-2}\sin^{s-i-1}(\pi\phi_i). \tag{B.29}$$

Of course, we may calculate $\det(\boldsymbol{T}\boldsymbol{T}')^{1/2}$ by induction. But the following way seems more simple. First let

$$\begin{cases} x_i = y_i, \qquad i = 1, \cdots, s-1, \\ x_s^2 = 1 - y_1^2 - \cdots - y_{s-1}^2 \end{cases}$$

and

$$\boldsymbol{H} = \left(\frac{\partial x_j}{\partial y_i}\right), \ i = 1, \cdots, s-1, \ j = 1, \cdots, s.$$

It is easy to conclude $\boldsymbol{H} = (\boldsymbol{I}_{s-1}, -\boldsymbol{y}/x_s)$ with $\boldsymbol{y} = (y_1, \cdots, y_{s-1})'$. Hence

$$\begin{aligned} \det(\boldsymbol{H}\boldsymbol{H}')^{1/2} &= \det(\boldsymbol{I}_{s-1} + \boldsymbol{y}\boldsymbol{y}'/x_s^2)^{1/2} \\ &= \det(\boldsymbol{P}\boldsymbol{P}' + \boldsymbol{P}\boldsymbol{y}(\boldsymbol{P}\boldsymbol{y})'/x_s^2)^{1/2}, \end{aligned}$$

where $\boldsymbol{P}$ is an orthogonal matrix of order $(s-1)$ and $\boldsymbol{P}\boldsymbol{y} = (\|\boldsymbol{y}\|, 0, \cdots, 0)'$, and

$$\begin{aligned} \det(\boldsymbol{H}\boldsymbol{H}')^{1/2} &= \det(\boldsymbol{I}_{s-1} + diag(\|\boldsymbol{y}\|^2/x_s^2, 0, \cdots, 0))^{1/2} \\ &= [1 + \boldsymbol{y}'\boldsymbol{y}/(1 - \boldsymbol{y}'\boldsymbol{y})]^{1/2} \ (\ as\ x_s^2 = 1 - \boldsymbol{y}'\boldsymbol{y}) \\ &= 1/(1 - \boldsymbol{y}'\boldsymbol{y})^{1/2} = 1/|x_s|. \end{aligned}$$

By (B.4) and (B.7) we have

$$I_3 = \int_{U_s} f(\boldsymbol{x})\,dv = \int_{B_{s-1}} f(\boldsymbol{y}, x_s)/|x_s| d\boldsymbol{y},$$
$$x_s = \pm\sqrt{1 - \boldsymbol{y}'\boldsymbol{y}}.$$

Take the first $s-1$ transformations of (B.28) for $\boldsymbol{y}$, which maps C^{s-1} into B_{s-1}, i.e.

$$\begin{cases} y_1 = \cos(\pi\phi_1), \\ y_2 = \sin(\pi\phi_1)\cos(\pi\phi_2), \\ \quad\cdots \\ y_{s-1} = \sin(\pi\phi_1)\cdots\sin(\pi\phi_{s-2})\sin(2\pi\phi_{s-1}), \end{cases}$$

where $\boldsymbol{\phi} = (\phi_1, \cdots, \phi_{s-1})' \in C^{s-1}$. The matrix $(\frac{\partial y_i}{\partial \phi_j})$ is triangular and the Jacobian of the transformation is

$$\begin{aligned} \det\left(\frac{\partial y_i}{\partial \phi_j}\right)_+ &= \prod_{i=1}^{s-1} \frac{\partial y_i}{\partial \phi_i} \\ &= 2\pi^{s-1} \prod_{j=1}^{s-2} \sin^{s-j}(\pi\phi_j)\sin(2\pi\phi_{s-1}) \\ &= J_1 \ (say). \end{aligned}$$

Therefore

$$I_3 = \int_{C^{s-1}} f(\boldsymbol{x}_3(\boldsymbol{\phi}))J_1/|x_s|\,d\boldsymbol{\phi}, \tag{B.30}$$

where

$$J_1/|x_s| = 2\pi^{s-1} \prod_{j=1}^{s-2} \sin^{s-j-1}(\pi\phi_j)$$

is the required $J_3(\boldsymbol{\phi})$. When $f \equiv 1$ we have from (B.30)

$$\begin{aligned} v(U_s) &= \int_{C^{s-1}} 2\pi^{s-1} \prod_{j=1}^{s-2} \sin^{s-j-1}(\pi\phi_j)\, d\phi_j \\ &= 2\pi \prod_{j=1}^{s-2} \left[\pi \int_0^1 \sin^{s-j-1}(\pi\phi)\, d\phi\right] \\ &= 2\pi \prod_{j=1}^{s-2} B\left(\frac{1}{2}, \frac{s-j}{2}\right) \\ &= \frac{2\pi^{s/2}}{\Gamma(s/2)}. \end{aligned} \tag{B.31}$$

Let $\boldsymbol{x} = (X_1, \cdots, X_s)$ be uniformly distributed on U_s, and $\boldsymbol{\phi} = (\phi_1, \cdots, \phi_{s-1})$ be defined by (B.28). By a similar argument to that for B_s, we can conclude that

(a) $\phi_1, \cdots, \phi_{s-1}$ are independent;
(b) $\phi_j (1 \le j \le s-1)$ has a p.d.f.

$$p_j(\phi) = \frac{\pi}{B(\frac{1}{2}, \frac{s-j}{2})} \sin^{s-j-1}(\pi\phi), \quad 0 \le \phi \le 1. \tag{B.32}$$

Before we finish this section we will discuss the following distributions which appear in (B.27) and (B.32).

Let X_m be a random variable with a p.d.f.

$$p_m(x) = \begin{cases} \dfrac{\pi}{B(\frac{m}{2}, \frac{1}{2})} \sin^{m-1}(\pi x), & 0 \le x \le 1, \\ 0, & \text{otherwise}, \end{cases} \tag{B.33}$$

when m is a positive integer. Then we have the following facts:

(a) When $m = 1, X_m \sim U(0,1)$, the uniform distribution.
(b) Let $Z = \sin^2(\pi X_m)$ be nonnegative random variable. Then $Z \sim Be(\frac{m}{2}, \frac{1}{2})$, the beta distribution with parameters $m/2$ and $1/2$.

In fact, the p.d.f. of X_m is given by (B.33). Take the transformation $Z = \sin^2(\pi X_m)$. Then the corresponding Jacobian is

$$\begin{aligned} J(z \to x_m) &= |2\pi \sin(\pi x_m) \cos(\pi x_m)| \\ &= 2\pi z^{1/2}(1-z)^{1/2} \end{aligned}$$

and the p.d.f. of Z becomes

$$\begin{aligned} &\frac{\pi}{B(\frac{m}{2}, \frac{1}{2})} z^{(m-1)/2} \frac{1}{2\pi z^{1/2}(1-z)^{1/2}} \\ &\quad = \frac{1}{B(\frac{m}{2}, \frac{1}{2})} z^{m/2-1}(1-z)^{1/2-1}, \quad 0 \le z \le 1, \end{aligned}$$

i.e. $Z \sim Be(\frac{m}{2}, \frac{1}{2})$.

(c) Let $F_m(x)$ be the c.d.f. of X_m. Then $F_m(x)$ can be expressed in terms of the incomplete beta function

$$\begin{aligned} B_x(\alpha, \beta) &= \int_0^x y^{\alpha-1}(1-y)^{\beta-1}\, dy, \\ &\alpha > 0,\ \beta > 0,\ 0 \le x \le 1 \end{aligned}$$

and the incomplete beta function ratio

$$I_x(\alpha, \beta) = B_x(\alpha, \beta)/B(\alpha, \beta).$$

Since for $0 \le x \le 1$

$$\begin{aligned} F_m(x) &= \frac{\pi}{B(\frac{m}{2}, \frac{1}{2})} \int_0^x \sin^{m-1}(\pi\phi)\, d\phi \\ &= \frac{1}{B(\frac{m}{2}, \frac{1}{2})} \int_0^{\pi x} \sin^{m-1}\theta d\theta \quad (\theta = \pi\phi), \end{aligned}$$

we have

(i) $0 \le x \le \frac{1}{2}$: Let $y = \sin\theta$. We have

$$dy = \cos\theta d\theta = (1-y^2)^{1/2} d\theta$$

and

$$
\begin{aligned}
F_m(x) &= \frac{1}{B(\frac{m}{2},\frac{1}{2})}\int_0^{\sin(\pi x)} y^{m-1}(1-y^2)^{-\frac{1}{2}}\,dy \quad (z=y^2)\\
&= \frac{1}{2B(\frac{m}{2},\frac{1}{2})}\int_0^{\sin^2(\pi x)} z^{\frac{m}{2}-1}(1-z)^{\frac{1}{2}-1}\,dz\\
&= \frac{1}{2}I_{\sin^2(\pi x)}\left(\frac{m}{2},\frac{1}{2}\right); \qquad \text{(B.34)}
\end{aligned}
$$

(ii) $\frac{1}{2} \le x \le 1$: By a similar way we obtain

$$
\begin{aligned}
F_m(x) &= 1-\frac{1}{2}I_{\cos^2(\pi x)}\left(\frac{1}{2},\frac{m}{2}\right)\\
&= 1-\frac{1}{2}I_{\sin^2(\pi x)}\left(\frac{m}{2},\frac{1}{2}\right) \qquad \text{(B.35)}
\end{aligned}
$$

because

$$I_x(\alpha,\beta) = 1 - I_{1-x}(\beta,\alpha).$$

Therefore with the inverse incomplete beta function we can find $F_m^{-1}(x)$ which is required in generating an NT-net on B_s and U_s by the spherical coordinate transformation. In fact we can avoid calculating F_m^{-1} if the TWWF method stated in Section 4.3 is used for generating an NT-net on B_s or U_s.

B.3 The domains V_s and T_s

Let V_s and T_s be the unit l_1-ball and the unit l_1-sphere of R^s, i.e.

$$V_s = \{(x_1,\cdots,x_s) : x_i \ge 0,\ i=1,\cdots,s, \sum_{i=1}^{s} x_i \le 1\} \qquad \text{(B.36)}$$

and

$$T_s = \{(x_1,\cdots,x_s) : x_i \ge 0,\ i=1,\cdots,s, \sum_{i=1}^{s} x_i = 1\}. \qquad \text{(B.37)}$$

Consider the integral

$$I_4 = \int_{V_s} f(\boldsymbol{x})\, d\boldsymbol{x}.$$

The transformation $\boldsymbol{x} = \boldsymbol{x}_4(\boldsymbol{\phi})$,

$$\begin{cases} x_i = \phi_1 \prod_{j=2}^{i} \sin^2\left(\frac{\pi\phi_j}{2}\right) \cos\left(\frac{\pi\phi_{i+1}}{2}\right), \\ \qquad i = 1, \cdots, s-1, \\ x_s = \phi_1 \prod_{j=2}^{s} \sin^2\left(\frac{\pi\phi_j}{2}\right), \end{cases} \tag{B.38}$$

where $\boldsymbol{\phi} = (\phi_1, \cdots, \phi_s) \in C^s$, maps C^s into V_s. In order to apply Lemma B.1 in (B.38), we use the following relation

$$\begin{cases} x_s = \phi_1 \prod_{j=2}^{s} \sin^2\left(\frac{\pi\phi_j}{2}\right), \\ x_{s-1} + x_s = \phi_1 \prod_{j=2}^{s-1} \sin^2\left(\frac{\pi\phi_j}{2}\right), \\ \quad \cdots \\ x_2 + \cdots + x_s = \phi_1 \sin^2\left(\frac{\pi\phi_2}{2}\right), \\ x_1 + \cdots + x_s = \phi_1. \end{cases}$$

The Jacobian of transformation (B.38) is

$$J_4(\boldsymbol{\phi}) = \prod_{i=1}^{s} J(x_i \to \phi_i),$$

where

$$J(x_i \to \phi_i) = \phi_1 \pi \sin\left(\frac{\pi\phi_i}{2}\right) \cos\left(\frac{\pi\phi_i}{2}\right) \prod_{j=2}^{i-1} \sin^2\left(\frac{\pi\phi_j}{2}\right)$$

for $2 \le i \le s$, and

$$J(x_1 \to \phi_1) = 1.$$

Therefore

$$J_4(\boldsymbol{\phi}) = (\pi\phi_1)^{s-1} \prod_{j=2}^{s} \sin^{2(s-j)+1}\left(\frac{\pi\phi_j}{2}\right) \cos\left(\frac{\pi\phi_j}{2}\right), \tag{B.39}$$

and

$$I_4 = \int_{C^s} f(\boldsymbol{x}_4(\boldsymbol{\phi})) J_4(\boldsymbol{\phi})\, d\boldsymbol{\phi}. \tag{B.40}$$

In particular,

$$\begin{aligned} v(V_s) =& \int_{C^s} J_4(\boldsymbol{\phi})\, d\boldsymbol{\phi} \\ =& \int_0^1 \phi_1^{s-1}\, d\phi_1 \\ & \times \prod_{j=2}^{s} \pi \int_0^1 \sin^{2(s-j)+1}\left(\frac{\pi\phi_j}{2}\right) \cos\left(\frac{\pi\phi_j}{2}\right) d\phi_j \\ =& 1/s!. \end{aligned} \tag{B.41}$$

Let $\boldsymbol{x} = (X_1, \cdots, X_s)$ be uniformly distributed on V_s. The p.d.f. of $\boldsymbol{x}$ is

$$p_4(\boldsymbol{x}) = \begin{cases} s!, & \text{if } \boldsymbol{x} \in V_s, \\ 0, & \text{otherwise.} \end{cases}$$

Take the transformation (B.38) for $\boldsymbol{x}$. Then the p.d.f. of $\phi_1, \cdots, \phi_s$ is given by

$$s! J_4(\boldsymbol{\phi}) = s\phi_1^{s-1} \prod_{j=2}^{s} \left[\pi(s-j+1) \sin^{2(s-j)+1}\left(\frac{\pi\phi_j}{2}\right) \cos\left(\frac{\pi\phi_j}{2}\right)\right]$$

which implies that

(a) $\phi_1, \cdots, \phi_s$ are independent;

(b) the p.d.f. and c.d.f. of ϕ_j are

$$p_j(x) = \begin{cases} s!\phi^{s-1}, & \text{if } j = 1, \\ \pi(s-j+1)\sin^{2(s-j)+1}\left(\frac{\pi\phi}{2}\right) & \\ \quad \times \cos\left(\frac{\pi\phi}{2}\right), & \text{if } 2 \le j \le s \end{cases} \tag{B.42}$$

and

$$F_j(\phi) = \begin{cases} \phi^s, & \text{if } j = 1, \\ \sin^{2(s-j+1)}\left(\frac{\pi\phi}{2}\right), & \text{if } 2 \le j \le s, \end{cases} \tag{B.43}$$

where $0 \le \phi \le 1$.

We now consider the integral

$$I_5 = \int_{T_s} f(\boldsymbol{x})\, dv,$$

where dv is the volume elements of T_s. Take the transformation $\boldsymbol{x} = \boldsymbol{x}_5(\boldsymbol{\phi})$, where

$$\begin{cases} x_i = \prod_{j=1}^{i-1} \sin^2\left(\frac{\pi\phi_j}{2}\right)\cos^2\left(\frac{\pi\phi_i}{2}\right), & i = 1, \cdots, s-1, \\ x_s = \prod_{j=1}^{s} \sin^2\left(\frac{\pi\phi_j}{2}\right). \end{cases} \tag{B.44}$$

We are going to show that the Jacobian $J_5(\boldsymbol{\phi})$ of the transformation is

$$J_5(\boldsymbol{\phi}) = \sqrt{s}\pi^{s-1} \prod_{j=1}^{s-1} \sin^{2(s-j)-1}\left(\frac{\pi\phi_j}{2}\right)\cos\left(\frac{\pi\phi_j}{2}\right) \tag{B.45}$$

and consequently

$$I_5 = \int_{C^{s-1}} f(\boldsymbol{x}_5(\boldsymbol{\phi})) J_5(\boldsymbol{\phi})\, d\boldsymbol{\phi}.$$

First we take the transformation which maps V_{s-1} into T_s as follows

$$\begin{cases} x_i = y_i, & i = 1, \cdots, s-1 \\ x_s = 1 - y_1 - \cdots - y_{s-1}. \end{cases}$$

It can be easily shown that

$$\boldsymbol{H} = \left(\frac{\partial x_j}{\partial y_i}\right) = (\boldsymbol{I}_{s-1}, -\boldsymbol{1}_{s-1}), \qquad \boldsymbol{1}_{s-1} = (1, \cdots, 1)'$$

and

$$\begin{aligned} \det(\boldsymbol{H}\boldsymbol{H}')^{1/2} &= \det(\boldsymbol{I}_{s-1} + \boldsymbol{1}_{s-1}\boldsymbol{1}'_{s-1})^{1/2} \\ &= (1 + \boldsymbol{1}'_{s-1}\boldsymbol{1}_{s-1})^{1/2} = \sqrt{s}. \end{aligned}$$

Therefore we have

$$I_5 = \sqrt{s} \int_{V_{s-1}} f(\boldsymbol{y}, 1 - \|\boldsymbol{y}\|) d\boldsymbol{y},$$

where $\|\boldsymbol{y}\| = \sum_1^{s-1} y_i$ is the l_1-norm of $\boldsymbol{y}$. Next, we take the first $s-1$ transformations of (B.44) which maps C^{s-1} into V_{s-1}. The matrix $(\frac{\partial y_j}{\partial \phi_i})$ is triangular and the Jacobian of the transformation is

$$\begin{aligned} \det\left(\frac{\partial y_j}{\partial \phi_i}\right)_+ &= \prod_{i=1}^{s-1} \frac{\partial y_i}{\partial \phi_i} \\ &= \prod_{i=1}^{s-1} 2\frac{\pi}{2} \prod_{j=2}^{i-1} \sin^2\left(\frac{\pi\phi_j}{2}\right) \sin\left(\frac{\pi\phi_i}{2}\right) \cos\left(\frac{\pi\phi_i}{2}\right) \\ &= \pi^{s-1} \prod_{j=1}^{s-1} \sin^{2(s-j)-1}\left(\frac{\pi\phi_j}{2}\right) \cos\left(\frac{\pi\phi_j}{2}\right) \end{aligned}$$

which implies (B.44) and (B.45).

The volume of T_s can be obtained by

$$\begin{aligned} v(T_s) &= \int_{C^{s-1}} J_5(\boldsymbol{\phi})\, d\boldsymbol{\phi} \\ &= \sqrt{s}\prod_{j=1}^{s-1} \pi \int_0^1 \sin^{2(s-j)-1}\left(\frac{\pi\phi_j}{2}\right)\cos\left(\frac{\pi\phi_j}{2}\right) d\phi_j \\ &= \sqrt{s}/(s-1)!. \end{aligned}$$

Let $\boldsymbol{x} = (X_1, \cdots, X_s)$ be uniformly distributed on T_s. The density function of $\boldsymbol{x}$ on T_s is

$$p_5(\boldsymbol{x}) = \begin{cases} (s-1)!/\sqrt{s}, & \text{if } \boldsymbol{x} \in T_s, \\ 0, & \text{otherwise.} \end{cases}$$

Take the transformation (B.44) for $\boldsymbol{x}$. Then the p.d.f. of $\phi_1, \cdots, \phi_{s-1}$ is

$$\begin{aligned} &p_5(\boldsymbol{x}_5(\boldsymbol{\phi}))J_5(\boldsymbol{\phi}) \\ &\quad = \frac{(s-1)!}{\sqrt{s}}\sqrt{s}\pi^{s-1}\prod_{j=1}^{s-1}\sin^{2(s-j)-1}\left(\frac{\pi\phi_j}{2}\right)\cos\left(\frac{\pi\phi_j}{2}\right) \end{aligned}$$

which implies that

(a) $\phi_1, \cdots, \phi_{s-1}$ are independent;
(b) the p.d.f. and c.d.f. of ϕ_j are

$$\begin{aligned} p_j(\phi) =& \pi(s-j)\sin^{2(s-j)-1}\left(\frac{\pi\phi}{2}\right)\cos\left(\frac{\pi\phi}{2}\right), \\ & 0 \le \phi \le 1 \end{aligned} \tag{B.46}$$

and

$$F_j(\phi) = \sin^{2(s-j)}\left(\frac{\pi\phi}{2}\right), \qquad 0 \le \phi \le 1, \tag{B.47}$$

respectively.

References

Abut, H., Gray, R.M. and Rebolledo, G. (1982) Vector quantization of speech and speech-like waveforms, *IEEE Trans. Acoust., Speech, Signal Processing,* **ASSP-30,** 423-435.

Aitchison, J. (1986) *The Statistical Analysis of Compositional Data,* Chapman and Hall, London/New York.

Allgower, E. and Georg, K. (1980) Simplicial and continuation methods for approximating fixed points and solutions to systems of equations, *SIAM Review,* **22,** 28-85.

Amari, S. -I. (1985) *Differential-Geometrical Methods in Statistics*, Lecture Notes in Statistics, **28**, Spring-Verlag, New York.

Amari, S. -I., Barndorff-Nielsen, O.E., Kass, R.E., Lauritzen, S.L. and Rao, C.R. (1987) *Differential Geometry in Statistical Inference*, IMS, Hayward, California.

Anderberg, M.R. (1973) Cluster Analysis for Applications, New York, Academic Press.

Anderson, E. (1935) The irises of the Gaspe Peninsula, *Bull. Am. Iric Soc.* **59**, 2-5.

Anderson, T.W. (1984) *An Introduction to Multivariate Statistical Analysis*, 2nd edition, Wiley, New York.

Anderson, T.W., Olkin, I. and Underhill, L.G. (1987) Generation of random orthogonal matrices, *SIAM J. Sci. Statist. Comput.*, **8**, 625-629.

Andrews, D. F. (1972) Plots of high-dimensional data, *Biometrics,* **28,** 125-136.

Arnold, B.C. and Balakrishnan, N. (1989) Relations, Bounds and Approximations for Order Statistics, Springer-Verlag, New York.

Avriel, *M.* (1976) *Nonlinear Programming, Analysis and Applications,* Prentice-Hall, Inc. Englewood Cliffs, New Jersey.

Baker, A. (1965) On some Diophantine inequalities involving the exponential function, *Canad. J. Math.,* **17,** 616-626.

Baringhaus, L. and Henze, N. (1988). A consistent test for multivariate normality based on the empirical characteristic function, *Metrika,* **35,** 339-348.

Bentler, P.M. (1980) Multivariate analysis with latent variables: causal modeling, *Annual Review of Psychology*, **31**, 419-456.

Bera, A. and John, S. (1983) Tests for multivariate normality with Pearson alternatives, *Commun. Statist.-Theor. Meth.*, **12**, 103-117.

Bernardo, M.C., Buck, R., Liu, L., Nazaret, W.A., Sacks, J., and Welch W.J. (1992), Integrated circuit design optimization using a sequential strategy, *IEEE Trans, on Computer – Aided Design*, **11**, 361–372.

Bertsimas, D. and Tsitsiklis, J. (1993) Simulated annealing, *Statistical Science*, **8**, 10-15.

Bickel, P.J. (1964) On some alternative estimates for shift in the p-variate one-sample problem, *Ann. Math. Statist.*, **35**, 1079-1090.

Bofinger, E. (1970) Maximizing the correlation of grouped observations, *J. Amer. Statist. Assoc.*, **65**, 1632-8.

Boot, J.C.G. (1964) *Quadratic Programming,* North-Holland Publishing Company, Amsterdam.

Bouleau, N. (1990) On effective computation of expectations in large or infinite dimensions, *J. Comput. Appl. Math.*, **31**, 23-34.

Braaten, E. and Weller, G. (1979) An improved low-discrepancy sequence for multidimensional Quasi-Monte-Carlo integration, *J. Comp. Physics,* **33**, 249-258.

Brent, R.P. (1973) Some efficient algorithms for solving systems of nonlinear equations, *SIAM, J. Numer. Anal.*, **10**, 327-344.

Brown, K.M. (1967) Solution of simultaneous nonlinear equations, *Comm. ACM*, **10**, 728-729.

Bundschuh, P. and Zhu, Y.C. (1993) A method for exact calculation of the discrepancy of low-dimensional finite point sets (I), Abhandlungen aus dem Math. Seminar der Univ. Hamburg, Bd. 63.

Cambanis, S. and Gerr, N.L. (1983) A simple class of asymptotically optimal quantizers, *IEEE Trans. Inform Theory,* **IT-29,** 664-676.

Cao, S.Y. (1992) Discussion on estimates in l_1-norm regression and number-theoretic method, *International Symposium on Multivariate Analysis and Its Applications*, Hong Kong.

Cheng, P. (1983) An open problem in steel rolling, *Mathematics in Practice and Theory,* **2,** 79-79.

Cheng, P., Zhu, L.X. and Wei, G. and Shi, P.D. (1991) On the life of a sphere roller, *Acta Math. Sci.* **11** 308-316

Chong, W.P. (1991) Representative points univariate continuous distributions – A number-theoretic method approach, *BSc thesis*, Hong Kong Baptist College, Hong Kong.

Chung, K.L. (1949) An estimate concerning the Kolmogoroff limit distribution, *Trans. Amer. Math. Soc.*, **67**, 36-50.

Cohen, A.C. (1965) Maximum likelihood estimation in the Weibull distribution based on complete and on censored samples, *Technometrics,* **7,** 579-88.

Cornell, J.A. (1973) Experiments with mixtures: A review, *Technometrics,* **15,** 437-455.

Cornell, J.A. (1975) Some comments on designs for Cox's mixture polynomial, *Technometrics,* **17,** 25-35.

Cornell, J.A. (1981) *Experiments with Mixtures, Designs, Models, and the Analysis of Mixture Data,* New York: Wiley.

Cox, D.R. (1957) Note on grouping, *J. Amer. Stat. Assoc.,* **52,** 543-547.

Cox, D.R. (1971) A note on polynomial response functions for mixtures, *Biometrika,* **58,** 155-159.

Cox, D.R. and Small, N.J.H. (1978) Testing multivariate normality, *Biometrika,* **65,** 263-272.

Cranley R. and Patterson T.N.L. (1976) Randomization of number theoretic methods for multiple integration, *Siam J. Numer. Anal.,* **13,** 904-914.

Csörgö, S. (1986) Testing for normality in arbitrary dimension, *Ann. Statist.*, **14**, 708-723.

Csörgö, S. (1989) Consistency of some tests for multivariate normality, *Metrika,* **36,** 107-116.

David, H.A. (1981) *Order Statistics*, 2nd Edition, Wiley, New York.

Davis P.J. and Rabinowitz P. (1984) *Methods of numerical integration,* Academic Press, New York.

Deng, N.Y. (1988) *Numerical Methods for Unconstrained Optimization,* Science Press, Beijing.

Dennis, J.E. and Moré, J.J. (1977) Quasi-Newton methods, motivation and theory, *SIAM Review,* **19,** 46-89.

Diaconis, P. (1989) Group Representations in Probability and Statistics, IMS Lecture Notes-Monograph Series; Vol. 11, Hayward, California.

Ding, Y. (1986) An exploration of optimality of uniform designs, *Chinese Journal of Applied Probability and Statistics*, **2**, 153–160.

Dixon, L.C.W. (1972) Quasi-Newton algorithms generate identical points, *Math. Prog.*, **2**, 383–387.

Dixon, L.C.W. and Szego, G.P. (eds) (1975). *Towards Global Optimisation*, North-Holland, Amsterdam.

Draper, N.R., and Lawrence, W.E. (1965) Mixture designs for three factors, *J. Royal Statist. Soc. B,* **27,** 450-465.

Draper, N.R., and Lin, D.K.J. (1990) Small response-surface designs, *Technometrics*, **32**, 187–194.

Dubey, S.D. (1967) Some percentile estimators for Weibull parameters, *Technometrics,* **9,** 119-29.

Eaton, M.L. (1989) *Group Invariance Applications in Statistics*, CBMS Regional Conference Series in Probability and Statistics, Vol. 1, IMS, Hayward, California.

Efron, B. (1975) Defining the curvature of a statistical problem with applications to second-order efficiency (with discussion), *Ann. Statist.* **3**, 1189-1242.

Fang, K.T. (1980) The uniform design: application of number theoretic methods in experimental design, *Acta Math. Appl. Sinica*, **3**, 363-372.

Fang, K.T. and Anderson, T.W. (1990) *Statistical Inference in Elliptically Contoured and Related Distributions*, Allerton Press, New York.

Fang, K.T. and Fang, B.Q. (1988) Some families of multivariate symmetric distributions related to exponential distribution, *J. Multivariate Analysis,* **24**, 109-122.

Fang, K.T. and He, S.D. (1982) The problem of selecting a given number of representative points in a normal population and a generalized Mill's ratio, *Technical Report No.5,* U.S. Army Research Office Contract DAAG 29-82-K-0156, Department of Statistics, Stanford University, California.

Fang, K.T. and He, S.D. (1984) The problem of selecting a specified number of representative points from a normal population, *Acta Math. Appl. Sinica,* **7**, 293-306.

Fang, K.T. and He, S.D. (1985) Regression models with linear constraints and nonnegative regression coefficients, *Math. Numer. Sinica,* **7**, 237-246.

Fang, K.T., Kotz, S. and Ng, K.W. (1990) *Symmetric Multivariate and Related Distributions,* Chapman and Hall, London.

Fang, K.T., Kotz, S. and Ng, K.W. (1992) On l_1-norm distributions, l_1-*Statistical Analysis and Related Methods*, (ed. by Y. Dodge), North-Holland, New York.

Fang, K.T. and Li, R.Z. (1992) An algorithm for generating a set of matrices uniformly scattered on the Stiefel manifold, to be submitted.

Fang, K.T., Quan, H. and Chen, Q.Y. (1988) *Applied Regression Analysis*, Science Press, Beijing (in Chinese).

Fang, K.T., Wang, D.Q., and Wu, G.F. (1982) A class of constrained regression-programing regression, *Math. Numer. Sinica,* **4**, 57-69.

Fang, K.T. and Wang, Y. (1990) A sequential algorithm for optimization and its applications to regression analysis, in *Lecture Notes in Contemporary Mathematics,* 1989 (eds by Yang, L. and Wang, Y.), Science Press, Beijing 17-28.

Fang, K.T. and Wang, Y. (1991) A sequential algorithm for solving a system of nonlinear equations, *J. Comp. Math.,* **9**, 9-16.

Fang, K.T., Wang, Y. and Bentler, P.M. (1992) Applications of number-theoretic method in multivariate statistics, invited talk in *the International Symposium on Multivariate Analysis and Its Applications,* Hong Kong.

Fang, K.T., Wang, Y. and Wong, K.L. (1992) A new method of generating an NT-net on the unit sphere, *Technical Report* **MATH-002**, Hong Kong Baptist College, Hong Kong.

Fang, K.T., Wang Y. and Yuan K.H. (1990) Quantizer and quasirandom F-number sequence of multivariate continuous distribution, *Technical Report,* **004,** Institute of Applied Mathematics, Academia Sinica, Beijing.

Fang, K.T. and Wei, G. (1989a) The distributions of a class of the first hitting time, to appear in *Acta Math. Appl. Sinica.*

Fang, K.T. and Wei, G. (1989b) The applications of number theoretic method to some geometric probability problems in statistics, *The Third Japan-China Symposium on Statistics,* Contributed Papers, Tokyo, 42-45.

Fang, K.T. and Wu, C.Y. (1979) The extreme value problem of some probability function, *Acta Math. Appl. Sinica,* **2,** 132-148.

Fang, K.T. and Yuan, K.H. (1990) A unified approach to maximum likelihood estimation, *Chinese J. Appl. Prob. Stat.,* **6,** 412-418.

Fang, K.T. Yuan, K.H. and Bentler, P.M. (1990) Quantizer and representative points of elliptically contoured distributions, invited talk in "The First Conference on Recent Developments in Statistical Research", Hong Kong.

Fang, K.T., Yuan, K.H. and Bentler, P.M. (1992) Applications of sets of points uniformly distributed on a sphere to testing multinormality and robust estimation, in *Probability and Statistics*, (eds Jiang, Z.P., Yan, S.J., Cheng, P. and Wu, R.), World Scientific, 56–73.

Fang, K.T. and Zhang, J.T. (1990) A new algorithm for evaluation of estimates of parameters of nonlinear regression modelings, to appear in *Acta Math. Appl. Sinica.*

Fang, K.T. and Zhang, J.T. (1992) Criteria for uniform design, *Technical Report* MATH–002, Hong Kong Baptist College, Hong Kong.

Fang, K.T. and Zhang, Y.T. (1990) *Generalized Multivariate Analysis,* Springer-Verlag and Science Press, Berlin and Beijing.

Fang, K.T. and Zhen, H.N. (1992) A maximum symmetric differences principle and its applications in uniform design, *Chinese J. Appl. Prob. Statist.,* **8,** 10-18.

Fang, K.T. and Zhu, L.X. (1993) On uniformly random design, *Technical Report* **MATH-020**, Hong Kong Baptist College.

Fang, K.T., Zhu, L.X. and Bentler, P.M. (1993), A nescessary test of goodness of fit for sphericity, to appear in *J. Multivariate Analysis.*

Faure, H. (1982) Discrépance associée à un Système de Numération (en dimensions), *Acta Arithmetica,* **XL1**, 337-361.

Feng, G.C. (1989) *Iterative Methods of A System of Nonlinear Equations,* Shanghai Press of Science and Technology, Shanghai.

Ferrari, P.A., Frigessi, A. and Schonmann, R.H. (1993) Convergence of some partially parallel Gibbs samplers with annealing, *Ann. Appl. Prob.*, **3**, 137-153.

Fisher, R.A. (1936) The use of multiple measurements in taxonomic problems, *Ann. Eugenics*, **VII**, 179-188.

Fischer, T.R. (1986) A pyramid vector quantizer, *IEEE Trans. Inform. Theory,* **IT-32,** 568-583.

Fischer, T.R. (1989) Geometric source coding and vector quantization, *IEEE Trans. Inform. Theory,* **IT-35,** 137-145.

Fischer, T.R. and Dicharry, R.M. (1984) Vector quantizer design for memoryless Gaussian, gamma, and Laplacian sources, *IEEE Trans. Commun.,* **COM-32,** 1065-1069.

Fletcher, R. (1987) *Practical Methods of Optimization*, 2nd Edition, Wiley, New York.

Fu, J.C. (1989) Method of Kim Zam – An algorithm for computing the maximum likelihood estimator, *Statistics & Probability letters,* **8,** 289-296.

Gasko, M. and Donoho, D. (1982) Influential observations in data analysis, Missing observations, in *Proceedings of the Business and Economic Statistics Section*, American Statistical Association, 104-109.

Gelfand, A.E., Hills, S.E., Racine-Poon, A., and Smith, A.F.M. (1990) Illustration of Bayesian inference in normal data models using Gibbs sampling, *J. Amer. Statist. Assoc.,* **85,** 972-985.

Gentleman, W.M. (1965) Robust estimation of multivariate location by minimizing pth power deviations, Ph.D thesis, Princeton University.

Gersho, A. (1979) Asymptotically optimal block quantization, *IEEE Trans. Inform. Theory,* **IT-25,** 373-380.

Gersho, A. (1982) On the structure of vector quantizers, *IEEE Trans, Inform. Theory,* **IT-28,** 157-166.

Gersho, A. and Cuperman, V. (1983) Vector quantization: A pattern-matching technique for speech coding, *IEEE Commun. Mag.,* **21,** 15-21.

Gill, P.E., Murray, W. and Wright, M.H. (1981) *Practical Optimization*, Academic Press, London.

Gnanadesikan, R. (1977) *Methods for Statistical Data Analysis of Multivariate Observations,* Wiley, New York.

Gnanadesikan, R., Pinkham, R.S., and Hughes, L.P. (1967) Maximum likelihood estimation of the parameters of the beta distribution from smallest order statistics, *Technometrics,* **9,** 607-20.

Goldberg, D.E. (1989) *Genetic Algorithms in Search, Optimization, and Machine Learning*, Addison-Wesley, Reading, Mass.

Gray, R.M. (1984) Vector quantization, *IEEE ASSP Mag.,* **1,** 4-29.

Gray, R.M. and Karnin, E.D. (1982) Multiple local optima in vector quantizers, *IEEE Trans. Inform. Theory,* **IT-28,** 256-261.

Gray, R.M. and Linde Y.L. (1982) Vector quantizers and predictive quantizers for Gauss-Markov sources, *IEEE Trans. Commun.,* **COM-30,** 381-389.

Gray, R.M., Kieffer. J. C. and Linde, Y. (1980) Locally optimal block quantizer design, *Inform. Contr.,* **45,** 178-198.

Gupta, R.D. and Richards, D. St. P. (1987) Multivariate Liouville distributions, *J. Multi. Anal.*, **23**, 233-56.

Gupta, S.S. (1963) Probability integrals of multivariate normal and multivariate *t*, *Ann. Math. Statist.*, **34**, 792-828.

Haber, S. (1970) Numerical evaluation of multiple integrals, *SIAM Rev.*, **12**, 481-526.

Haber, S. (1972) Experiments on optimal coefficients, *Applications of Number Theory to Numerical Analysis* (S.K. Zaremba, ed.), Academic Press, New York, 11-37.

Halton, J.H. (1960) On the efficiency of certian quasi-random sequences of points in evaluating multi-dimensional integrals, *Numer. Math.*, **2**, 84-90.

Hammersley, J.M. (1960) Monte Carlo methods for solving multivariable problems, *Ann. New York Acad. Sci.*, **86**, 844-874.

Hampel, F.R., Ronchetti, E.M., Rousseeuw, P.J., and Stahel, W.A. (1986) *Robust Statistics: The Approach Based on Influence Functions*, Wiley, New York.

Harter, H.L. and Moore, A.H. (1965) Maximum likelihood estimation of the parameters of gamma and Weibull populations from complete and from censored samples, *Technometrics*, **7**, 639-43.

Hartley, H.O. (1961) The modified Gauss-Newton method for the fitting nonlinear regression function by least squares, *Technometrics*, **3**, 269-280.

Haselgrove, C.B. (1961) A method for numerical integration, *Math. Comp.*, **15**, 323-337.

Hensler, G.L., Mehrotra, K.G. and Michalek, J.E. (1977) A goodness of fit test for multivariate normality, *Commun. Statist.-Theor. Math.*, **6**, 33-41.

Hlawka, E. (1961) Funktionen von beschränkter variation in der Theorie Gleichverteilung, *Ann Mat. pure Appl.*, **54**, 325-333.

Hlawka, E. (1962) Zur angenäherten Berechnung mehrfacher Integrale, *Monatsh, Math.*, **66**, 140-151.

Hlawka, E. (1964) Uniform distribution modulo 1 and numerical analysis, *Compositio Math.*, **16**, 92-105.

Hlawka, E. (1990) *Selecta*, Springer-Verlag, Berlin.

Hlawka, E. and Muck, R. (1972) A transformation of equidistributed sequences, *in "Applications of Number Theory to Numerical Analysis"* (Zaremba, S.K. ed), Academic Press, New York, 371-388.

Horst, R. and Tuy, H. (1990) *Global Optimization: Deterministic Approaches*, Springer-Verlag, Berlin.

Horswell, R.L. and Looney, S.W. (1992) A comparison of tests for multivariate normality that are based on measures of multivariate skewness and kurtosis, *J. Statist. Comput. Simul.*, **42**, 21-38.

Hsu, L.Z. and Zhou, Y.S. (1980) *Numerical Evaluation of Multiple Integrals*, Science Press, Beijing.

Hua, L.K. (1956) *Introduction to Number Theory,* Science Press, Beijing.

Hua, L.K. (1984) *Introduction to High Mathematics, Supplement*, Science Press, Beijing.

Hua, L.K. and Wang, Y. (1960) Remarks concerning numerical integration, *Sci. Record, (N.S.),* **4,** 8-11.

Hua, L.K. and Wang, Y. (1961) *Numerical Evaluatuion of Integrals,* Science Press, Beijing.

Hua, L.K. and Wang, Y. (1963) *Numerical Integration and Its Applications,* Science Press, Beijing.

Hua, L.K. and Wang, Y. (1964) On Diophantine approximations and numerical integrations, *(I) Sci. Sin.,* **13,** 1007-1008 *(II) Sci. Sin.,* **13,** 1009-1010.

Hua, L.K. and Wang, Y. (1965) On numerical integration of periodic functions of several variables, *Sci. Sin.,* **14,** 964-978.

Hua, L.K. and Wang, Y. (1981) *Applications of Number Theory to Numerical Analysis,* Springer-Verlag and Science Press, Berlin and Beijing.

Huber, P.J. (1964) Robust estimation of a location parameter, *Ann. Math. Statist.,* **35,** 73-101.

Huber, P.J. (1972) Robust statistics: a review, *Ann. Math. Statist.,* **43,** 1041-1067.

Huber, P.J. (1973) Robust regression: asymptotics, conjectures and Monte Carlo, *Ann. Statist.,* **1,** 799-821.

Huber, P.J. (1985) Projection pursuit, *Ann. Statist.*, **13**, 435–475.

Iman, R.L. and Conover W.J. (1980) Small sample sensitivity analysis techniques for computer models, with an application to risk assessment (with discussion), *Communications in Statist, Part A – Theory and Methods*, **9**, 1749–1874.

Jarrett, R.G. and Morgan, B.J.T. (1984) A critical reappraisal of the Patch-Gap model, *Biometrics*, **40**, 203-207.

Jiang, S. and Chen, R.C. (1987) Homogeneous designs of latin square type, *Applied Math., J. Chinese Universities*, **2**, 532–541.

Johnson, M.E. Moore, L.M. and Ylvisaker, D. (1990) Minimax and maximin distance designs, *J. Statist. Planning and Inference*, **26**, 131-148.

Johnson, N.L. and Kotz, S. (1972) *Distributions in Statistics: Continuous Multivariate Distributions*, Wiley, New York.

Kahaner, D. Moler, C. and Nash, S. (1989) *Numerical Methods and Software*, Prentice Hall, Englewood Cliffs, N.J.

Kendall, M.G. and Moran, P.A.P. (1963) *Geometrical Probability*, Griffin, London.

Kiefer, J. (1961) On large deviations of the empiric d.f. of vector chance variables and a law of the iterated logarithm, *Pacific J. Math.*, **11**, 649-660.

Kirkpatrick, S., Gelatt, C.D., Jr., and Vecchi, M.D. (1983) Optimization by simulated annealing, *Science,* **220,** 671-680.

Korobov, N.M. (1957) Approximate calculation of multiple integrals with the aid of methods in the theory of numbers, *Dokl. Akad. Nauk. SSSR,* **115,** 1062-1065.

Korobov, N.M. (1959a) The approximate computation of multiple integrals, *Dokl. Akad. Nauk. SSSR,* **124,** 1207-1210.

Korobov, N.M. (1959b) Computation of multiple integrals by the method of optimal coefficients, *Vestnik Moskow Univ. Ser. Mat. Astr. Fiz. Him.,* **4,** 19-25.

Korobov, N.M. (1960) Properties and calculation of optimal coefficients, *Dokl. Akad. Nauk. SSSR,* **132,** 1009-1012.

Korobov, N.M. (1963) *Number Theoretic Methods in Approximate Analysis,* Fizmatigiz, Moscow.

Korobov, N.M. (1989) *Trigonometric Sums and Their Applications,* Nauk, Moscow.

Koziol, J.A. (1982) A class of invariant procedures for assessing multivariate normality, *Biometrika*, **69**, 423-427.

Kuipers, L. and Niederreiter, H. (1974) *Uniform Distribution of Sequences,* Wiley, New York.

Kurker, C.M. Paulos, J.J., Gyurcsik, R.S. and Lu, J.C. (1993), Hierarchical yield estimation of large analog integrated-circuits, *IEEE J. Solid State Circuits* **28** 203-209.

Laarhoven, van P.J.M. (1987) *Simulated Annealing: Theory and Applications*, D. Reidlt, Amsterdam.

Lai, W.Y. (1992) A hybrid Newton-like/number-theoretic optimization algorithm, *BSc thesis*, Hong Kong Baptist college.

Lapeyre, B. and Pagès, G. (1989) Familles de snits à discrépance faible obtenues par itérations de transformations de [0,1], *CRAS*, **308, Série** I, 507-509.

Lapeyre, B., Pagès, G. and Sab, K. (1990) Sequence with low discrepancy, generalization and application to Robbins-Monro algorithm, *Statistics,* **21**, 251-272.

Li, G.Y. (1985) Robust Regression, in *Exploring Data Tables, Trends and Shapes*, eds by D.C. Hoaglin, F. Mosteller, and J.W. Tukey, Wiley, New York.

Li, G.Y. and Zha, W.X. (1992) PP Neyman tests for multivariate goodness-of-fit, in *The Development of Statistics: Recent Contributions from China: Achivevement and Progress*, eds by Chen X. R., Fang K.T. and Yang C.C., Longman, 261-274.

Linde, Y.L., Buzo, A. and Gray, R.M.(1980) An algorithm for vector quantizer design, *IEEE Trans. Commun.,* **COM-28,** 84-95.

Lloyd, S.P. (1982) Least squares quantization in *PCM, IEEE Trans. Inform. Theory,* **IT-28,** 129-137.

Machado, S.G. (1983) Two statistics for testing for multivariate normality, *Biometrika*, **70**, 713-718.

MacQueen, J. (1967) Some method for classification and analysis of multivariate observations, *in Proc. 5th Berkeley Symposium on Math. Statist. and Prob.*, **1,** 281-296.

Malkovich, J.F. and Afifi, A.A. (1973) On tests for multivariate normality, *J. Am. Statist. Assoc.*, **68,** 176-179.

Mardia, K.V. (1970) Measures of multivariate skewness and kurtosis with applications, *Biometrika,* **57,** 519-530.

Mardia, K.V. (1971) The effect of nonnormality on some multivariate tests and robustness to nonnormality in the linear model, *Biometrika,* **58,** 105-121.

Mardia, K.V. (1972) *Statistics of Directional Data,* Academic Press, New York.

Mardia, K.V. (1974) Applications of some measures of multivariate skewness and kurtosis for testing normality and robustness studies, *Sankhya,* **A36,** 115-128.

Mardia, K.V. (1975) Assessment of multinormality and the robustness of Hotelling's T test, *J. Roy. Statist. Soc.*, **C24,** 163-171.

Maisonneuve, D. (1972) Recherche et utillisation des "bons treillis". Programmation et résultats numériques, *Applications of Number Theory to Numerical Analysis* (S.K. Zaremba, ed.), Academic Press, New York, 121-201.

Malone, K.T. and Fisher T.R. (1988) Contour-gain vector quantization, *IEEE Trans. Acoust., Speech, Signal Processing,* **ASSP-36,** 862-870.

Marsaglia, G. (1972) Choosing a point from the surface of a sphere, *Annalys. Math. Statist.*, **43,** 645-646.

Max, J. (1960) Quantizing for minimum distortion, *IRE Trans. Inform. Theory,* **IT-6,** 7-12.

McKay, M.D., Beckman, R.J. and Conover, W.J. (1979), A comparison of three methods for selecting values of input variables in the analysis of output from a computer code, *Technometrics*, **21**, 239–245.

Menon, M.V. (1963) Estimation of the shape and scale parameters of the Weibull distribution, *Technometrics,* **5,** 175-82.

Moon, A.M. (1941) On the joint distrution of the median in samples from a multivariate population, *Ann. Math. Statist.*, **12,** 268-278.

Moon, Y.S. (1974) Some numerical experiments on number-theoretic methods in the approximation of multi-dimensional integrals, *Technical Report,* **72,** Dept. of Computer Science, Toronto Univ., Canada.

Morris, M.D. (1991) Factorial sampling plans for preliminary computational experiments, *Technometrics*, **33**, 161–174.

Müller, P. (1991) A generic approach to posterior integration and Gibbs sampling, *Technical Report* **91-09**, Purdue University.

Nash, J. C. and Walker-Smith, M. (1987) *Nonlinear Parameter Estimation: An Integrated System in BASIC,* Marcel Dekker Inc., New York.

Naylor, J.C. and Smith, A. F. M. (1982) Applications of a method for the efficient computation of posterior distributions, *Appl. Statist.*, **31**, 214-225.

Neyman, J. (1937) "Smooth" test for goodness of fit, *Skandinavisk Aktuanetidskrift*, **20**, 149-199. (Reprinted in *A Selection of Early Statistical Papers of J. Neyman*, Cambridge, Cambridge Univ. Press, 1967).

Niederreiter, H. (1973a) Metric theorems on the distribution of sequences, *Proc. Symp. Pure Math.*, **24, AMS Pro. R.I.,** 195-212.

Niederreiter, H. (1973b) *Application of Diophantine Approximations to Numerical Integration, Diophantine Approximation and Its Applications* (ed by C.F. Osgood), Academic Press, New York, 129-199.

Niederreiter, H. (1977) Pseudo-random numbers and optimal coefficients, *Advances in Math.,* **26,** 99-181.

Niederreiter, H. (1978a) Existence of good lattice points in the sense of Hlawka, *Monatsh. Math.,* **86,** 203-219.

Niederreiter, H. (1978b) Quasi-Monte Carlo methods and pseudo-random numbers, *Bull. Amer. Math. Soc.,* **84,** 957-1041.

Niederreiter, H. (1983) A quasi-Monte Carlo method for the approximate computation of the extreme values of a function, *Studies in Pure Math,* Birkhauser, Basel, 523-529.

Niederreiter, H. (1987) Point sets and sequences with small discrepancy, *Monatsh. Math.,* **104,** 273-337.

Niederreiter, H. (1988a) Quasi-Monte Carlo methods for multi-dimensional numerical integration, *International Series of Numerical Mathematics,* Birk. Springer-Verlag.

Niederreiter, H. (1988b) Low-discrepancy and low-dispersion sequences, *J. Number Theory,* **30,** 51-70.

Niederreiter, H. (1992) *Random Number Generation and Quasi-Monte Carlo Methods*, Soc. Industr. Appl. Math. (SIAM), Philadelphia.

Niederreiter, H. and McCurley, K. (1979) Optimization of functions by quasi-random search methods, *Computing,* **22,** 119-123.

Niederreiter, H. and Peart, P. (1986) Localization of search in quasi-Monte Carlo methods for global optimization, *SIAM J. Sci. Statist. Computing*, **7,** 660-664.

Owen, A. (1991), controlling correlations in latin hypercube samples, *J. Amer. Statist. Assoc.*, to appear.

Owen, A. (1992) Orthogonal arrays for computer experiments, integration and visualization, *Statistica Sinica,* to appear.

Pagès, G. and Xiao, Y. J. (1991) Sequences with low discrepancy and Pseudo-random numbers, *Technical Report,* LaboratoireDe Mathematiques ET Modelisation, Paris.

Patel, J.K., Kapadia, C.H. and Owen, D.B. (1976) *Handbook of Statistical Distributions*, Marcel Dekker, Inc., Hew York.

Pearson, E.S. and Hartley, H.O. (1956) Biometrika Tables for Statisticians, Cambridge University Press, Cambridge.

Pollard, D. (1981) Strong consistency of k-means clustering, *Ann. Statist.,* **9,** 135-140.

Pollard, D. (1982) Quantization and the method of k-means, *IEEE Trans Inform. Theory,* **IT-28,** 199-205.

Ratkowsky, D.A. (1983) *Nonlinear Regression Modeling,* Marcel Dekker, Inc., New York.

Ratkowsky, D.A. (1990) *Handbook of Nonlinear Regression Models*, Marcel Dekker, Inc., New York.

Reilly, P.M. (1976) The numerical computation of posterior distributions, *Appl. Statist.*, **25**, 201-209.

Rincon-Gallardo, S., Quesenberry, C. P. and O'Reilly-Federico (1979) Conditional probability integral transformation and goodness-of-fit tests for multivariate distribution, *Ann. Statist.*, **7**, 1052-1057.

Rosenblatt, M. (1952) Remarks on a multivariate transformation, *Ann. Math. Statist.*, **23**, 470-472.

Roth, K.F. (1954) On irregularities distribution, *Mathematika,* **1,** 73-79.

Rousseeuw, P.J. and Leroy, A. (1987) *Robust Regression and Outlier Detection*, Wiley, New York.

Rousseeuw, P.J. and van Zomeren, B.C. (1990) Unmasking multivariate outliers and leverage points, *J. American Statist. Assoc.*, **85**, 633-639.

Roy, S.N. (1975) *Some Aspects of Multivariate Analysis,* Wiley, New York.

Royston, J.P. (1983) Some techniques for assessing multivariate normality based on the Shapiro-Wilk W., *Appl. Statist.*, **32**, 121-133.

Rubinstein, R.Y. (1981) *Simulation and the Monte Carlo Method,* Wiley, New York.

Rubinstein, R.Y. (1986) *Monte Carlo Optimization, Simulation and Sensitivity of Queuing Networks*, Wiley, New York.

Sacks, J., Schiller, S.B. and Welch, W.J. (1989), Designs for computer experiments, *Technometrics*, **31**, 41–47.

Sacks, J., welch, W.J., Mitchell, T.J., and Wynn, H.P. (1989), Design and analysis of computer experiments (with discussion), *Statist. Sci.*, **4**, 409–435.

Saltykov, A.I. (1963) Tables for computing multiple integrals by the method of optimal coefficients, *Z. Vycisl. Mat. i Mat. Fiz.*, **3,** 181-186.

Scheffè, H. (1958) Experiments with mixtures, *J. Royal Statist. Soc. B,* **20,** 344-360.

Scheffè, H. (1963) The simplex-centroid design for experiments with mixtures, *J. Royal Statist. Soc. B,* **25,** 235-263.

Schmidt, W.M. (1970) Simultaneous approximation to algebraic numbers by rationals, *Acta Math.,* **125,** 189-201.

Schmidt, W.M. (1972) Irregularities of distribution, VII, *Acta Arith.*, **21,** 45-50.

Seber, G. A.F. (1977) *Linear Regression Analysis*, Wiley, New York.

Serfling, R.J. (1980) *Approximation Theorems of Mathematical Statistics*, Wiley, New York.

Shannon, C.E. (1948) A mathematical theory of communication, *Bell Syst. Tech. J.,* **27,** 379-423 and 623-656.

Sharma, D.K. (1978) Design of absolutely optimal quantizers for a wide class of distortion measures, IEEE Trans. Inform. Theory, **IT-24,** 693-702.

Shaw, J.E.H. (1988) A quasirandom approach to intergration in Bayesian statistics, *Ann. Statist.,* **16,** 895-914.

Shih, S.C. (1980) Une generalization des $\ll$ bons trellis $\gg$, *C. R. Acad. Sc. Paris,* **290,** 527-530.

Sibuya, M. (1962) A method for generating uniformly distributed points on n-dimensional spheres, *Ann. Inst. Statist. Math.*, **14**, 81-85.

Sirvanci, M. and Yang, G. (1984) Estimation of the Weibull parameters under Type I censoring, *J. Amer. Stat. Assoc.,* **79,** 183-187.

Small, N.J.H. (1980) Marginal skewness and kurtosis in testing multivariate normality, *Appl. Statist.*, **29**, 85-87.

Smith, A.F.M. et al (1985) The implementation of the Bayesian paradign, *Commun. Statist.-Theor. Meth.*, **14**, 1079-1102.

Smith, H. and Dubey, S.D. (1964) Some reliability problems in the chemical industry, *Industrial Quality Control,* **21,** 64-70.

Snee, R.D. (1973) Techniques for the analysis of mixture data, *Technometrics,* **15,** 517-528.

Sobol, I.M. (1967) The distribution of points in a cube and the approximate evaluation of integrals, *U.S.S.R. Comput. Math. and Math. Phys.*, **7**, 86-112.

Steck, G.P. (1962) Orthant probabilities for the equicorrelated multivariate normal distribution, *Biometrika*, **49**. 433-445.

Stein, M. (1987) Large sample properties of simulations using latin hypercube sampling, *Technometrics*, **29**, 143-151.

Stroud, A.H. (1967) Some seventh degree integration formulas for the surface of an n-sphere, *Num. Math.*, **11**, 273-276.

Stroud, A.H. (1971) *Appproximate Calculation of Multiple Integrals,* Prentice-Hall, Inc. Englewood Cliffs, New Jersey.

Tang, B.X. (1991) OA-based Latin hypercubes, Ph.D. Proposal, Department of Statistics and Actuarial Science, University of Waterloo.

Tang, B.X. (1992) OA–based latin hypercubes, to appear in *J. Amer. Stat. Assoc.*

Tang, M.C. (1992) Mixtures of number-theoretic methods and Monte Carlo methods for optimization, *BSc thesis*, Hong Kong Baptist College.

Tashiro, Y. (1977) On methods for generating uniform points on the surface of a sphere, *Annals of Institute of Statistical Mathematics,* **29,** 295-300.

Teichroew, D. (1956) Tables of expected values of order statistics samples of size twenty and less from the normal distribution, *Ann. Math. Statist.*, **27**, 410-420.

Thompson, W.O. and Myers, R.H. (1968) Response surface design for experiments with mixtures, *Technometrics,* **10,** 739-756.

Todd, M. (1976) *The Computation of Fixed Points and Applications,* Springer Lecture Note in Econ. and Math. Systems, Springer-Verlag, Berlin.

Toth, L.F. (1959) Sur la representation d'une population infinie par un nombre fini d'elements, *Acta Math. Acad. Scient. Hung.,* **10,** 299-304.

Tseng, H.C. and Fisher, T.R. (1987) Transform and hybrid transform/ DPCM coding of images using pyramid vector quantization, *IEEE Trans. Commun.,* **COM-35,** 79-86.

Tong, Y.L. (1990) *The Multivariate Normal Distribution*, Springer- Verlag, Berlin.

Tukey, J.W. (1960) A survey of sampling from contaminated distributions, In *Contributions to Probability and Statistics: Essays in Honour of Harold Hotelling* (Olkin, I., Ghurye, S.G., Hoeffding, W., Madow, W.G., and Mann, H.B., eds), Stanford University Press, Stanford, California 448-485.

Van der Corput, J.G. (1953) Verteilungs funktionen, I, *Proc. Akad. Amsterdam*, **38**, 813–821.

Vasicek, O. (1976) A test for normality based on sample entropy, *J. Royal. Statist. Soc. B,* **38,** 54-59.

Wang, X.R. (1982) *Multivariate Statistical Methods for Geological data*, Science Press, Beijing.

Wang, Y. and Fang, K.T. (1981) A note on uniform distribution and experimental design, *Kexue Tongba,* **26,** 485-489.

Wang, Y. and Fang, K.T. (1990a) Number theoretic methods in applied statistics, *Chinese Ann. Math. Ser. B,* **11,** 41-55.

Wang, Y. and Fang, K.T. (1990b) Number theoretic methods in applied statistics (II), *Chinese Ann. Math. Ser. B,* **11,** 859-914.

Wang, Y. and Fang, K.T. (1992) A sequential number-theoretic method for optimization and its applications in statistics, in *The Development of Statistics: Recent Contributions from China, eds by* Chen X.R., Fang, K.T. and Yang, C.C., Longman Academic Scientific and Technical, London, 139-156.

Wang, Y., Xu, G.S. and Zhang, R.X. (1978-1982) On number-theoretic method of numerical integration in multi-dimensional space, I, *Acta Math. Appl. Sin.*, **2**, 106-114, II, ibid. 412-417.

Wang, Z.K. (1986) *Basic Algorithm for Fixed Points,* Publishing House of Zhongshang University, Guangzhou, China.

Wang, Z.K. (1987) *Algorithms for Fixed Points,* Shanghai Press of Science and Technology, Shanghai.

Waterman, M.S. (1974) A restricted least squares program, *Technometrics,* **16,** 135-136.

Watson, G.S. (1983) *Statistics on Spheres,* Wiley, New York.

Wei, B.C. (1989) *Modern Nonlinear Regression Analysis,* Southeast University Press, Nanjing.

Welch, W.J. (1983), A mean squared error criterion for the design of experiments, *Biometrika,* **70**, 205–213.

Welch, W.J., Buck, R.J., Sack, J., Wynn, H.P., Mitchell, T.J. and Morris, M.D. (1992) Screening, predicting, and computer experiments, *Technometrics* **34** 15–25.

Welch, W.J., Yu, T.K., Kang, S.M., and Sacks, J. (1990), Computer experiments for quality control by parameter design, *J. Quality Technol.*, **22**, 15–22.

Weyl, H. (1916) Über die Gleichverteilung der Zahlem mod Eins, *Math. Ann.,* **77,** 313-352.

Wilson, S.G. (1980) Magnitude/phase quantization of independent Gaussian variates, *IEEE Trans. Commun.*, **COM–28**, 1924–1929.

Wingo, D.R. (1972) Maximum likelihood estimation of the parameters of the Weibull distribution by modified quasilinearization, *IEEE Trans. Rel.,* **R-21,** 89-93.

Wolfe, P. (1969) Convergence conditions for ascent methods, *SIAM Review*, **11**, 226–235.

Wozniakowski, H. (1991) Average case complexity of multivariate integration, *Bull. Amer. Math. Soc.* **24,** 185-194.

Yamada, Y., Tazaki, S. and Gray, R.M. (1980) Asymptotic performance of block quantizers with difference distortion measures, *IEEE Trans. Inform. Theory,* **IT-26,** 6-14.

Yamauti, Z. (1972) Statistical Tables and Formulae with computer application, JSA, Tokyo.

Ylvisaker, D. (1975) Designs on random fields, in *A Survey of Statistical Design and Linear Models*, ed by J.N. Srivastava, North-Holland, Amsterdam, 593–607.

Ylvisaker, D. (1987), Prediction and design, *The Annals of Statistics,* **15**, 1-19.

Zador, P.L. (1982) Asymptotic quantization error of continuous signals and quantization dimension, *IEEE Trans. Inform. Theory,* **IT-28,** 139-149.

Zaremba, S.K. (1972) *Applications of Number Theory to Numerical Analysis,* Academic Press, New York.

Zhang, J.T. (1990) Computation of moments of order statistics, *J. Graduate School, Academia Sinica* **7,** 1-29.

Zhang, R.C. (1991) Uniform design in the Stiefel Manifold and its application, *Technical Report*, University of Waterloo, Canada.

Zhang, R.C. (1993) On design of column-orthogonal matrix with with uniform distribution, *Technical Report*, Series No. **MATH-023**, HKBC, Hong Kong.

Zhang, Y.T. and Fang, K.T. (1982) *An Introduction to Multivariate Analysis,* Science Press, Beijing.

Zinterhof, P. (1976) Über einige Abschätzungen bei der Approximation von Funktionen unit Gleich Verteilungs merhoden, *S. B. Akad. Wiss. Math. Nat. ABt. II,* **185,** 121-132.

Zhu, L.X., Wong, H.L., and Fang, K.T. (1993). A test for multinormality based on sample entropy and projection pursuit, *Technical Report*, Series No. **MATH-022**, HKBC, Hong Kong.

Zhu, Y.C. (1990) On some multi-dimensional quadrature formulas with number-theoretic nets, *Technical Report No. 047*, Institute of Applied Math., Academia Sinica, Beijing.

Author index

Subject index

ERRATUM

pp. 334–5 insert.

The author index is completed as follows:

Number-theoretic Methods in Statistics. K.T. Fang and Y. Wang. Published in 1994 by Chapman & Hall, London. ISBN 0 412 46520 5.